John of the Mountains
The Unpublished Journals of John Muir

Edited by Linnie Marsh Wolfe

THE UNIVERSITY OF WISCONSIN PRESS

The University of Wisconsin Press
114 North Murray Street
Madison, Wisconsin 53715

3 Henrietta Street
London WC2E 8LU, England

Printed in the United States of America

ISBN 0-299-07880-9 cloth, 0-299-07884-1 paper
LC 79-83859

This edition is reprinted by arrangement with Houghton Mifflin Co.

TO

WANDA MUIR HANNA

THIS, HER FATHER'S BOOK,

IS DEDICATED

ACKNOWLEDGMENTS

To DOCTOR WILLIS LINN JEPSON, a very special acknowledgment is due for his labor so generously given to the task of surveying all botanical terms used in selected portions of the journals, such a survey being made necessary by the dimness of the writing and Mr. Muir's habit of abbreviating scientific names. Footnotes signed W. L. J. have been supplied by him.

To Mr. and Mrs. Thomas R. Hanna I am deeply grateful for the privilege of doing this work, and for their unfailing help throughout.

To Charles Keeler, Beloved Poet, so recently passed from our midst, is owing a debt not to be put into words, for courage to undertake a task of such import. And to Ormeida Keeler, his wife, thanks are due for permission to quote from her husband's prospective book, 'Friends Bearing Torches.'

To Mrs. William Frederic Badè I pay my tribute for the gracious spirit with which she turned over the mass of Muiriana gathered and organized by her husband.

To Miss Mary Barmby, Librarian of the Alameda County Free Library, thanks are due for the temporary storage of the Muiriana and the facilities for working therewith.

To Doctor John Wright Buckham I am indebted for valuable suggestions upon the manuscript, and to Doctor Stanley Armstrong Hunter for encouragement and advice.

Others to whom I wish to express gratitude for information as to place-names, persons, and park history are: Mr. Francis B. Farquhar; Mr. William E. Colby, co-worker in the cause of forests and parks; Mr. C. A. Harwell, Park Naturalist, and Mr. M. E. Beatty, Assistant Naturalist, both of the Yosemite National

Park; Mr. John R. White, Superintendent of the Sequoia National Park; Mr. Guy Hopping, Superintendent of the General Grant National Park; Mr. John C. Preston, Superintendent of the Lassen Volcanic National Park; and Miss Mabel Gillis, Librarian, California State Library.

Recognition of courtesy is due the D. Appleton-Century Company for permission to use certain paragraphs from Muir's 'Rival of the Yosemite,' *Century Magazine*, November, 1891; to Harper & Brothers for passages from Muir's 'New Sequoia Forests of California,' *Harper's Magazine*, November, 1878; and to Mr. Lee Ettleson of the San Francisco *Call-Bulletin* for several brief extracts from Muir's 'Letters,' published in the San Francisco *Bulletin* in the years 1874–75.

I am glad to acknowledge the aid of Professor August Vollmer, of the University of California, in reading faded script; and I wish to express appreciation to Mrs. Isabel Longbourne for her efficient and enthusiastic labor in transcribing journals and preparing manuscripts.

Especially do I thank Houghton Mifflin Company for that friendly spirit of co-operation in which we have worked together in the making of this book.   Footnotes signed F. H. A. were supplied by Mr. Francis H. Allen of their staff, an ornithologist.

In grateful memory of my husband, Roy N. Wolfe, who shared my early interest in John Muir, I have done this work.

L. M. W.

CONTENTS

Introduction   ix

Milestones   xvii

Chapter I
    Note   1
    At Smoky Jack's Sheep Camp   2

Chapter II
    Note   34
    1. My First Winter, Spring, and Summer in the Yosemite Valley   36
    2. Sierra Fragments   56
    3. Explorations in the Great Tuolumne Canyon   68
    4. Trip to Mount Clark and the Illilouette Basin   79
    5. Sierra Fragments   86
    6. Mountain Thoughts   92

Chapter III
    Note   100
    1. Sunnyside Observations   101
    2. Trip to the Upper Canyons of the Middle and North Forks of the San Joaquin River, Including an Ascent to the Minarets of South Mount Ritter and the Summit of the 'Matterhorn'   141
    3. Sierra Fragments   163
    4. Tuolumne Days and Nights   166
    5. An Autumn Saunter   172
    6. Down the Sierra to Mount Whitney   173

Chapter IV
    Note   189
    1. Going Home to the Mountains   190
    2. Hunting Shasta Wild Sheep   193
    3. A June Storm   201

4. Exploring the Sequoia Belt from the Yosemite to the White
   River, Tulare County, 1875   209
5. A Night Scene at Lake Tenaya   235
6. Canoeing Down Three Rivers   236

Chapter V
   Note   245
1. First Journey to Alaska, 1879   246
2. Memories of the Dog Stickeen   275

Chapter VI
   Note   281
1. 'Fixing' Nevada Falls   283
2. Camping at Lake Tahoe with Charles C. Parry   285
3. A Journey to Mount Rainier   289

Chapter VII
   Note   298
1. A Voyage to Alaska   299
2. Alaska Fragments, June – July, 1890   311
3. A Trip to Kings River, Yosemite   322

Chapter VIII
   Note   334
1. At Home on the Ranch   335
2. A Mountain Ramble   342
3. Thoughts Upon National Parks   350
4. At Home on the Ranch   354
5. With the National Forestry Commission   356
6. Rambling Through the Southern States   364

Chapter IX
   Note   378
1. Cruising with the Harriman-Alaska Expedition, 1899   379
2. The True Story of J. B. and Behring Sea   422

Chapter X
   Note   427
1. Trees   428
2. Arizona Notes   431
3. Thoughts Written on the Birthday of Robert Burns   433
4. Yesterday, Today, and Tomorrow   436

Index   441

INTRODUCTION

THE sixty extant journals of John Muir were written over a period of forty-four years, from 1867 to 1911. Closely allied with them is a mass of notes scribbled upon loose sheets and bits of paper of all shapes and sizes. The journals cover nearly the whole of his career as a naturalist, narrating his wanderings, describing what he saw, recording his scientific researches. The contemporary fragments are largely devoted to an expression of his inmost thoughts upon Nature and her transcendental meanings.

Taken in their order the journals have considerable biographical interest. But since in the earliest of them he already reveals a surprising mastery of poetic prose, one must go back of them to discover how he learned to write. Apparently he had no wish to become a writer, hence there was no conscious playing of 'the sedulous ape.' His boyish ambition was to invent machines, for which he had unique ability. A few years later he planned to study medicine. Still later he decided to be a botanist. Finally in mature life, with the utmost reluctance and distrust of his own powers, he took up the 'making of books.'

However, from childhood he was sensitive to the beauty of rhythmic language. Of a certain ballad he said, 'It was one of my earliest memories, and it will ring in my ears when I am dying.' Moreover, the pungent folk-speech of his native Scotland, 'so rich in love-words,' became part of his fiber, as did the poetry of Burns and Milton and the King James English of the Bible which he had to learn by heart as a child. All these lent an unconscious music to his spoken words. Rhymed jingles rose spontaneously to his lips, in his teens he was writing metrical letters to his friends, and in 1860 he composed a fairly long poem in blank verse, entitled 'The Old Log Schoolhouse.' All this was simply the effervescence of high spirits moulded by an innate law of rhythm. In later years when he wrote down his thoughts in prose sentences, they still had to conform to this law. Members of his family recall his habit of beating time with uplifted forefinger as he wrote.

No journals survive from his boyhood, nor from his four years at the University of Wisconsin, if, indeed, he wrote any. And none records those early botanical ramblings in Canada. However, during this period, 1864–66, he began a remarkable series of letters to Mrs. Jeanne C. Carr, wife of Doctor Ezra Slocum Carr of the University faculty. Mrs. Carr, herself a botanist, persuaded him to write to her of his wanderings. His efforts began in the somewhat stilted diction of a shy youth unaccustomed to put down his thoughts for others to read, but quickly developed into a style of freedom and beauty.

With this brief preliminary practice in the art of self-expression he wrote his first journal in 1867, recording what he saw as he tramped south through Kentucky, Tennessee, Georgia, and Florida, and adorning his pages with sketches of trees, scenery, and a humorous drawing of himself sleeping among the tombs of Buenaventura Cemetery in Savannah, Georgia, watched over by owls and prowling animals.

On the flyleaf he encircled with a flourish his address as 'John Muir, Earth-Planet, Universe.' This was his declaration of freedom to wander. Coming with his family from Scotland in 1849, when he was eleven years old, and settling on the Wisconsin frontier, he absorbed with the air he breathed the venturing, mystical spirit of the pioneer, the spirit that went seeking for 'Something lost behind the ranges.' 'When I first left home to go to school,' he said, 'I thought of fortune as an inventor, but the glimpse I got of the Cosmos at the University, put all the cams and wheels and levers out of my head.' And so without money, but gloriously happy, he journeyed on from the Gulf States to the 'wild side of the continent' on his life's business of studying the Cosmos.

His first California journal, written in 1869, forms Chapter I of this book, and the succeeding chapters contain most of the important journals that followed. Regretfully his foreign travel journals have been omitted from this book. Rich and flowing abundantly out of his undimmed joy in Nature, although some of them were written near the end of life, they form in themselves a unit too large to be included here.

Having shared the hardships of his wanderings, his notes, mostly written in pencil, are not easy to read after the lapse of years. Many were scribbled by flickering campfires when his body was numb with fatigue; or in the dark lee of some boulder or tree while the storm raged without; or tramping over a vast glacier, his fingers stiff with the cold, and his eyes blinded by the snow glare. Often before the notes could be carried home to camp or hut, they were smudged by ferns and flowers pressed between the pages, or water-soaked in bogs that had to be waded through.

John Muir apparently took with him on his longer excursions two or more notebooks, and when starting off on each day's trip, tied to his belt whichever one he happened to pick up. Into this he wrote his notes, sometimes in the front, sometimes in the back, often not dating the entries. The task of the editor has been to arrange these in chronological order as nearly as that may be determined from contemporary correspondence, and from other clues. Needless to say, only an approximate accuracy can be attained.

Botanists reading the journals will remember that most of these field notes were unrevised by Muir, and do not always represent his final judgment in the matter of plant names. Moreover, place names used by Muir are not always to be found on maps. He frequently gave his own names to features of the landscape that he described.

Having written his journals under such conditions, not until late in life did he make much use of them as source material for articles and books. His memories were fresh, and these he wrote and rewrote into language that never satisfied him. The 'infinite shortcoming of words' filled him with despair. How describe in 'muddy' English beauty 'so filled with warm God?' He hailed William Keith's ability 'to paint the poems of the mountains,' but as for the books people were urging him to write, they would be only 'dead bone-heaps' of the Living Reality!

His Scotch reticence also complicated the task of writing. 'The best things and thoughts we get from Nature we dare not tell,' he said to a friend. To lay out all his delicate treasures for the coldly critical eye of an unimaginative world seemed a kind of betrayal. So he found it impossible to write freely for the printed page the high poetic fancy and transcendental philosophy he poured white-hot into his journals.

But his journals share one quality with all he wrote for publication. They are impersonal, especially in the early years. To a Scotland cousin who begged him to write more about himself, he replied: 'As to putting more of myself into these sketches, I never had the heart to spoil their symmetry with mere personal trials and adventures. It looks too much like having to say, "Here is the Lord, and here is Me!"'

In the journals is only a chance word now and then of his many warm friendships, and of his devotion to wife and children. There is nothing at all of the loyal care with which he anticipated every need of his parents, and watched over his brothers and sisters and their growing families, and of course not a word of the amazing generosities shown toward friends and people who had no claim upon him but that of sympathy. Indeed, there is surprisingly little of the quarter-of-a-century battle for national parks and forest reserves. For all these human and public phases of his many-sided life, one must go to the vast correspondence written by him and to him, and assembled after his death, forming in itself a rich biography. For when John Muir went into the wilderness, he went in absolute surrender of self and all the concerns of self, to become 'like a flake of glass through which the light passes.'

Yet, in spite of their impersonality, his journals record most intimately his development as a naturalist and his progress in the humanities. During those early four years when he made the Yosemite Valley his headquarters, his notes present glacial action as the dominating theme. The peaks and domes and canyons were his laboratories, where he found a wealth of evidence to prove that a vast ice-sheet once swept over the entire region, sculpturing the Sierra into the present endless variety of forms. Trees, flowers, and ouzel songs delighted him, but they were not allowed to interfere with the main quest. During this period he remained aloof from human associations, becoming almost a hermit. 'I always feel glad,' he said, 'that ... my lessons are simple rocks and waters and plants and humble beasts, all pure and in their places, the Man Beast with all his complications being laid upon stronger shoulders.' On his rare visits to the Valley to get bread, strangers looked askance at him, this tall, queer fellow with the 'rapt look' and unkempt hair and tattered clothes. Some of his friends became alarmed about him. Emerson, whom he had met in the Yosemite, begged him to 'hurry done with mountains, come east ... and teach young men in colleges.' Mrs. Carr, by this time living in California, troubled lest he become too unpolished and wild, tried to persuade him to come down from the mountains to live in Oakland, thence to make short study trips into the Coast Range. To all of which he replied: 'I will not be done here for years. I am in no hurry.... My horse and bread are ready for upward.... I will fuse in spirit skies!'

So until December, 1873, this man who was no 'mush of concession' continued to scale the untouched summits, notebook at belt, in pursuit of glaciers living and dead, to answer his own last questions and establish for other scientists his theory of geological origins.

At the close of this period came those drear ten months of literary labor in Oakland, ended by his precipitate 'escape' to the mountains, his sudden mystic realization that his intensive study of the earth in past ages was finished, and his utter surrender to the living present, rejoicing in storms, flowers, squirrels, birds, and trees, and finally the declaration of his new purpose: 'I care to live only to entice people to look at Nature's loveliness!'

Released from his preoccupation with the glaciers and rocks

of the Yosemite region, he went on wider journeys to Mount
Shasta, Utah, Nevada, the Columbia River, and Alaska, making
between-whiles an intensive study of sequoia trees in the Sierra.
The winter months he now spent less rebelliously writing for
publication, and even began to lecture up and down the State in
defense of the forests. Friendships arose wherever he went.
John Swett, the beloved schoolmaster of California, opened his
San Francisco home to him in 1875, and the gracious family life
he there became a part of, fastened 'the hooks of civilization'
into him for good and all.

His marriage in 1880, to the daughter of a large landowner,
brought a train of new responsibilities that threatened to engulf
his whole future.   To publishers and nature-lovers all over the
country the middle eighties seemed to be 'lost years' in the life
of John Muir, scientist and writer. But when one considers the
great work of the ensuing quarter-century, during which, largely
owing to his leadership, the national park and forest reserve
systems were established in America, it is apparent that the six
or seven years of withdrawal contributed their necessary share
to the pattern and purpose of his life. For it was then he learned
to live and work with men and women, and to understand and
utilize social institutions. When he emerged in 1889, to take up
his public work, he was no longer 'an unknown nobody in the
woods,' but a shrewd, practical man of the world, and a lover
of his fellow man.

During the travels of later years he wrote many notebooks.
As might be expected, they reflect the changes that had come
with the years. They contain only occasional flights into the
empyrean of transcendental poetry so characteristic of his
writings during the seventies, but they contain more humor, and
are mellowed by his concern for human rights, and his desire to
lead men to the sanity and joy of wild nature.

However, these too joined the earlier journals in collecting
dust upon the crowded shelves — and floor — of John Muir's
'den.' During the last dozen years these accumulations came to
lie heavily upon his conscience, especially since John Burroughs

and Walter Hines Page were prodding him to write up 'all those bags of notes.' In rare hours of leisure he tried to sort them out, with the following results:

> I've been reading old musty dusty Yosemite notes until I'm tired and blinky blind, trying to arrange them in something like lateral, medial, and terminal moraines on my den floor. I never imagined I had accumulated so vast a number. The long trains and embankments and heaped-up piles are truly appalling.... I'm beginning to see that I'll have to pick out only a moderate-sized bagful,... and abandon the bulk of it to waste away like a snowbank or grow into other forms as time and chance may determine.

Revising the 'moderate-sized bagful,' he published, in 1911, 'My First Summer in the Sierra,' and the following year, 'The Yosemite.' Then he began to prepare his earlier Alaskan journals for a book, and was nearing the end of that task when he died in 1914. Mrs. Marian Randall Parsons, who had assisted him, ably completed the labor of editing the book entitled 'Travels in Alaska.'

Doctor William Frederic Badè, appointed in 1915 to serve as literary executor and biographer of his friend John Muir, published from the journals the 'Thousand-Mile Walk to the Gulf,' and 'The Cruise of the Corwin,' and from published articles, 'Steep Trails.' Then, after assembling from all the world the correspondence of John Muir, and a vast mass of other material pertaining to his life and public service, Doctor Badè wrote with rare scholarship and insight 'The Life and Letters of John Muir,' using as his main sources the naturalist's unpublished autobiography, and the enormous correspondence. It was Doctor Badè's intention, when he had finished his archeological labors in Palestine, to bring out one or two additional Muir books. Death cut short his plans.

It has been the desire of the present editor to salvage and give to the world, in this centenary year of his birth, the best of John Muir's writings yet remaining in his unpublished notes, and thus make available that intimate understanding of his life and thought which may be gleaned from following his journals down through the years, from early mountain raptures and painstaking scientific investigations to the mature reflections of a man dwelling on the plain and laboring among his fellow men to bring to them the Reality he has known.

LINNIE MARSH WOLFE

BERKELEY, CALIFORNIA
January 3, 1938

MILESTONES

1838 John Muir born in Dunbar, Scotland, April 21.
1849 Emigrated with family to Wisconsin frontier.
1860 Won prize for inventions at State Agricultural Fair. Entered University of Wisconsin.
1864–66 Botanized in Canada.
1867 Walked from Kentucky to Florida. Wrote first journal.
1868 Arrived in California, March 28.
1869–73 Made headquarters in Yosemite Valley. Explored Sierra for evidence of glacial action.
1874–76 Ascended Mount Shasta. Began intensive study of trees. Launched movement for Federal control of forests.
1877–78 Explored the Great Basin.
1879 First trip to Alaska.
1880 Married Louie Wanda Strentzel, April. Second trip to Alaska.
1881 Cruised in Alaskan and Arctic waters.
1882–87 Home and fruit-raising.
1888 Ascended Mount Rainier. Resumed writing.
1889 Worked for the creation of the Yosemite National Park.
1890 Yosemite National Park Bill passed by Congress. Explored Muir Glacier in Alaska.
1891–92 Founded Sierra Club. Worked for recession of Yosemite Valley.
1893–94 Visited Europe. 'Mountains of California' published.
1896–98 Received degree of M.A. from Harvard. Worked with National Forest Commission. Received degree of LL.D. from University of Wisconsin.
1899 Joined the Harriman-Alaska Expedition.
1901–02 'Our National Parks' published. Began fight to save the Hetch Hetchy Valley.
1903–04 Guided President Roosevelt through the Yosemite. Made world tour.
1905 Yosemite Valley receded to Federal control. Death of Mrs. Muir, August.

1906   Explored Arizona.
1909   'Stickeen' published.
1911   'My First Summer in the Sierra.'   Received degree of
       Litt.D. from Yale.
1912   'Yosemite.'
1913   'Story of My Boyhood and Youth.'   Degree of LL.D re-
       ceived from University of California.   Lost fight to save
       Hetch Hetchy Valley.
1914   Died in Los Angeles, December 24.

John of the Mountains

Chapter I : 1868-1869

'A strong butterfly full of sunshine settles not long
at any place. . . . Such a life has been mine.'

JOHN MUIR wrote the earliest of his extant journals while
tramping through the Southern States in 1867. This has been
published under the title, 'A Thousand-Mile Walk to the
Gulf.'

His second journal was written in California in the winter and
spring of 1868 and 1869. Traveling steerage via Panama, he had
arrived in San Francisco March 28, 1868, and had sauntered away,
'in the bloomtime of the year,' through the San Joaquin Valley and
on to the High Sierra, visiting the Mariposa Big Trees, and spend-
ing eight or ten days in the Yosemite. But being without money
to supply his simple wants, he had to return to the foothill ranches
in the vicinity of La Grange to work in summer harvests, and
later to break mustangs and herd sheep. It was while he was
shepherding the flock of an Irishman named John Connel, alias
'Smoky Jack,' that he wrote the journal here published for the
first time. A letter written November 1, 1868, to Mrs. Carr[1]
announced:

> I am engaged at present in the very important and patriarchal
> business of sheep. I am a gentle shepherd. The gray box in which
> I reside is distant about seven miles northwest from Hopeton,
> two miles north of Snelling's. The Merced pours past me on the
> south, from the Yosemite; smooth, domy hills and the tree
> fringe of the Tuolumne bound me on the north; the lordly Sierras
> join sky and plain on the east; and the far coast mountains on

[1] Mrs. Ezra S. Carr. See the Introduction.

the west. My mutton family of eighteen hundred range over about ten square miles, and I have abundant opportunities for reading and botanizing. . . . I shall be in California until next November, when I mean to start for South America.

The introductory paragraphs of the sheep-camp journal, reviewing the months of November and December, 1868, were probably written about the time the young shepherd made his first dated entry, January 1, 1869.

The events of the journal immediately precede those recorded in Muir's 'My First Summer in the Sierra.'

- - - - -

At Smoky Jack's Sheep Camp[1]

[*December*, 1868.]

Since coming to this Pacific land of flowers I have walked with Nature on the sheeted plains, along the broidered foothills of the great Sierra Nevada, and up in the higher piney, balsam-scented forests of the cool mountains. In these walks there has been no human method — no law — no rule. A strong butterfly full of sunshine settles not long at any place. It goes by crooked unanticipated paths from flower to flower. Sometimes leaving blossoms of every taste, it alights in the mud of a stream, or glances up into the shadows of high trees, or settles on loose sand or bare rock. Such a life has been mine, every day and night of last summer spent beneath the open sky; but last month brought California winter and rain, so a roof became necessary, and the question came, What shall I do? Where shall I go?

I thought of the palmy islands of the Pacific, of the plains of Mexico, and of the Andes of Peru, but the attractions of California were yet stronger than all others, and I decided to stay another year or two.

About the middle of last month, after warning him that I knew

[1] See Badè's *Life and Letters of John Muir*, vol. I, pp. 190–95. See also Muir's *A Thousand-Mile Walk to the Gulf*, chap. IX, 'Twenty-Hill Hollow.'

nothing of the business, I sold my labor to an opaque little Irishman to herd sheep. I was then at Pat Delany's on the Tuolumne. 'Your camp,' said he, 'is about five miles from here, over on Dry Creek. You had better go right over tonight, for the man that is there wants to "quit." You will have no trouble in finding the way. Just take the Snelling road and the first shanty you see on a hill to the right — that's *the place.*' Well, I found *the place,* and a remarkably dirty and dingy old misshapen box of a place it was — like the poor Briton's dwelling which the King durst not enter; the rains entered it and all the winds of heaven whistled through it.

The shepherd that I was to relieve was busy when I knocked at the shanty door, in the work of supper, which, as in most pastoral cabins here, consisted of an uneasy-looking liquid called tea, brown beans, and speckled flapjacks. This supper affair accomplished, the young shepherd began the recital of grievances, domestic and pastoral, mixed with much admonition for my profit. He was about eighteen years of age, and had tried the work of shepherd because he thought it would be easy, but the sheep, finding little to eat, roamed over the hills and levels of their pasture at a discontented pace, and in most disorderly and widespread companies, making the limbs of the gentle shepherd weary in their pursuit. 'I have tried many kinds of work,' said he, 'but this of chasing a band of starving sheep is the worst of all'; and he cursed the whole flock and the entire sheep business with cordial energy, and assured me that I had chosen a bad job.

I asked him if he would stop with me one day to point out the range and to instruct me in dealing with the sheep. 'Oh! No need for that,' said he. 'It would do you no good. I am going away tonight. All that you have to do is just to open the corral in the morning and run after them like a coyote all day and try to keep in sight of them. They will soon show you the range.' And thus I was left to my own resources, out on the unfielded plains with eighteen hundred sheep to feed, besides the work of keeping house.

On examining my cabin premises, the winter home of my adop-

tion, I found one and a half benches for seats, one teapot black as
its contents from smoking on the fire, one pot for boiling beans,
another larger cast-iron pot with cast-iron lid, called a Dutch
oven, for baking bread, one tin and one earthenware cup, and a
wooden water bucket. There was a sort of a rough rickety shelf
in one corner, which I surmised might be intended for a bed. The
sheep corral joined the shanty on the east side. At a late hour I
sought a sleeping-place. None in the cabin looked innocent, and
I thought of sleeping on the ground outside, but scattered everywhere there were ashes, old shoes, sheep skeletons with tough
tendons and ligaments still attached to them, old jaws and
craniums, and rams' horns, etc. And besides these dead evils,
a good many wild hogs prowled about, so I had to submit to the
perhaps lesser evils of the black cabin. I laid myself very doubt-
fully down on the bed-shelf, and after speculating half-repentingly
on the morrow's duties and star-gazing through the faulty roof,
I drifted off into merciful sleep.

Next morning ere I awoke, long sheets of light came streaming
through the seams of the cabin that reached from the gable to the
floor. I was late and jumped hastily from the shelf, made tea,
fried some watered flour, and hastened to the sheep, who were by
this time crowding against the panel at which they were accustomed to get out. Opening the gate, they came crowding out,
gushing and squeezing like water escaping from a broken flume,
crossed Dry Creek, and scattered over a dozen hills and rocky
banks, and I followed with an impression that like spilt water
they would hardly be gathered into one flock again. About a
hundred old lean ones, together with a dozen or so of halt and
lame, were inclined to stop behind and eat any kind of dried
weeds or grass. About a thousand or so wanted to scatter until
each had about an acre of ground to itself, while the remaining
four or five hundred were 'leaders,' or rather secessionists, that ran
convulsively at a gallop full of very short stops, as if determined
to leave the flock for dreamed-of new pastures, but too hungry
to fully carry out their plans.

Of course, under the action of so many disintegrating forces the flock was soon widely scattered, the stragglers, called the 'tail end,' were soon far in the rear, torn off the 'main body' with ragged lines of rupture, the 'leaders' were mostly out of sight, and the 'main body' was sprinkled and dotted over the plains halfway to Snelling's. After a long chase I headed off the leaders and drove them back on the main body. The sky was cloudless, and the blazing sun inclined the woolly runners to halt for rest. All became moderate in their motions and nibbled as best they could at the short, dry grass, and I had time to round them into one flock, while a few lay down on the warm hillsides and chewed contentedly upon what they had already gathered.

In the evening, when the sun was still two hours high, I headed the flock homewards. The cabin was about two miles distant, and, feeding towards it, they reached the rocky banks of Dry Creek about sunset, when, of their own accord and much to my surprise, they formed themselves into long parallel columns and marched willingly across Dry Creek, up the bank, and into the corral. Thus my first day of shepherd life began and ended.

I had to make as well as earn my daily bread, both of which operations were attended with difficulty. I filled the big cylindrical pot with dough and applied hot coals on the hearth, trusting the result might be bread, but the sticky compost, innocent of yeast or any patent inflating mixture, remained as passive beneath the fire as an Indian martyr, and upon being pried out of the pot next morning was found to be black and hard over all its surface, and perfectly solid. It became extremely hard in cooling and looked like a cartwheel, and on attempting to cut out a section of it with a butcher knife it broke with a glassy fracture and I began to hope that like Goodyear I had discovered a new article of manufacture. My teeth were good, and I rasped on a block of it like a squirrel on a nut. I told my troubles to a neighboring shepherd, and he made me wise about sour-dough ferment, and henceforth my bread was good.

The grass of 'the plains' is thoroughly dried in May and the young green grass does not sprout until the rains set in, usually in November or December. During all these months between

May and January, either the dry grass, which, on account of the absence of dew or rain, still retains most of its nourishment like hay, has to be relied on for pasture, or the sheep and cattle are driven from fifty to a hundred miles into the Sierra mountains. Here 'the plains' means all of the immense valley of the San Joaquin and Sacramento Rivers — a fertile, treeless tract of country, measuring about forty or fifty miles in width by five hundred in length.

This year, the first general heavy rain of the season fell on the eighteenth of December, a few days after I began life as a shepherd. It is not often the rains are delayed so long, and farmers feared crop failure, and sheep-owners were disturbed by visions of starving sheep. Every cloud-flake was watched, as were the changes of the wind and moon. 'Will it never, never rain?' they wailed in despair. At length the generous rain came to the thirsty sun-baked plains. Every waiting seed was bathed for ten hours and on the following morning each dead stalk and leaf was given a row of gem-like drops. Myriads of seeds beneath them sprang to life, while insects appearing suddenly played in the balmy air with the hum of a harvest noon. The plains, ere this dappled yellow and purple and brown, soon began to show tintings of green, and the ever-beautiful and glorious mountains became yet more divinely delectable. The nearest foothills are now rosy purple, which forms the lowest of four bands of color — the next higher, blackish purple, the next, deep blue, and beyond and above all, the summit peaks, clustered and spiry, are pure deep white — and all of these magnificent level bands are joined to each other, and to the sky above, and to the plains below as harmoniously as are the color bands of the rainbow. The first shower of the eighteenth belonged to a whole company which came trooping in close succession in loud-pouring earnest.

I begin to know a little more of sheep nature, but do not think that the study or aught else connected with it will ever be interesting to me. Five days ago I had difficulty in getting the flock home. About four o'clock they were scattered far and wide over five or six square miles of hills, feeding quietly. I was reading on a hill that commanded a full view of them, not watching them much because they were so quiet, when suddenly I heard the storm-bleat amongst them, and all at once the whole flock came pelting over hill and dale as if every mutton-eating animal in the world was at their heels. Every one of the terror-stricken wretches ran for the middle point of the flock, and in a minute or so all were solidly massed. The cause of their alarm was not far to seek. A black threatening storm-cloud covered half the sky and was coming grandly on from the west, more like one of the summer storms of Wisconsin than any I had yet seen. When it broke upon the poor silly sheep, they turned their heads from it with piteous looks and crowded still closer together until they did not in number appear to exceed two or three hundred.

The direction of the corral was at right angles to the path of the storm. I wanted to get them home, as night was approaching, but they would not stir a single step. I ran in front of them, shouting till I was hoarse, but they were as immovable as the hillside they stood on, and I might as well have tried to speak to the storm. In the meantime my two dogs were crouching together and making a yet more piteous appearance than the sheep, and trying vainly to find a hillock where the sheet of flowing water coming from the hill would not reach them, observing me, however, all the time. Just as I stopped my noisy exhortations in despair and began to consider what to try next, the more experienced of the dogs, seeing the desperate condition of affairs, dashed upon their conglomerate ranks, made vigorous use of teeth and voice, and soon compelled the felted phalanx to break homeward with ordinary life and sheepishness. I succeeded in getting all corralled before dark, excepting a few that drifted off like chaff before the wind, but these found shelter back of some hills and I found most of them the next day.

After making tea in the black pot I ventured out in the dark to find the hiding-place of some weak sheep and to hear the storm. The night was what is usually called wild and dismal. The sky was one shapeless mass of blackness and gave rain in torrents, and the harmonious powers of the storm were glorious to feel and hear. Not winter winds among the solemn pines, nor the storm-blasts of ocean I used to hear at Dunbar where the shore is rockiest, are more impressively glorious than the black night-storms of these broad happy plains.

The last day of 1868 was a season of rich light and shade — of clear sky and heavy clouds. Cloud-shadows drifted heavily over the brown plains like islands of solid darkness in a sea of light. The air was balmy, making springtime for the flowers and inspiring lark congregations with unmeasured joy. Like an ardent life this day was full of very bright and very dark places, meeting grandly and goldly like deep experiences in a noble character.

January 1, 1869.

The New Year was ushered in with rain, a black day without a single sunbeam. The purple and brown colors are fast fading from the plains, the bright youthful plant green is deepening with astonishing rapidity. Every groove and hollow, however shallow, has its stream — living water is sounding everywhere. 'Tumbling brown, the burn comes down, and roars from bank to brae.' I celebrated the Happy New Year crossing countless streams, running 'ower moor and mire through gude and gide' in full chase of the wretched sheep.

Everything is governed by laws. I used to imagine that our Sabbath days were recognized by Nature, and that, apart from the moods and feelings in which we learn to move, there was a more or less clearly defined correspondence between the laws of Nature and our own. But out here in the free unplanted fields there is no rectilineal sectioning of times and seasons. All things *flow* here in indivisible, measureless currents.

January 2.

Rain with equal cloudiness over the whole sky continued unbroken until the afternoon, when sunshine gushed through jagged openings in the black heavens, and that portion of it which fell into the breasts of the blessed larks [1] was transformed to sweetest song.

[1] The 'larks' of this journal are Western meadowlarks, not the horned larks more strictly so called. — F. H. A.

My insatiable sheep ran today faster and farther than ever, seeming to be assured that better grass was always just beyond them. Sheep are more restlessly discontented at the time of the first coming of green grass than at any other time of the year, for then they will no longer eat the old decaying grass, while there is not enough of the new to fill them. About the end of May most of the flocks in this and the adjoining counties are driven to the mountains, where they fatten in the sweetest and most luxuriant of wild pastures. On the approach of winter in October, they are driven back to their old long-dried pastures on the plains, where they discontentedly and famishingly roam the smooth bare hills until they die, or, if strong enough to pass this dead point of sheep life, gradually fatten and grow strong on the new year's grasses.

This forenoon I saw a gray eagle come swooping down past the edge of the flock and alight on a bare sandy hillside. At first I could not guess what could bring this strong sailor of the sky to the ground. His great wings, seven or eight feet broad, were folded as he stood, dim, motionless, and clod-like as if all excelsior instincts — all of cloud and sky — were forgotten, and that, lark-like, he meant to walk about in the grass hereafter, or stand on a hillock like the little burrowing owl. But as I watched him attentively, the cause of his earthiness became apparent. Food, which brings both eagles and men from the clouds, was a-wanting — he hungered and came down for a hare. The hare he was after stood at the top of his burrow, erect and motionless as the eagle, and with his great beautiful eyes stared the bird king full in the face. They were but eight or ten feet apart, and well did each appreciate the other's thoughts. If the eagle should try to strike the hare, he would instantly dive into his den, and perfectly well did the eagle understand this. His only hope was that his long-eared game, weary at last of inaction, might venture to skim the hillside to some neighboring burrow, when he might sweep above him and strike him dead at a single blow from his pinions, then, checking his impetuous flight, wheel back, pick him up, and bear him to some favorite table rock, satisfy his hunger, wipe off all marks of grossness, and go again to the skies.

I cannot remember having seen any specimen of leg locomotion so swift, so graceful, and so effortless as that of the long-eared California hare. When disturbed by a dog, he scorns to hide in the ground, and bounds over the softly curved hills and hollows like the shadow of a flying bird, knowing, I suppose, the dog might dig him out. They are exceedingly abundant on the plains and smooth plain-like foothills, but do not go high into the pine regions of the Sierra. No butterfly North or in the sunny South lives a more flowery life than he — he lives and moves in one vast flower garden.

January 3.

Sky half cloudy. Great warm, soothing, magnificent mist-sheets are upon the mountains this morning, bathing the tree buds and myriads of quickening seeds with a gentleness of gesture and touch that has no word symbol on earth. How indescribable in texture they are, and how finely do they conform to the sloping and waving topography of the hill-bands!

January 4.

Clouds cumulus. A warm, balmy, bright creation is this day. The purple and yellow of the soil and of the old plant stems are rapidly fading in the deepening green of young life. The little triangular rock fern, *Gymnogramme triangularis,* is unrolling its tiny fronds in sweetly arranged knots and mantlings along the rocks of Cascade Creek. I do not know of any fern that has so wide a vertical range as this hardy and contented gold-powdered fellow. I have met it on the lower Joaquin and at all altitudes on the Sierra as far as Yosemite. Sunset-sky purple of the most refined quality.

Dry Creek, on whose happy bank my cabin stands, is subject to sudden swellings, and overflows in the rainy season. Then it becomes a majestic stream, almost a river, with serious and confident gestures curving about its jutting banks and horseshoe bends, carrying fence and bridge timbers, logs and houses within reach of its ephemeral power. In the course of a few hours after the close of a rain, it will retire within its banks, leaving many flat, smooth fresh sheets of sand. I like to watch the first writings upon these fresh new-made leaflets of Nature's own making. One of these pages was made last night and was already written upon when I saw it this morning. It is made from a pulpy mass of ground lavas and slate and old ocean sands, beautifully smoothed and wavily shaded like a high cloud at rest. The first apparent writing was done by a mollusc, the valves of which were about two inches in diameter. I found one belated specimen in his tracks. They set themselves on edge with the valves slightly opened to allow the worm-like motion of their muscle foot. Thus they slide along like a topheavy ship in handsome and inimitable crossing curves. A great blue crane had also printed this virgin sheet with footprints eight inches in length, and some other smaller birds and beasts had left their mark before I came to make mine — all easily read at present, but soon writing above writing in countless characters will be inscribed on this beautiful sheet, making it yet more beautiful, but also carrying it far beyond our analysis. There are no unwritten pages in Nature, but everywhere line upon line. In like manner every human heart and mind is written upon as soon as created, and in all lives there are periods of change when by various floods their pages are smoothed like these sand-sheets, preparing them for a series of new impressions, and many an agent is at once set in motion printing and picturing. Happy is the man who is so engraved that when he reaches the calm days of reflection he may rejoice in following the forms of both his upper and under lines.

January 5.

Clouds — a few filmy touches through which the sun is shining as through glass. Balminess and springiness from the same fountain as yesterday. The blessed larks are aglee with sunshine and fresh grass.

January 6.

Black cirrus cloud, cold. A few flossy isolated parcels of mist are floating about loose in the foothills. When the sun was a half-hour high in the morning, it seemed as if made of silver in a condition a little more liquid and glowing than the silver of the moon, but still without any indistinctness of outline or glare. The sun's edge was not smooth, but had long unequal leg-like projections as if glowing silver had fallen upon a hard sky and spattered out in ragged rays.

This afternoon all magnitudes were greatly exaggerated by the peculiar condition of the atmosphere. Sheep at a short distance looked like oxen, and hills at a mile's distance appeared to be at least ten miles away. These conditions are oftentimes reversed. One day I observed in looking over a plain a few moving dots on a hillock, and after watching their motions attentively for some minutes fully decided that they were not sheep from my flock but rambling hogs. On approaching them nearly, they proved to be oxen! Variations, mirages, etc., of all kinds are common upon these plains, but I seldom have seen any landscape so wonderfully disturbed as this one. If the Creator were to bestow a new set of senses upon us, or slightly remodel the present ones, leaving all the rest of nature unchanged, we should never doubt we were in another world, and so in strict reality we should be, just as if all the world besides our senses were changed.

January 7.

A magnificently lighted cloud rested for a long time upon one of the foothills, a few miles away. Trees peered above it very clearly. It appeared to have become too heavy in some way and sunk — foundered from its fellows to the bottom of the great ocean of purple air. The sunset is most resplendent — by far the most glorious in color of all the California sunsets I have beheld. The clear ordinary sunsets of these plains are exceedingly beautiful, and when one in full repose opens himself to their influences he will find himself happier than he dare tell; but blurred as we commonly are, the gorgeousness of an evening like this is far more impressive. This is like a Wisconsin summer sunset, only the clouds are upon a smaller scale.[1] The flakes and hillocks and tufts and crumpled furrows of cloud are saturated in purple and gold, and all the edges are on fire. Below the clouds float delicate films of yellow upon pale green and blue — a great many dots also and islets and bars have place on the blue, and their crimson is rich and faultless as that of the main continents. A most glorious day, for which I thank God.

January 8.

Dim black cloud. Discovered myriads of Hepaticæ[2] today when lying on the ground looking for germinating seeds.

January 9.

Black cloud — cool. Larks — ever-blessed — rabbits, mosses, and liverworts in throbbing joyful life.

January 10.

Light winds. Clouds silvery and translucent, few at the sunset, but brilliant crimson and gold. A most soothing, bland, warm day. Larks, the blessed, and the insect people overjoyed. In the forenoon when the sheep were quiet, I counted five hundred and fifty mosses upon one quarter of a square inch of rock by a creek-side. I saw longer light and shade splices upon the mountains than ever before. The blending overlaps were wondrous fine.

Visited a pair of Chinamen rocking for gold in one of my pasture gulches. They are patient fellows, easily satisfied.

January 11.

High north wind, variable beyond measure. A smooth, clear, open day, bounded by a misty morning and a rainy night.... No clouds. Discovered an exceedingly small black mushroom

[1] The reader will remember that much of John Muir's boyhood and youth was spent in Wisconsin.

[2] Liverworts, plants of a low order resembling the mosses, not to be confused with the flowering plant *Hepatica*.

on the ground, the head but one eighth of an inch in diameter. Great numbers of very handsome purple mushrooms half an inch in diameter are coming to light everywhere. They are about one inch in length and concave on top.

The plants of these plains may naturally be divided into two classes, viz., water-loving, semi-aquatic plants, flourishing now in this raininess when all the ground is covered by a film of water — mosses, liverworts, fungi, cresses, etc.; and the drier plants maturing their seeds in April and May, not in moisture and beneath cold clouds, but in unbroken, unflecked sunshiny days and dewless nights.

January 13.

Light and filmy clouds. This afternoon my sheep were inclined to be contented over their short mixed nibbling of plants, new and old, and towards night all were tranquilly outspread upon the wrinkled gulchy banks of Dry Creek just opposite my cabin. I always carry a book and usually find a few moments for reading. At this time I was luxuriating in Shakespeare, looking up occasionally to see that my leaders were not running off. Suddenly I heard a rushing sound which was followed by a kind of low frightened bleating. I looked over the flock and saw what appeared to be two handsome shepherd dogs, with very bushy tails and erect ears. They were about one hundred yards from me, and I saw that they had killed and were eating a lamb. I set Fanny after them, but after chasing them up the hill she abandoned the hopeless pursuit. I knew they must be coyotes — California wolves, abundant everywhere on the plains and nearly to the summits of the Sierra. They are small, about the size of an English shepherd dog, but with longer, stronger legs. They are yellowish dark gray, and have a keen, spiky, whistly, musical bark. They are the greatest of all the enemies of the California sheep-raiser, destroying many of his lambs despite his watchfulness and industry in killing them off with poison, etc. They are beautiful animals, and, although cursed of man, are loved of God. Their sole fault is that they are fond of mutton.

Forenoon rainy, foggy, and opaque. Afternoon glistering and transparent. Sunset common purple.

Wind about equally divided in time from all points of the compass. Slight frost. A dense black mist came down from the mountains, lifting repeatedly to let the light through, and again bathing all the hills and their plants in dampness and shade. About one hour before sunset the whole landscape was splendidly illumined.

A danger real or imaginary rattled the silly nerves of my sheep, and they galloped with a noise like thunder from the slopes and gulches, driving headlong for the center of the flock, and in a few seconds the whole eighteen hundred were squeezed and felted into a solid circular cake of mutton and wool. A sheep scarcely possesses a separate existence — the whole flock is required to make an individual. The body of the flock contracts and expands or bends like the body of a worm, but like the legs of a centipede all motions have reference to one common object.

Wind from all points and very light. Clouds — a few dim coastlike streaks over the mountains. In the morning all the plains were white with hoar-frost crystals, and a slim grating of ice lances was shining upon the calm shallows. A bright, balmy, genial day. Read in my shirt-sleeves, or lay with closed eyes, when my sheep also became meditative and lay soaking and steeping in the sunbeams, which reached to the joints and marrow. The ordinary tranquil purple of morning and evening divinely laid on.

A few films of clouds formed over the Coast Mountains, to which they appeared to belong rather than to the sky. Hoar-frost and dense mist in morning, and when the sun broke through,

a beautiful double white rain, or rather mist, bow was visible for
fifteen or twenty minutes. When the mist was rolling off, that
portion which was seen against the hills looked like an immense
black wall with a perpendicular front rising within a few hundred
yards of me. Another glorious day, full of light and joy and
life. A purple evening.

January 19.

Clouds in transparent flakes. Warm, balmy life in every sunbeam. Perfect harmony in all things here.

January 20.

Purple morning and evening. The evening lark song is 'Queed-
lix boodle.' Today is Tuesday with us mortals, but it must be
Sabbath day in the lark calendar, for they have been holding
meetings extraordinary, and their songs were sweet and pure as
the light which inspired them. Lark song is very absorbable by
human hearts. It is about the only bird song of these plains that
has been made with reference to our ears. Yet how grand must
be the one general harmony of all nature's voices here — winds,
waters, insects, and animals. Music belongs to all matter. There
is not a silent, songless particle in the Lord's creation. A little
fragment of wind is broken off from the main ocean, specialized
and made to eddy and gurgle in the bosom of a lark, and that is
made into music, all precious sweet. Wind also gurgles and
vibrates about the angles and hollows of every surface grain of
sand — each sand grain making a perfect song, but not for us.
How spiritual must be the tunes that are born in the groves of
these golden daisies! How the wind will pulse among the curves
and points of these lovely corollas and among the pistils and the
stamens with their sculptured pollen, but not one note is for
mortals! But thank God for this arrangement of the wind beneath the feathers of a lark, and for every wind vibration that
our ears can read.

January 21.

Light southeast wind. Clouds transparent veils. Hoar-frost. The larks this morning sang to the words 'We'-ero spe'-ero we'-eo we'-erlo we'-eit.' I wish I could understand lark language.

Dreamed in the sunbeams, when the sheep were calm, the plan of a hermitage: walls of pure white quartz, doors and windows edged with quartz crystals, windows of thin smooth sheets of water with ruffling apparatus to answer for curtains. The door a slate flake with brown and purple and yellow lichens. And oh, could not I find furniture! My table would be a grooved and shining slab of granite from the bed of the old mountain glaciers, my stool a mossy stump or tree bracket of the big dry, stout kind, and a bed of the spicy boughs of the spruce, etc., *ad infinitum.*

January 22.

Warm and balmy — the most exquisitely refined of all that is lovely in bright hushed springtime.

Almost every bundle of plant life has been unrolled and Mother Earth is brooding her beautiful children with loving appreciation of their coming glory. Lark song this evening is 'Chee, cheel cheedildy choodildy.' Perched upon a stone or old post, they will repeat these words to music that is invariable for an hour at a time. Sweet humanic song is this of 'the holy lark,' but doubtless there are ears to whom the small peepings and chapperings of the other little feathered people are equally sweet.

January 23.

Rained three hours. Yesterday was the Saturday of a glorious weather-week. Its first day, the fifteenth, was between light and dark. With heat and frost, wind and calm, the week developed from day to day like a flower in atmospheres of purple and gold — a sweet, bright, balmy cluster of radiant January days.

January 24.

Clouds light, of large pattern. There are a great many beautiful snowy mist-lakes on the mountain hills, and evening comes grandly on.

This morning, while observing the movements of my flock from a round hill towards Snelling's, I saw a coyote stealing from a thicket of dead weeds and earnestly watching an opportunity for a lamb. I did not make any allowance for his morning hunger, but almost wished I had not seen him, that he might have had a lamb in peace. The flower hungers and watches for the sunshine, the sparrow for the grass seeds, and the wolf for the sheep.

January 25.

A yellow flower of the Umbelliferæ came to the warm side of the sandy hills today. I notice also that one or two of the mosses have adjusted their cowls.

January 26.

A strong southeast wind with clouds of all species, mostly of large pattern. It rained one hour towards evening from finely developed cumuli. Two plants are now in flower, an umbellifer and a crucifer. The little white cress is the first, perhaps, of all, and is eagerly sought by sheep. It occurs in patches of miles in almost unbroken extent. The size, governed by soil, is from seven to fifteen or twenty inches high. The old grasses do not now rule the landscape color; they appear as if separate from the ground, floating on the surface of the green. A magnificent double rainbow bends over the foothills.

January 27.

Had a heavy rain during the night, perhaps hail also. Clouds — grand islands and rocks of the finest textured cirrus — in some places are resting upon the ground. At one time in the morning I observed a well-defined colorless bow from the sun shining through one of these foundering mist-clouds. Dry Creek is booming full now, and behaves like a river. I have to pasture my muttons upon the north side of the creek now. Found another flower today, making a beautiful trio in all.

January 28.

High winds. Magnificent cloud curtains in massive folds trailing over plain and mountain. Rain two hours in the afternoon. A beautiful black circle is about the moon this night, forty degrees in diameter, and the pure white moon shows grandly in so comprehensive and dark a setting.

January 29.

I found a few small companies of a splendid yellow starry Composita, the first of the glorious sheet-gold soon to cover all these hills and hollows and rocky banks like a sea. The heads are daisy-like and very simple, with rays yellow golden above and purple beneath; discs of the flowers are yellow, the stamens united only by tips, leaving the lower portions to bulge out in a crown-like form.

This mongrel, manufactured, misarranged mass of mutton and wool called a sheep band, which I have tended, lo, these six weeks with a shepherd's care, are now rapidly being increased in number by little thick-legged, wrinkled duplicates — unhappy lambs born to wretchedness and unmitigated degradation.

My master, Jack Smoky, alias Smoky Jack, called the other day and asked me whether I had ever 'lambed' a band of sheep. I answered that I had not, and that I had not the slightest idea of the duties of an *accoucheur pastoralibus*. 'Well, thin,' said he, 'you better sind this band over to the ranch, and I will sind over Mike wid the wedder band for you to herd.' Here followed a long discourse upon the difficulties of lamb-raising, the profundity of the skill required, of his own wonderful attainments in raising the largest number of lambs from the smallest number of mothers, etc. In the afternoon Mike appeared on a neighboring hill and I ran to him to ask about my new flock of males — their habits and general behavior as compared with my congregation

of mothers. 'Och, shure,' said Mike, who like his lord and master
was from the green, green isle, 'och, shure, all I can tell ye is
that they'll run from marnin' till night a full gallop, and if ye
keep in sight of them you'll do better than I've done. They have
scattered over a hundred hills every day in as many different
flocks, and they have mixed with every other band in the country,
and now I'm plaguey glad to git rid of them, and my best advice
to you is to rub plenty of mutton tallow and swate ile on your
knee-joints every marnin' before you open the corral, for you bet
they'll "run" — they'll run like h — l.' 'Never mind, Mike,'
said I, 'I'll set my dogs after them.' 'Oh, they can outrun your
dogs. Not a mustang in California can kape up wid 'em.' 'Well,
at least I will not be bothered with bleating lambs.' 'Oh, yes,
you will. There's ewes scattered all through the band; but I
must go,' and away the fellow went with my ewes, exultant over
the change he had made, only stopping once and shouting back,
'They'll run, they'll *run* like h — l!'

I studied the disposition of the brutes, and found that their
greatest fault was a tendency towards total disintegration.
First a split of the whole flock, one half going on one side of a
hill, the other half choosing the other side, these again bifurcating,
until they finally deliquesced like an oak into branches uncountable. This fault was soon cured by constantly setting the dogs
upon every branch as soon as sprouted, and they soon learned
to stop and turn back when I whistled, and the whole fifteen
hundred became a unit in every way and as manageable as a
trained dog.

January 30.

High southeast wind and clouds. One storm-cloud blue like
tempered steel. Rain fell in great quantities during last night,
also three hours today. The whole face of the plains is brilliantly mirrored with pondlets and netted with songful silvery
rills. A cold bleak day. A good many of my sheep are dying of
cold.

January 31.

Nature's fields are fully green from mountain range to mountain range over all the plains. The late rains have made many shallow pools, and this evening they are noisy with a musical population. An immense crop of batrachians have come to life more suddenly than mushrooms. Where have all these frogs been during the long dry season?

My sheep are long-legged and long-tailed, and come in gallant style from the hillsides when pursued by my two dogs. Today I observed one of the sheep inquisitively smelling and examining a large hare. The hare allowed the sheep to touch him with his nose.

Without question this has been the most enjoyed of all the Januarys of my life.

February 1.

The sheep galloped along the rim of Twenty-Hill Hollow. Warm; the grass increasing rapidly.

February 3.

Large cumulus clouds. The sheep jumped excitedly from bank to bank over the creek. The excitement of each appears to be the united excitement of all. Heavy rain fell last night from the southeast. The river is higher than before in the season.

February 4.

Found a fragrant yellow violet [1] near Twenty-Hill Hollow, perennial, thick-rooted.

February 5.

Many a plant is tasting life now.

February 7.

A slight rain sprinkle. Flowers in rich plats are opening fresh and pure as ever did the best of uncursed Eden. Golden Composita, rays eight or ten, horizontal, purple outside, calyx-like

[1] Gold violet (*Viola douglasii*). — W. L. J.

involucre single, constantly one leaf to each ray. Disc flowers twenty to thirty, of a rich waxy yellow. Stamens and pistils beautifully united, white pistils united in a head or dome. Leaves pinnatifid. A most delightful plant.

February 8.

It rained six hours, also in the night. All streams are full again, and weak sheep dying. Mosses, liverworts, and cresses in full prime of health and stature; violets coming in deeper ranks. I observed a small plant on a rock with a pair of undivided cotyledon leaves. Next leaf higher, three-parted; the next five-parted, next seven-parted with great regularity. Parents sometimes lecture children for snipping paper into fanciful shapes. How busily the Creator is at work today upon ornamental flower tissue! Those terribly correct parents ought to consider their Creator and learn of Him, or, to be consistent, include Him in their fault-finding lectures forbidding waste of time on frivolous fancy. These grasses and flowers would make as good and as much mutton without such great pains of nicking and printing.

February 9.

Clouds are curtainy around the horizon. Rain half an hour. I found a rich purple bed of Alfilaria, and explored a rocky tributary of Niagara Creek. The smoothest, whitest of mist is laid evenly over the foothills, conforming to every hollow and curve like a downy mantle of infinite fineness and lightness. We speak of mist as touching the mountains or lying upon the plains, but no name, no word is fine enough for the fair expression of mist contact with everything.

February 10.

Clouds form one equal sheet. Rain twelve hours — a cold, stormy day. All streams are full, the plain in soak — one shallow lake. My sheep are too cold and wet to feed, they run from shelter to shelter behind the hills. There is a wonderful charm about so rainy and dark a day.

February 11.

A light dusting of snow fell last night. The snow-line of the mountains for the last month has been thirty or forty miles distant upon the upper foothills, but this morning it reached down below the last foothills to the very plains. About a hundred of my sheep died during the night. They lay in groups of three or four in the wind shelter of hollows and rocks. Behind my cabin, close against the wall, there were piled this morning five or six dead sheep, a few living sheep, some live hogs, and two or three dead hogs, all in one heap. Away out on the open plain I found a pair of little black pigs five or six months of age, exactly alike, lying dead in a small bed which they had dug in the sand just the length of their bodies. They had died without a struggle side by side in the same position. Poor unfriended creatures. Man has injured every animal he has touched.

February 12.

The snow-line has retired about ten miles up the mountains. The little early Composita is in small companies, bright and cheerful as ever — one of the most delightful flowers I ever met. Nature's three great colors — yellow, purple, and green — it has in perfection. Rays yellow above, purple beneath, involucre scales also purple tipped with white hairs. Stamen and pistil column white — all colors in one.

Discovered on a hillside, near Twenty-Hill Hollow, a shaggy bush of the Compositæ, namely, Baccharis. The smallest bush of any kind is a marvel upon these plains, and my very dog knew this to be a strange specimen. He stiffened his hairs and ran around it at a cautious distance, barking at it as if it had been a bear. I saw several bushy Compositæ, six or seven feet high, in the Coast Range at Pacheco Pass.

Found a large piece of silicified wood near Niagara Creek — the season rings and knots very distinct.

Warm and unwintry again. The blood of plants and birds, chilled in the sleety cold of past days, is cheerfully warm and pulses once more with summer heat. The snow-line has taken another long step backwards towards the great strongholds of the summit canyons.

I was singing bits of 'Highland Mary' among the hills of Twenty-Hill Hollow, often repeating the lines:

> 'There simmer first unfaulds her robes,
> And there they longest tarry,'

when I chanced to look over into the Hollow and discovered patches of the first golden summer robe of the plains. I left my sheep on the rim hills and went down into the Hollow to meet the lovely visitors in their robes of gold. They numbered about one million souls in five or six companies. I welcomed them to the world, congratulated them upon the goodness of their home, and blessed them for their beauty, leaving them a happy flock in keeping of the Great Shepherd, while I turned to the misshapen half-manufactured creatures of mine.

The Coast Mountains look grand in their new snow, and the mist-mountains over the San Joaquin plain appear like a central snow-clad range as rocky as the Sierra.

Perhaps I do not understand the request of Moses, 'Show me thy glory,' but if he were here I would like to take him to one of my Twenty-Hill Hollow observatories, and after allowing him time to drink the glories of flower, mountain, and sky I would ask him how they compared with those of the Valley of the Nile and of Mount Pisgah, and then I would inquire how he had the conscience to ask for more glory when such oceans and atmospheres were about him. King David was a better observer: 'The whole earth is full of thy glory.'

I think that if a revivalist, intoxicated with religion of too high a temperature for his weak nerves, were to awake from his exhaustion and find himself upon the rim of Twenty-Hill Hollow, he would, above such sheets of plant gold and beneath such a sky, fancy himself in heaven. Especially if a camp-meeting were going on at the time. He would say as he gazed upon the sheeted Compositæ filling the flat spaces, paving every straight or winding ravine: 'Here are the streets of gold, and there are Wesley's never-withering flowers. I hear the blessed multitude, I breathe the atmosphere of angels, I am beneath the cloudless sky of New Jerusalem.'

February 14.

The go-quicks are out of their holes. Birds, squirrels, hares, and flowers are full of springtime joy. Winter made but one invasion of these plains, when he came down from his mountains with sleet and icy rain. Perhaps some birds died, many sheep and hogs, but not a single plant. Flower suns are rising in every quarter of the valley sky and mingle their gold with the silvery gleaming crystal stars of quartz so abundant here.

February 17.

Clouds of a filmy dimness on the summit. Black ants are out and at work. Many still in dull heaps dressing and getting fully awakened.

February 18.

A most lovely day. 'The ant his labor hath begun.' The bee is on the wing. The spiders are busily engaged in building and weaving. Nature's song is daily enriched with the singing wings of flies, and the air is sweeter with the fragrance of added flowers. *Nemophila maculata* came today. Its stigmas are small and black like spiders' eyes, and the corolla very delicate. Only the fingers of God are sufficiently gentle and tender for the folding and unfolding of petaled bundles of flowers. A gorgeous sunset sky.

February 19.

Clouds none, or only a few scarcely discernible touches. Earth's green mantle is rapidly deepening. Three flowers today

— Dodecatheon, Ranunculus, and a crucifer. The construction of a very fine quality of spider web is going on rapidly at present. A species of spider making an elegant circular web upon the dead branches of Eriogonum has the power of invisibility by rapid swinging in the center of its web.

All the bird-music labor of the plains is at present laid upon the larks, but their hearts are in the work, and it is easy for them. The hills about Castle Creek, Lily Hollow, and Twenty-Hill Hollow are composed of porous, stratified lavas and quartz and slaty sands mixed and coarsely conglomerated and easily acted upon by rain. The high winds of winter, armed with rain globes, cut rapidly into the soil which is not protected by binding grasses. These hills are depressed about from one-eighth to one-quarter inch per season, as shown by the naked tops of perennial roots which have to strike deeper and deeper every year. It is interesting to observe how the Lord removes mountains without destroying a single plant inhabitant. Marine floating plants rise and sink upon the waves of the sea; so also sink and rise these land flowers upon the hill waves of the plain.

February 20.

Mist in the morning. I was lost in it with my sheep. Saw a small beautiful gray bird among the rocks of Niagara Creek.

February 22.

Squirrels, ants, and lizards are engaged in mending their dwellings. In walking the plains now one sees a sort of doubtful motion at the mouths of holes. These uncertain twinklings proceed from crickets as they suddenly retreat to the bottom of home; also from squirrels and skim-fields — go-quicks, as the small swift lizards are sometimes called.

February 24.

Saw a wonderful effect of the setting sun in a sheep's eyes. The eyes appeared like green blazing gems. . . .

February 25.

Four flowers were born today — the coiled fragrant Myosotis, a yellow Composita in Twenty-Hill Hollow, violet-like Scroph,[1] and a very minute rock flower.

February 26.

Clouds absolutely none in the forenoon. The ground squirrels, having found easy tunneling in a soft stratum of one of the hills of Twenty-Hill Hollow, have bored it round and round. It is curious to observe them standing bolt upright in a row watching the movements of my dog. They shout excitedly, 'Seekit seek, seek seekit,' of which I don't know the meaning, or whether meant for one another or for the dog.

February 27.

Very hot. Sunset glorious in purple clouds melted into films. This forenoon, while sitting upon the brow of Twenty-Hill Hollow, I was startled by the swish of wings and by the sudden outcry of my dog Fanny. An eagle had swooped upon her, though she was close to my heels, I suppose mistaking her for a hare.

March 1.

Eight plants are now in bloom in Twenty-Hill Hollow, colored in sections like a map.

March 2

Three or four new flowers. Serrated ridge of cloud upon the Coast Mountains at sunset. Air transparent and delicious.

March 3.

Saw a horned toad today. Butterflies and polished beetles are coming out daily.

March 4.

Flower ranks deepening. Poppies and sweet Portulacæ have come, and multitudes of insects.

[1] Muir has abbreviated Scrophularia to Scroph, which is a sort of botanical slang used still. — W. L. J.

March 6.

Ants labor in earnest. All the plain is in full gush and blaze of glory.

March 7.

Myriads of insects are trying their new-made wings. Flowers in full, glad, rejoicing holiday attire.

March 10.

A cumulus bank over the Coast Range. Three flowers today — Leptosiphon, Gilia, Polytrichum. Hot. Caught a young hare; it had from instinct all the gait, tricks, and deceptions of old ones.

March 13.

Canary music today.

March 14.

Killed a rattlesnake that was tranquilly sunning himself in coiled ease about a bunch of grass. After dislodging him by throwing dirt, I killed him by jumping upon him, because no stones or sticks were near. He defended himself bravely, and I ought to have been bitten. He was innocent and deserved life.

March 15.

The first grass panicle is open.

March 16.

It rained two hours. Found a little cress with beautifully embroidered silicle.[1]

March 17.

Rain ten hours. Flowers are rising into life in glorious array. Observed a company of small waders, very rapid on wing, about the size of swallows and as swift. Ash-colored belly, white and shining like steel, graceful, and they have many gestures of the killdeer.

[1] One of the species of fringe-pod (*Thysanocarpus*). — W. L. J.

March 20.

After an experience of a hundred days, I cannot find the poetry of a shepherd's life apart from Nature. If ancient shepherds were so intelligent and lute-voiced, why are modern ones in the Lord's grandest gardens usually so muddy and degraded? California shepherds become sheepish. They will not commit great crimes, but only for lack of courage, not because of the possession of greater virtue. The whole business, with all of its tendencies, exerts a positively degrading influence. Milton in his darkness bewailed the absence of 'flocks and herds,' but I am sure that if all the flocks and herds, together with all the other mongrel victims of civilization, were hidden from me, I should rejoice beyond the possibility of any note of wail.

March 21.

A rainstorm from the northwest came over our golden hills swiftly and grandly as those of sultry summer in the Mississippi Valley. It lasted about one minute, but was the most sudden downpour from the clouds, the most magnificent cataract of the sky mountains I have seen. A portion of the sky to the east was brushed smooth with thin white clouds, and the rain torrents showed clearly against them to a great height. This cloud waterfall, like those of Yosemite rocks, was neither spray, rain, nor solid water. How glorious a baptism did our flowers receive, and how sweet their breath! . . .

The fields are lovely. The warm rains of the last week have doubled the depth of plant life and opened whole lakes and seas of color.

There is a small Plantago, a silky gray, half-silvery plant which has done more during the last few days for the general beauty of the plains than all others combined. It is now long enough to wave and show ripples of shade. It gives its peculiar gray plushy color to all of the lower shallows of gravelly hillsides. A happy, modest plant.

March 22.

Bright, blowy, and cool. One of the Eschscholtzias has stamens and pistils nearly alike, and both resemble leaves. There is a circle of pods arranged like a vase. The filaments show, not as answering to the stalk of the leaf, but as the leaf itself. The apparent separation of the anther is only a line drawn to show that something else is to be written. The organs of plants form a circle, and, considering the cotyledon as the primary, least developed organ, the leaves exactly opposite, or halfway to the pistil, would be the most perfect organ. In plants of distinctness of variation this point of greatest variation is always about midway betwixt root and flower.

The simpler plants may be considered as a series of concentric cylinders. The leaves form one or many cylinders, of which only a small portion of the upper end is free. The whorls of the floral organs are fringed ends of other cylinders. It is only where many individual plant existences mass into groups of greater or lesser complexity upon one root that the simplicity of this arrangement of plant bodies is hidden.

March 23.

Vegetation is deepening with inconceivable rapidity.

March 27.

The wavy hills are mantled in abundant, divine, gushing, living plant gold, forming the most glowing landscape the eye of man can behold. The light-colored radiate yellow Composita is upon warm banks. The frothy white plants of ravines and dimple cups show like patches of unmelted snow, and purple gilias and lilies and heliotropes blend with the grand hill waves of heavenly gold.

March 31.

Sun-gold to sun-gold with not a cloud to separate.

April 5.

Four cloudless April days filled in every pore and chink with unsoftened undiluted sunshine.

April 9.

Saw two splendid bird waders in a shallow. They were more than half crimson.

April 15.

The foothill deciduous *Quercus douglasii* oaks are in full leaf. Their leaf fall is about December.

April 20.

Clouds. Scattering of hail, cool.

April 24.

Warm. The plains are dim and brown already. The light of the whole sky of plant suns has faded. The glory has departed, but another glory has come, scarcely less obvious to those who look and wait for it.

April 25.

Warm. The plains are colored a dim purple and gray and yellow, with faint fading patches of green. The purple is from the stems of Compositæ and cresses, and the bare soil. The gray is from a species of Gnaphalium and withered corollas of other Compositæ and the second scale-corollas of Compositæ. The yellow is from grass stems and a few lingering flowers.

May 2.

The plants of the plains are rapidly finishing their course. Many have sown their seeds, but, still undecayed, assist the landscape with color of stem or persistent calyx and involucre. Some have dry chaffy stars of five rays like a flower composed of the withered involucres. Others have the seeds with calyx and corolla attached, spread in silvery shining heads, thus giving the charm

of a second flowering. The death of flowers in this garden is only
a change from one form of beauty to another.

May 6.

Clouds. Thunder in the mountains.

May 9.

Hot. At a ranch just out of sight of the plains in the first oak-
planted foothills a great number of handsome birds have made
homes in a piazza in hanging nests. I never saw birds so elegant
and beautiful choose the confusion of a human home — also the
safety and scenery of an old barn — before the pure purple and
green of the forest or plain. I observed one that had a nest beneath a dirty mop which accidentally had been left against a
fence. Nevertheless, with all this depravity, they, the *Carpodacus*, or house finch, are sweet singers and are well dressed.

At Snelling's Ranch.
Undated.

The sun is a few degrees above the ground and is shining dimly
through fleecy clouds that drift about the Sierra. All the rest of
the sky-dome is lighted, and a little of the morning purple is
still left unburied by the stronger light of the growing day. This
place is neither mountain nor plain. It belongs to the band of
hills that skirt the higher mountains. The Coast Range is not
visible, neither are the high peaks so constantly seen from the
level plains.

The landscape is open, with flowing undulating hills, colored in
autumn yellow and purple, dotted freely with the low branchy
foothill white oaks and lichen-colored groups of slate slabs erect
or leaning like decaying tombstones. Nearly every flower has
received its latest portion of life blessings, but continues to make
the world beautiful. A few lilies, Lilium, Brodiæa, Calochortus, etc., that have a store of food in deep-buried bulbs, continue
to bloom with a few humble scrophs and clovers along water-
courses.

May 16.

Warm. The plains are finely dappled in yellow from grass stems in hollows. The only plants in full bloom are a species of deep blue liliaceous flower, the Brodiæa, and a purple larkspur.

May 27.

Hazy and hot at Hopetown on the Merced. The mountain summits are scarce visible. The last rains have hastened the decomposition of the parched grasses of the plains, which have now permanently assumed the ordinary purple and yellow of the winter of dry heat. Living plants are at the lakes and streams, and up in the cool, shadowy, watered mountains.

May 28.

Hazy and hot. All of the larger animals seek for cool shadows. Insect life is in full gush of vigor.

May 30.

Hot. Dreamy, hazy, glowing days.

May 31.

Oveny and oppressive.

Chapter II : 1869-1872

'Patient observation and constant brooding above the
rocks, lying upon them for years as the ice did, is the
way to arrive at the truths which are graven so lavishly upon them.'

EARLY in June, 1869, the tall auburn-haired young shepherd
John Muir took general charge of the sheep of another Irishman,
Pat Delaney, and went with them in quest of high green pastures. Assisting him were two dogs and the sub-shepherd Billy,
so he had leisure to explore much of the Divide between the
Tuolumne and Merced Basins, climb Mount Hoffman and Mount
Dana, and penetrate Bloody Canyon to Mono Lake, which lay on
the ashen plain 'like a burnished disk.' The journal of this first
summer in the Sierra, so well known to Muir readers, reveals him
already arriving by observation and intuition at his conviction
that a vast ice mantle had once covered all this mountain region,
and in its slow descent, following rock cleavages and faults in
mountain structure, had ground and sculptured the landscape
into the forms we know today. In September, he drove his
woolly charges down to their home pastures, grateful for what
he had seen, and praying he might return.

His following several journals, published here in part, begin
with the narrative of that return in November to take up his
abode in the Yosemite 'as a convenient and grand vestibule of the
Sierra.' His companion on this adventure was Harry Randall, of
Rhode Island. Muir, the elder, took the lead in their 'partner-
ship,' and 'sold' their labor to J. M. Hutchings, the hotel-keeper.
Harry's job was to do chores around the hotel, to milk Buttercup,
the cow, and drive the oxen, Paddy and Duke. Muir's contract

called for the construction and running of a sawmill for fallen timber. Soon the two friends built for themselves a snug cabin, and boarded at Hutchings's cabin, where they devoured quantities of Mrs. Sproat's 'memorable muffins.' Several long, chatty letters, exchanged thirty-five years later, give pictures of their life to-gether — of the evenings by the firelight, when the elder partner tried to teach the younger arithmetic, or sat 'leaned back' in his 'home-made chair covered with old sheepskins, reading Humboldt.' The letters also hint of many 'a good Sabbath day's journey' when the two 'hired men' explored the nearer canyons and divides, arriving home Monday morning in time to do the chores and start the saws.

From the early days of his Yosemite sojourn, John Muir had two favorite 'observatories' — Sentinel Dome, from which high point he studied the whole Tuolumne–Merced system, and Sunny-side Bench, a ledge on the north wall just east of Yosemite Falls. From this latter lofty camping spot he surveyed the main valley and its tributaries 'spread out like a map' before him, and here he found much to confirm and enlarge his theories of glacial action and of the influence of shadows upon landscape formation. Many of his most important journal notes were written during days and nights at Sunnyside.

Such progress did he make during these first months in scientific investigation that on August 8, 1870, Professor Joseph Le Conte, starting with Muir on a ten-day mountain ramble, wrote in his diary: 'Muir believes that the Valley has been wholly formed by causes still in operation in the Sierra — that the Merced Glacier and the Merced River and its branches . . . have done the whole work.'

John Muir, wearing blue jeans or 'tough old clothes, gray like the rocks,' and willing to subsist on raw oatmeal or dry bread and tea, could get along on three dollars a month. So when he had saved up enough from his wages to last him 'for years to come,' he left the employ of J. M. Hutchings in September, 1871, and began his 'glorious toil,' with 'unmeasured time and independent of companions and scientific association.'

The same month, he explored for the first time the Great
Tuolumne Canyon known as Hetch Hetchy. The following
month he discovered at the head of the Illilouette Basin his first
'living glacier.' From there he went farther afield to find the
glaciers of Lyell, Maclure, and Hoffman, equally 'alive.' Hot
on the trail now of the whole grand story, he spent the next two
years tirelessly working among the remote and almost inaccessible
canyons of the Tuolumne and Merced system.

- - - - - - - - - - - - -

1. *My First Winter, Spring, and Summer in the
Yosemite Valley*

Yosemite. *December*, 1869.

I left the foothills at French Bar November 16, 1869, with
H[arry] Randall, for Yosemite in particular, and the Sierra in
general. We advanced slowly by unforced marches, making
every principal mountain bench which we encountered the labor
of a day — generally eating breakfast at the foot, dinner halfway up, and supper at the summit. We were armed with a very
formidable stabbing instrument, a kind of halfway thing betwixt
a sword and a butcher knife. At Coulterville we further armed
and burdened ourselves with a shotgun and projectiles suited to
the killing of all kinds of mountain life. The scenery was enchanting, and we allowed our bodies ample time to sponge it up.
We found no snow, and in many places, upon sunny stream-
banks, some late flowers were lingering like flies after frost. The
magnificent conifer and King of Pines *Pinus lambertiana* warmed
us with its beauty all the way from Coulterville. The lovely fir
Abies magnifica towered in exact and measured grandeur in its
unchangeable green. Maples and the black oaks were casting
their brown and yellow leaves, and in their crumpling and rustling beneath our feet touched many a memory of the sweet Indian-
summer days of the old West.

As we ascended the mountains, the sky became thinner and

bluer, and though filled with free sunshine, lost gradually the richness and balmy purple cream of the smooth plains. In one week we reached our goal, the great mountain trough, Yosemite Valley. We thought of a visit to Hetch Hetchy, and to the grand valley upon the South Fork of Kings River, also to the settlements of Sequoia in Fresno, but after a week of resting, sketching, and general climbing we sold ourselves to J. M. H[utchings] to feed sows and turkeys, build henroosts, laying-boxes, etc. Also to take charge of the ladies and to build a sawmill.

I had long lived in bright flowery summer, and I wished to see the snow and ice, the divine jewelry of winter once more, and to hear the storm-winds among the trees and rocks, and behold the thin azure of the mountains, and their clouds. I expected to find all these pleasures in all their moods and unions as soon as I should reach the altitude of the firs. But week after week passed with sunshine and blue sky. Winds spoke not, or so feebly that they could not be understood but by ears better than ours. A few droppings of rain would come from changing clouds, but generally the sky of our rockery was bright by night and day. Frost at from sixteen to thirty degrees at sunrise crusted calm shallows stiffly, and hung slim crystals upon the grass instead of dew. But the long-looked-for pearly snow came at last.

December 16.

South Dome [1] is the barometer of our valley, perhaps the grandest in the world. The scale is divided into an unreadable number of degrees, but only four or five points are understood by us, and the scale is nearly a mile in length. The weight of the atmosphere is indicated by the rising and falling of a cloud upon its face. On the morning of December 15, the barometer touched the snow point. Long fleecy rolls of mist came softly down the canyons, or floated about in masses in the open valley, like icebergs in the sea. 'I do not like these clouds hanging about the rocks so far

[1] South Dome is now commonly known as Half Dome. Tissiack, the Indian name of the great rock that heads the valley, is also frequently used by John Muir.

down,' said our weather prophetess, Mrs. Sproat, as she gazed wistfully and fearfully from the doorsill of our log cabin, reading the signs of coming storm. I was not accustomed to the Tissiack barometer, but hoped and prayed that the storm would come, corresponding in all its forms with our glorious mountain mansion, and it did so come with winter beauty unspeakable.

Just as the dark of evening fell, the clouds gave rain and sleet which soon changed to snow, coming in large matted flakes. All through the long dark night it fell. Every peak and dome, every bench and tablet had their share of winter jewelry, as we our daily bread; and the nodding tassels of the pine, and the plumes of the Libocedrus, and the naked oaks shared in the first precious gift of the winter sky, and blessed are the eyes that saw morning open upon the glory accomplished in that first storm night. In vain as I crossed the open meadow did I search for some special palpable piece of beauty on which to rest my gaze, and on which, too, my mind might stop to gather ideas as from a text. But no such resting-place appeared in the completed heaven of winter perfection. Tissiack stood like a god, a real living creature of power and glory, awful, incomprehensible, yet not as clad in vasty, inseparable masses of dim sublimity, but in ten thousand things of equal, separate, outspoken loveliness — feathery pines so shaded, so dazzled by snow, cavelet wreathings and archings of the same, masses of the Ceanothus and manzanita bushes glorious with a brighter than their own summer bloom. Never have I beheld so great and so gentle and so divine a piece of ornamental work as this grand gray dome in its first winter mantle woven and jeweled in a night.

I rode to the very end of the valley gazing from side to side, thrilled almost to pain with the glorious feast of snowy diamond loveliness. The walls of our temple were decorated with exact reference to each other. The eye mounted step by step upon the dazzling tablets and flowering shrubs which ruled the mountain walls like a sheet of paper, and published the measure of their sublime height. The sky was mostly clear, but here and there a fleece of muffling cloud lingered about the canyons and larger tablets, or floated up the perpendicular steep like a balloon, as if taking a last look at the night's work accomplished, to see if it could spy a place where a flake or two more was needed. A filmy veiling, wondrous fine in texture, hid the massive front of Capitan. Only in little spots with melted edges did the solemn gray of the rock appear. 'El Capitan smokes mucha este dia,' said a Spaniard who rode with me.

From end to end of the temple, from the shrubs and half-buried ferns of the floor to the topmost ranks of jeweled pine spires, it is all one finished unit of divine beauty, weighed in the celestial balances and found perfect.

Christmas brought us a cordial, gentle, soothing snowstorm — a thing of plain, palpable, innocent beauty that the frailest child would love. The myriad diamonds of the sky came gracefully in great congregational flakes, not falling or floating, but just coming to their appointed places upon rock or leaf in a loving, living way of their own — snow-gems, flowers of the mountain clouds in whose folds and fields all rivers take their rise. The floral stars of the fields above are planted upon the fields below. The pines, the naked oaks, the bushes, the mosses too and crumpled ferns are all in equal bloom, and belong to the same one great icy order....

Our little meadow, lying like a patch of brown cloth, suddenly stretches away into a vast and boundless plain, for the mighty ramparts of the valley are hidden, and we feel as if in a wide plain. Occasionally a thin place in the storm recalls our rocks in dim, uncertain patches, wonderfully smoothed and softened.

But now the last sky blossom has fallen, the clouds depart in separate companies, leaving the valley open to other influences and communions. Every tree seems to be possessed with a new kind of life — in sounds and gestures they are new creatures, born again. The whole valley, sparkling in the late sunlight, looks like a trim, polished, perfect existence. The dome Tissiack looks down the valley like the most living being of all the rocks and mountains; one would fancy that there were brains in that lofty brow. How grandly comes the gloaming over this pearly beauty! What praise songs pour from the white chambers of the falls! Surely the Lord loves this new creation, and His angels are now looking down at this new thing that His hands have wrought.

During the night preceding this blessed day a great quantity of rain fell — rain to the valley, snow to the mountains. Yosemite Fall was living again with the full throbbing glory of prime — louder sprayed the divine shafts of song, deeper grew the intense pressed whiteness of the clustered meteors. The wind went fearlessly into the black-walled chambers of the waters — now softly bearing away only the light outer sprays, now seizing the whole column, swaying it back and forth as if seeking to direct its efforts to some particular rocks, or taking up the whole ponderous mass and bending it upward, turning the whole stream up the mountain and back to its sources in a magnificent bow. How wonderful that so irresistible a thing as a descending torrent can be checked and turned backward like a film of mist, by so ordinary an agent as the air we breathe.

January 1, 1870.

The Yosemite sky is calm and blue without a cloud, and most of our Christmas snow has been taken back. The thin sunshine is dissolved into a grand display of waving irised colors in the spray of Yosemite Falls. The whole stream for a considerable length seems to be transformed from snow white to the substance of a rainbow.

I celebrated this day by thinking of my friends and climbing into heaven's blue on the crown of El Capitan by a new route. The valley with its rocks, streams, and shadows lay sublimely at my feet. To the right waved the timbered billows of the Sierra, and beyond these I could see the yellow, hazy San Joaquin plains, and still farther on the rim of the sky the blue mountains of the coast. To the east the spiry peaks of the Sierra crown were bathed with equal light, each mantled reposingly in its first fresh snow.

... I enjoyed gazing over the verge without becoming giddy and started for the nether world by a still newer route than that by which I ascended. Got on bravely until halfway down, when I found myself suddenly halted by a sheer descent of five or six hundred feet. Had to march back to the very summit and find my old route. I had two hours scrambling in the dark, but by running, jumping, sliding, besides other modes of locomotion terrestrial and aquatic too numerous to mention, I accomplished my craggy task in half time and escaped a long fast and a cold night on the mountain.

January 4.

The second, third, and fourth belonged to a genus of days slightly cloudy and calm — rest days after all heavy storm work is done, when the very torrents are soothed to gentleness, and Sabbath rest and the confidence of settled tranquillity beam from every feature of rock and sky.

January 5.

Gray and cold, with a red sky at sunset gilding the upper spires and domes in rare beauty.

January 6.

Calm and warm. The valley contains many mansions for the winds and clouds. During the glowing hours of this resting day, inert or slow-moving bodies of air slept in rock chambers and in the shelter of steep walls like lakes and potholes of air. Each atmospheric lake and rivulet had a temperature of its own, but at the coming of night they floated about in companies, and in walking half a mile my face was bathed with air of every temperature and quality — hot, cold, and balmy in startling suddenness of succession. I seemed to walk through narrow zones of summer, winter, and spring.

January 14.

The seventh, eighth, and ninth formed a golden trio of the sweetest winter children of the sun. Tenth, eleventh, twelfth,

thirteenth, and fourteenth — one storm has possessed all the
hours of these five days and nights. Faint lines of division are,
indeed, seen at times. A gleam of sunshine shoots occasionally
from the clefts of cloud and granite, but everything in our magnificent mountain temple during these blessed days has formed
one grand storm. Snow and rain, rocks and clouds are mingled
together; rain driven by high winds seems to form one continuous
waterfall having a lateral motion equal in rapidity and grandeur
to the perpendicular. Finely moulded cloud-patches float about
among the pines and rock seats or span from canyon to canyon
in arches scarcely less splendid than those of the rainbow. At
times these clouds dilute and soften to thinnest textured mist,
shutting out the last dull outlines of the valley ramparts —
making our patch of Carex meadow spread away to boundless
level plains. At other moods and stages of progress the whole
central gorge lies open, but the walls on either side are veiled
in a cloud fabric which for clearness, together with deftness of
outline and divine delicacy of texture, is unlike anything I ever
beheld elsewhere. When robed thus the rocks seem to rise to
their real stature, and they do look lonesomely down from the
clouds. Patches of pillared pines show themselves at times
through the thin cloud garments half pulled apart.... Again the
whole of our temple throughout all of its cells and halls is filled
smoothly full of uniform mist-cloud, but this cloud mass at the
same time reaches only to within fifty or sixty feet of the bottom,
and on looking upward one feels as if at the bottom of a sea.

In the blending of cloud and rock, to say that the cloud *rested*
on the rock or touched the rock conveys no idea of the fineness
and gentleness of their meetings and unions....

The light of the full moon makes most startling, uncommon-
place effects among these mists.

January 16.

The fifteenth and sixteenth were bright and clear, sunshine two
days in breadth....

January 18.

Today the falls were in terrible power. I gazed upon the mighty torrent of snowy, cometized water, whether in or out of the body I can hardly tell — such overwhelming displays of power and beauty almost bring the life out of our feeble tabernacle. I shouted until I was exhausted and sore with excitement. Down came the infuriate waters chafed among combative buttresses of unflinching granite until they roared like ten thousand furies, screaming, hissing, surging like the maddened onset of all the wild spirits of the mountain sky — a perfect hell of conflicting demons.

But I speak after the manner of men, for there was no look nor syllable of fury among all the songs and gestures of these living waters. No thought of war, no complaining discord, not the faintest breath of confusion. One stupendous unit of light and song, perfect and harmonious as any in heaven.

This gathering of mountain waters, on reaching this Yosemite portion of their lives, is carefully prepared for this rock display and rock music. Before reaching the brow of their falls, they are deflected from side to side upon granite instruments of proper angle, whirled and gurgled in eddies and potholes, carefully mixed with measured portions of air, calmed in pool basins, and finally moved over the brink with songs that go farther into the substance of our being than ever was touched by man-made harmonies — songs that bear pure heaven in every note. The fleecy, spiritualized waters take the form of mashed and woven comets, going with a grace that casts poor mortals into an agony of joy.

If my soul could get away from this so-called prison, be granted all the list of attributes generally bestowed on spirits, my first ramble on spirit wings would not be among the volcanoes of the moon. Nor should I follow the sunbeams to their sources in the sun. I should hover about the beauty of our own good star. I should not go moping among the tombs, nor around the artificial desolation of men. I should study Nature's laws in all their crossings and unions; I should follow magnetic streams to their source, and follow the shores of our magnetic oceans. I should go among the rays of the aurora, and follow them to their beginnings, and study their dealings and communions with other powers and expressions of matter. And I should go to the very center of our globe and read the whole splendid page from the beginning.

But my first journeys would be into the inner substance of flowers, and among the folds and mazes of Yosemite's falls. How grand to move about in the very tissue of falling columns, and in the very birthplace of their heavenly harmonies, looking outward as from windows of ever-varying transparency and staining!

Alas, how little of the world is subject to human senses!

January 23.

The clump of days bounded by the sixteenth and twenty-third was pictured with the sublimest forms of mountain tempest. The mountains received snow during all the nights and days between these two dates. The six days may be compared with a plant whose root is in the sky. The storm-plant sprouted on the night of the sixteenth, reached full stature and flowered on the eighteenth — such petals of cloud and crystals, such columns of snowy pistil falls! How plant-like did the organs of this storm crumple up and decay and return to the common fund of the weather resources to be moulded and worked up into other fabrics.

January 24–26.

Warm and bright, the valley was spanned by fibrous bows of white cloud, heated masses of air from currentless ovens of chambered and bushy rocks lifted by newborn winds and borne whole or in fragments about the open gulf of the valley.... The richness of the light of these days recalls our best mellow autumns and springs.

January 27.

Cloudy, hushed, and pensive.

January 28.

Fine-grained soft clouds.

January 29.

Winged ants have arrived and are happy in new existence. . . .

January 30–31.

Warm, bright, creamy light is laid on Tissiack. . . . Finely mingled iris matter of falls. . . .

February 1 and 2.

A deep blue atmospheric sea without a tinging shadow or island or barrier of continental cloud — coastless, waveless, tideless. . . .

February 9.

Sky equally obscured — half-lighted. The alder aments ripe. Mistletoe spheres becoming more yellow, with a beautiful effect upon the boiling, swelling green of the live-oaks.

February 10.

A storm commenced today — the largest and most complicated of storms. It was made up of all kinds of cloud, all kinds of light and water in all forms — snow, hail, mist, and coarse rain coming slowly in separate globes or in long, slant rods of water. Night darkness would come on suddenly at times only to be instantly removed by thick glowing light poured from a jagged flaw in the black cloud mantle. The snow-line would descend suddenly to the meadow, then rise in a few minutes to the top of the walls. Not only were these storm elements broadly mixed and blended, but the rocks and clouds seemed also to mix. Isolated peaks would loom to sublime heights, and appear to be drawn forward over the very middle of the valley.

February 12.

A night of misty, foggy moonlight.

February 13.

Constant floods — thousands of torrents and slipping stream-lets — come curving and leaping over the rocks, fine and white

like warped and crinkled strips of silver. The falls in glorious
dress and voice.

February 16.

The whole valley in full pomp of snow jewelry.

February 17.

A multitude of patient frogs sing lustily in the shallow ponds on
the meadow, though up to the throat in blue sludge.

February 18–21.

Devoted to clouds and their treasures, rain and snow.

February 22.

The day opened in dazzling splendor upon the variously put-on
garments of snow and ice that enveloped and mantled all the
earthly forms of our valley....

March 1.

Cumulus banks like additions to mountains. Butterflies like
flowers drifting in the wind.

March 2.

South Dome grandly capped with cloud. Warm.

March 3.

Misty masses of white upon the rocks. Sunny.

March 10.

A full hearty day of calm snowstorm.

March 12.

Snow all day. Separated from storm of tenth by a block of
wintry sunshine on the eleventh.

March 14.

Bright and balmy. Tissiack is ruffed by a huge cloud upon
which her split dome seems to float gay and warm with yellow
light.

March 15.

Throbbing with all that's bright and sweet in mountain transition weather from winter to bland spring.

March 17.

Dim all day with floating cloud islands. Snow in evening.

March 18.

The snow of last evening was soft and abundant. Trees were copiously supplied with this winter fruit. Never before beheld I such grandeur of snow-clad forest and rock. Pines with branches closely drooped appear like barbed arrows aimed at the sky, each tree far and near very distinct. Blue of sky after sunrise towards the west, divinely beautiful.

March 21.

Calm and dim. God's glory is over all His works, written upon every field and sky, but here it is in larger letters — magnificent capitals. These winter days are like the broken heaps of granite rocks at the foot of the falls on a bright day. Light and shade in clear masses lie among the larger tablets and blocks; smaller wedges and gravel are like the mixed days of checkered clouds and light....

March 25.

Birds, bats, and butterflies in fervent motion. Cloud on Tissiack. Snow-slides from Dome like falls. Waterfalls striking louder notes. Spring!...

March 27.

Warm. Purple gooseberry in flower with two other species. Falls in high dress and music, glistening with beads extending down curve of brow.... Cloudy light on Dome, light streaming from seams in the clouds.

March 28.

Winds are speaking the language of flowers. They are coming. Four or five have arrived.

March 29.

Flowers and flower winds and spring balm. Falls in spring costume — web-like mists bent and bordered with meteors.

March 31.

Cumuli about Dome and Starr King.

April 1.

Falls gloriously outspread and bent by winds. Rain at night.

April 4.

Snow fell all day — ten inches — in cold, copious winter earnest. Now at evening there stands on rock and meadow, tree and weed, as ripe and full and perfect a harvest of snow-flowers in this April as I ever beheld in the stormiest and most opaque days of midwinter. The streams are full of sludge, dim blue like cirrus clouds.

April 5.

Snow mostly removed from the valley floor, but untouched above at a low, well-defined height on the walls.

April 6.

The bright meeting of sky-flowers and earth-flowers ended. Snow-flowers, withered, have returned to the sky, as shortly will the violets and lilies return to the earth. I discovered the handsome long-spurred blue violet, cheerfully glinting forth upon a hillock in the meadow from decayed snow and weeds.

April 7.

Warm, bright, springy. Cushions of white cloud cumulating upon slope of Starr King.

April 9.

Flowers are born every hour.

April 10.

Spring, genial, gushing, balmy, in all the sky and in every sound and song. Falls in glory that excelleth. Caverns and arched openings in the falling column. At the base of the upper falls I saw and heard things of unwritable grandeur. Plant life is rapidly maturing.

April 11.

A magnificent craggy cloud on the Dome.

April 12.

Snow this morning is drifting in frozen dust in the high winds. Ten inches at noon. All summer distinctions of flowering and flowerless plants are leveled. All are equally polypetalous. Complete, unmitigated winter has returned, although not less than twenty plants are in flower. Brier rose is in leaf, also wild cherry. Willows are in flower, and two violets... gooseberry, three crucifers, etc.

In the afternoon, the upper falls pour out of a cloud at the head into a cloud at the bottom. A splendid array of dark and white cumulo-cirrus clouds crown the rocks from Three Brothers to Indian Canyon.

April 13.

Spring again. Frogs in a deep gurgling chorus.

April 14.

All spring influences arrive in one homogeneous mass. Early flowers uninjured by the snow burial.

April 15.

Eighty degrees in shade at noon. Water increasing.... Full moon, looking over the shoulder of South Dome, had three pine trees and a rock outlined upon its disk with startling distinctness. No other tree, however faint, was at all visible. It looked like a medal. The trees seemed to be melted into the white silver of the disk....

April 18.

Water much increased. A strong wind at the bottom of the falls is waving the pines to their own music in divine harmony. I could not approach within two hundred yards of the lower falls.

April 19.

Poplar leaves open this forenoon. Falls in glorious array — comets more matted and rapid in movements, not a quiet drop in the whole column.

April 20.

Falls in growing magnificence. Fifty thousand horsepower at present. Oak buds are expanding. The fragrance of warm joyous spring is in every sight and sound conjoined. The river much swollen.

April 21.

Water higher than ever. Poplar grove is in green leaf, long enough to tremble. The column of falls is not greater in appearance, but has increased ten times in volume, no longer floating down softly and downily, but striking at once for the bottom with determined energy in close concentrated columns. The Lower Fall, uttering deepest bass, drives out a heavy wind-blast which waves the trees that stand in waiting, making them keep time to its song.

April 22.

Spring is working at high pressure among plants and insects. Masses of cumuli tower about the Dome and Starr King.

April 23.

Windy at night. Temperature at sunrise forty-four degrees, at noon seventy-six degrees. Trees are opening buds. Poplars in a good covering of green. Big ferns uncoiling....

April 24.

Warm, hazy light. Nature palpitating with prime vigor. Oaks unfolding leaves. Waterfalls in matured majesty....

April 28.

A light shower sending forth the fragrance of opening buds.

April 29.

Light warm showers fall upon the beauty of young leaves and the deep green of the meadow. A splendid display of cumulo-cirrus cloud-heaps. South Dome is muffled from bottom to top of its perpendicular face, clouds ever varying to new forms. A cable of white cloud like the falls in substance stretches in a grand curve from the top of Glacier Point to the shoulder of the Dome.

May 1.

A cool, bright day of healthful, hopeful spring. Oaks are unfolding leaves and maturing their flowers with astonishing rapidity.

May 2.

Black, dense clouds are laid equally down upon all the valley walls, making darkness like night.

May 4.

Sweet spring. Oaks are in full flower and their leaves are just so far open as to resemble petals. . . .

May 5.

Very warm. Spring advances her work among flowers and buds with rapid strides. The oaks are gloriously tasseled and the young leaves are colored in many shades of purple and yellow. Much that is tranquil and joyous blends into the vast sublimity of rocks and waters. In what is termed awful, stupendous, terrific, furious, raging, savage, ungovernable, etc., there is order and tranquillity as among the smoothest lowland landscapes. . . .

May 6.

Falls in yet deeper power and glory.

May 7.

Flowers come in troops. ... Waterfalls louder than ever. The meadow is like a Florida swamp. Magical reflections of the valley walls shine in the meadow pools after sunset.

May 8–10.

Eighty degrees after sunrise. Water high. Falls booming. Visitors coming to see....

May 11.

Hot. Beetle, fly, and flower life in fervid, throbbing activity. The snow on the mountains is melting fast, and the river is over its banks.

May 12.

Magnificent cumuli rise above the upper rocks. Oak leaves of full size droop and wave gracefully.... Flowers troop on in uncountable ranks....

May 15.

Snow! Cold, copious driven snow is hurled at times through the gaps of the valley with the velocity and noise of the loudest storms of winter, but is mostly falling slowly in big flakes that daintily take their places on the trees and brilliant companies of shrinking flowers. I never before witnessed so square and flat a meeting of winter and summer flowers. It is not a last hesitating flurry upon sprouting grass and early violets, but a steady, heavy, majestic, high-mannered snowstorm falling into summer verdure of the richest luxuriance of fern-fields miles in extent and shoulder high, banks of Ceanothus in full flower and leaf, wild cherry also and a hundred other flowers, and meadows of sedges and grasses gone to seed.

May 16–18.

Snow; cold and cloudy with occasional showers. Many ferns and tender leaves frost-killed....

May 19–20.

A magnificent display of white clouds in the valley and over it. Small showers all day. Snow is falling upon the mountains, the snow-line being about two thousand feet above the valley....

May 21.

A grand cloud-cap settles on the Dome, encircling it like a collar, the bald gray top clear in the bright light. The falls spread in indescribable magnificence, allowing the light to penetrate to the deepest recesses of the ever-changing long-tailed meteors. The pulse of plant life is beating high as ever; flowers already seem to have forgotten the snow and frost — perhaps are all the happier for it.

These beautiful days must enrich all my life. They do not exist as mere pictures — maps hung upon the walls of memory to brighten at times when touched by association or will, only to sink again like a landscape in the dark; but they saturate themselves into every part of the body and live always.

May 22.

A sunny day, 'so sweet, so calm, so bright.' Nature is always lovelier on Sabbaths, I suppose, because of leisure to feel and enjoy....

May 23

Warmer and the river is rising. The falls are at the stage of greatest palpable display.

May 25.

Warm, genial days of transition from spring to summer Magnificent groups of cumuli rise almost daily from the summits at eleven o'clock A.M., making a rich background for the fringing trees of the valley rim.

May 26.

Cumuli arrived at the regular hour and a few raindrops fell. These summer days go on in quiet magnificence, without the

presence of any very obtrusive phenomena. Yet no period of the
year or part of any of the seasons is more delightful and instructive. ...

May 29.

Early in the forenoon the sky was covered with black clouds,
blue seen only through narrow, unsteady seams. Soon a lively
fall of hailstones came pattering down among flowers and birds
and summer insects. The sun would break forth at times as if
trying to make peace, but the storm was determined, and snow
followed thick and fast at eleven o'clock, unconscious of cruelty
to the plant and insect people. It fell steadily for three hours,
whitening the terraced bands of dwarf oaks on the sides of the
valley, and so loading the willows and briers they were bent to
the ground.

These balmy, generous summer days are split into clumps and
abruptly cut into sections by cold, icy, snowy days of harsh-
favored storm. Men will say the Lord manages but heartlessly
with his mountain gardens, but where in the most level, equable
climates is there more of tender, healthy, celestial beauty? Where
in all the world can plant creatures be seen of purer spirituality
and cheerier temper? These storm-days are instruments by
which immortal loveliness is accomplished. ...

June 6.

Here summer seems to be content, full of ripe beauty and joy.
Young birds are tasting life in every grove. Happy myriads of
flying insects shake all the air into music. Hundreds of flowers
have ripened and planted their seeds, and all seem ready for
the sere and yellow leaf. But the end is not yet, for remaining
segments of sunny June are developing a glory that excelleth —
a glory seldom seen by mortal eyes. The deep dense banks of
green grow yet deeper and denser. The frost-nipped brackens call
up their reserves, that soon overtop the tallest man and form a
perfect forest of fronds that high overarch and embower groves of
tall leaning grasses, and many with feathery dainty spikelets

arise and bloom in the commonest places. Whole kingdoms of Ceanothus rise high above sedges and ferns like yellow stars on a green sky. Spikes of purple mints eight feet in height shoot high above the common green. Family groups of starry Compositæ glow on the meadow, and every other feature of plant beauty joins in the late summer glory.

June 20.

I walked to the top of Nevada Fall with William Pratt. We made many a halt by the rapids, which are a labyrinth of music and water forms. The Vernal is the greenest, most rectangular fall in the valley. It is a sheet composed of five or six united parallel falls. The South Canyon fall is composed of three principal ribs of water, but at the top there is a compound web of them.

June 27.

Arose betimes and walked to the top of Sentinel Dome. The Coast Range was blue and distinct on the sky, nearer the plains were of yellow, and still nearer the Sierra foothills were also blue, making three bands of beautiful color. Had a fine view of the ever-glorious Sierra crest. I walked to Starr King, passing the old moraine on the banks of Illilouette Fall. Also I walked above Nevada Fall into Little Yosemite. I made a memorable descent of the Nevada cliff on the left.

June 29.

Went up the mountain to Sentinel Dome with Judge Colby.[1] Camped in a spruce grove. Sunrise from the Dome.

July 6–7.

Mighty rainstorm. Glorious hail. Streams everywhere. Yosemite Falls muddy.

[1] Judge Colby was probably Gilbert Winslow Colby, an attorney of Benicia, California, the father of William E. Colby, long associated with John Muir in conservation work.

2. *Sierra Fragments*

Merced Canyon. 1870 (?).

This glorious canyon is a Yosemite the second....

It is the most jagged river pathway I ever saw.... The river is now at earnest work with all the beauty and poetry of unmarred nature. In the middle of the stream are immense rock islands hewn from the grandly sculptured walls. I wished to camp in midstream on a flat, smooth rock, but H.[1] said he would have nightmare....

Here is material from all the valleys and all the mountains beyond, chips from the Lord's manufactory — quartz, granite of all shades and texture, feldspar of diorite, fragmentary chippings and rubbish, each bit a perfect creation in itself, each a volume written within and without. In factories we can read upon the chips the story of the tools that made them. So here we read of lightning and frost and rain and wind and the silent dew, and the hammer-blows of the driven hail.

We are in the shade of the canyon wall, beneath the shelter of a chestnut oak. Near-by are two silver firs, a cedar, maple, flowering Cornus, and Calycanthus... and a glossy black pool in which bushes and a few starry Compositæ are reflected as in a lens.

A mild wind comes up the canyon playing upon all kinds of leaves, upon each with a music for itself, and for the falls that are white in the rocks and for every creature that makes its home here. A willow shows endless shades of black, green, and silver white as the wind exposes the under side of the leaf — as beautiful and compound a shade as ever was cast over mortals. I am giddy with the intoxication of it. Here and there are small patches of sand upon which are written the footprints of ducks and blue cranes and water-ouzels, ever-present in loving romantic dells where water blooms in whiteness.... I saw a little bluish wagtail, and abundant signs of grizzlies....

---

[1] Harry Randall may have been Mr. Muir's companion on this tramp.

The Singing Merced. Thousands of joyous streams are born in the snowy range, but not a poet among them all can sing like Merced. Men are not born equal, neither are rivers. The Merced was born a poet, a perfect seraph among its fellows. The first utterances of its childhood are sweet, uncommon song, and in its glorious harmonies of manhood it excels all the vocal waters of the world, and when its days of mountain sublimity are past, in quiet age down on the plain amid land-waves of purple and gold, its life-blood throbs with poetic emotion, and with a smooth sheet of soft music it hushes and tinkles and goes to its death in a maze of dripping willows and broad green oaks — an Amazon of thoughtfulness and majesty. Ah, who shall describe these golden plains of age, these grand green forests of spruce and boundless tangles of blooming shrubs, and the level velvet glacial daisy-gentian meadows in which the gleaming arteries of this most noble of rivers lie?

Memories of the Thousand-Mile Walk. One day, when walking along the coral shores of Cuba gathering shells, I found a tiny fragile purple flower with its circlet of petals confidingly open to the bright tropic sun. It lived in coral rocks that were washed by the heavy whitecapped waves of every storm from the north. In these 'northers,' the dread of seamen, wave after wave rolled over it — tons in weight — sufficient to crush a ship, but the little purple plant, tended by its Maker, closed its petals, crouched low in its crevice of a home, and enjoyed the storm in safety.

Once I was very hungry and lonely in Tennessee. I had been walking most of the day in the Cumberland Mountains without coming to a single house, but in crossing a dark-shaded stream whose border trees closed over it like a leafy sky I found the frail Dicksonia that I had looked for so long, and the first magnolia, too, that I had ever seen. I sat down and reveled in the glory of my discoveries. A mysterious breathing of wind moved in the trees, and the stream sang cheerily at every ripple. There is no place so impressively solitary as a dense forest with a stream passing over a rocky bed at a moderate inclination.

Feelings of isolation soon caught me again among these hushed sounds, but one of the Lord's smallest birds came out to me from some bushes at the side of a moss-clad rock. It had a wonderfully expressive eye, and in one moment that cheerful, confiding bird preached me the most effectual sermon on heavenly trust that I had ever heard through all the measured hours of Sabbath, and I went on not half so heart-sick, nor half so weary.

California Flora. The flora of California has fewer vines than that of any country I have seen, with the exception of a few leguminous plants in the meadows. I can think of but two in all these mountains, and amid the grand ocean of plants on the Joaquin plains I remember but one — one of the Cucurbitaceæ. The happy beings who belong to the plant kingdom of Florida dwell together in gorgeous heaps and twistings and tangles, but California plants rise side by side with scarce a prickle or tendril of attachment, looking skyward and proper, like good people at church.

Yosemite Valley. Undated.

The Falls. Yosemite is crowned with glory above its fellows. Only Nevada approaches it in the attributes of grandeur and power; these two belong to the same variety of blooming, vocal waters, and are comparable. In the snowy Nevada Fall there is a more sudden and overwhelming development of glory, but at the same time it is the more fathomable. Yosemite comes to us as an endless revelation — mysterious, unreadable, immeasurable, yet holding its hearers spellbound with the divine majesty and loveliness of its forms and voices. Pohono, the Vernal, and Illilouette belong to another variety; the Ribbon, Royal Arch Falls, and the countless snowy threads that streak and adorn all the rocks in the grand days of storm, to another; and the joyous cascades that sing unseen in the canyons of Tenaya and Dome Creek, to still another kind well marked. The cheap adjectives

— charming, lovely, pretty, etc. — are bestowed upon Pohono, and the Vernal, with abundance of proper squeaking emphasis, because these falls are more easily approached and their external beauty is more easily observed....

Yosemite. 1871 (?).

Glacial Landscape. Standing on Sentinel Dome, the position of Yosemite Valley during the Glacial Period is indicated only by a gentle depression. But around the edge of it we discover huge yawning fissures where the ice-sheet is broken in crossing diagonally over the valley trough. Looking northward, over what is now the basin of Yosemite Creek, no rock is seen to interrupt the snowy expanse, which waves gently, heaving and rising in one huge general sheet against the flank of the Hoffman Range. In the distance rise the black peaks of the summit, pushing up as short, sharp points out of the all-embracing mantle of ice-covered snow, and the tops of the Merced group are seen more to the right, casting dark blue shadows athwart the white expanse.

In winter storms the sky is dark at noonday with thick clustering flakes or wind-driven snow-dust.

When summer shines, the surface of the icy sea resounds with the song of running waters. A thousand glittering rills, glancing in the light, glide and glint in channels of ice and vanish in yawning crevasses with loud rumble and roar.

Here and there long trains of boulders, black, gray, and red, are seen in curving lines, reaching away from the summits of Dana and Lyell and Hoffman, and rose-tinted granite in larger block boulders from the quarries of Mount Clark.

Later, as the long icy centuries circle away, the first of the valley rocks is seen bare and bright on the north wall. It is the highest of the Three Brothers. Next the massive brow of El Capitan appears, glowing like burnished silver. And after that the crest of Clouds' Rest and the great Half Dome seem to rise higher and higher as the ice mantle is divided into separate glaciers that melt, shallower and shallower, until all the wall rocks of the glorious temple come to the light, fashioned and finished.

Winters and summers come and go with their storms and clouds and balmy sun-days, like those of the present time, and the main Yosemite Glacier shrinks slowly until its wasting snout reaches no farther than the base of El Capitan, where it lingers long enough to build a terminal moraine across the valley from the foot of El Capitan to the Cathedral Rocks, while, from the main side canyons separating the headlands, ice-falls and cascades pour in as tributaries to the main valley trunk glacier.

These shorten and vanish in succession and are followed by waterfalls and cascades — first those of the north wall, then those of the south, which linger centuries in the shadows after the disappearance of the sun-beaten northern tributaries. The main trunk, gradually receding, leaves a lake in its place, into which the shortened and shallow glacier discharges icebergs, which drift to and fro beneath lofty walls. At length the ice-work is done and the valley lies open to the day. A lake extending from the terminal moraine to the head of the valley stretches from wall to wall, reflecting the beauty of the new rocks. This endures for centuries, until, filled by avalanches and detritus and the slow and steady work of the inflowing streams descending from a thousand fountains, a sedgy meadow takes the place of the lake. Then in-washing gravel and sand gradually cover the margins of the mead, until at length the valley floor becomes dry and is planted with all the summer beauty we find today.

Meanwhile, the summit peaks rise higher and higher, casting blue spiky shadows over ice and névé. Dome after dome emerges from the prairie of ice until, in the fullness of time, the glaciers, retiring into their upper strongholds of shadowy peaks, leave all the lower landscapes smiling in the sunshine.

Flowers flock to their places in many a garden spot. The trees take up their line of march to the moraine soil prepared for them, and all the flood of happy life rejoices in new-made homes, and a thousand lakes and falling, foaming streams begin to sing and shine together as if created simultaneously in all their present glory.

Sunnyside Bench.
April, 1871.

The sun is a few minutes below the domes from my standpoint, but all the upper walls are lighted with the purest of silver morning light I ever beheld. Glorious is the coming of morning on the lowlands in spring when the vital golden beams steep the flowery fields throbbing with life. Truly golden are those mornings, but this is a silver morning. The light now raying over Tissiack seems a sort of pausing Sabbath light, as if all the quickening, energizing principles were absent from it, as if only the soul of the sun were shining now, and marvelous are the radiations of that sun-soul.

The spired pines and spruces hung with brown cones, and the warm green plume-clad Libocedrus on the bank of Sunnyside, stand about me, transfigured in the ethereal light, and above the base of the upper fall beyond the trees the rock is covered with irised spray, and all the waters of the fall in their snowy robes, comets, smooth fibrous sheets, and atoms of whirling spray illumined by this silver light, seem full of spiritual life, while every thunder-tone echoes from cliff to cliff. The sun of last week melted the upper snow and gave the grandest notes of all the yearly anthem. . . .

Midnight, *April* 3.

At night, the lunar bows in the spray make a most impressive picture. There was the huge dark cavern of the gorge filled with tempestuous foam and scud and roar of many storms. The fall above hung white, ghostlike, and indistinct. Slowly the moon coming round the domes sent her white beams into the wild uproar, and lo, among the tremendous blasts and surges at the foot of the pit, five hundred feet below the ledge on which I stood, there appeared a rainbow set on end, colored like the solar bow only fainter, strangely peaceful and still, in the midst of roaring, surging tempestuous power. Also a still fainter secondary bow.

I had intended to stay all night, but an hour ago I crept out on a narrow ledge [1] that extends back of the fall, and as the wind swayed the mighty column at times a little forward from the face of the precipice, I thought it would be a fine thing to get back of the down-rushing waters and see them in all their glory with the moonlight sifting through them. I got out safely, though the ledge is only about six inches wide in one place, and was gazing up and out through the thin half-translucent edge of the fall, when some heavy plashes striking the wall above me caught my attention; then suddenly all was dark, and down came a dash of outside gauze tissue made of spent comets, thin and harmless to look at a mile off, but desperately solid and stony when they strike one's shoulders. It seemed as if I was being pelted with a mixture of choking spray and gravel. I grasped an angle of the ledge and held hard with my knees, and submitted to my frightful baptism with but little faith. When I dared to look up after the pelting had nearly ceased, and the column swaying back admitted the light, I hastily pounced back of a block of ice that was frozen to the ledge, squeezing myself in between the ice and the wall, and no longer feared being washed off.

When the moonbeams again slanted past the ever-changing edge of the torrent, I took courage to make a dash for freedom and escaped, made a fire and partially warmed my benumbed limbs, then ran down home to my cabin, reached it sometime towards morning, changed my clothing, got an hour or two of sleep, and awoke sane and comfortable, some of the earthiness washed out of me and Yosemite virtue washed in, better, not worse, for my wild bath in lunar bows, spent comet-tails, ice, drizzle, and moonshine.... Wonderful that Nature can do such wild passionate work without seeming extravagant, or that she will allow poor mortals so near her while doing it.

Undated.

Spring. Thick clouds of fragrance are wafting from azaleas, welling from every corolla like water from a fountain. Yosemite

[1] See Badè's *Life and Letters of John Muir*, vol. I, pp. 249–52; also Muir's *Mountains of California*, vol. II, pp. 163–65.

syringa is leaning its wealth of white fragrant bloom upon the rugged granite boulders about the Bridal Veil.

It is delightful walking in velvety bogs that make no sound to the foot, and in glacial meady gardens. But the clear ringing clank of clean glacial pavements is always pleasant, too.

Undated.

Unity. Vegetable cells of every form and color are units of perfect beauty, and so are their combinations in flowers — units also of beauty composed of many units. And flowers builded into structures of racemes and clusters and starry heads are also units of beauty; and so are the patches and tangles of each, and the grand sheets of all combined spread in purple and gold over smooth hills and plains; and these plain sheets united with the shaggy, spiry woods of the mountains and the dun blanks of summits and deserts are still units. . . .

Canyon[1] west of
Lower Yosemite Fall.
Undated.

This is a delightful nook, or recess, running back of the foot of the fall about a hundred yards on the west side, its walls well fringed with maidenhair and spiræa and tufts of live-oak. The head of the nook is occupied by the largest laurel I have yet seen in the valley — diameter of trunk near base, eighteen inches. Grapevines depend in graceful festoons from its spicy branches and from the small oaks that have found anchorage in crevices of the walls.

The purple Leuchera with its airy panicles is abundant here, and near the fall is a ledge thickly fronded with Polypodium and Gymnogramme.

The view in front is measured by the stern black jaws of the canyon embracing the shapely cone of Starr King, rising into the blue sky to a height of nine thousand feet from a dark forest of fir. Sentinel Dome in the middle of the view is seen above

[1] This is really a nook, as Muir calls it in the first sentence below, rather than a canyon.

Glacier Point, forming one of the most symmetrical fronts presented to the valley. The fall itself, seen and heard from this point, exhibits water-forms and sings songs never to be anticipated by any amount of imaginings. The bottom of the valley, seen from here past the fall, contains a mile of river with its attendant groves of poplar and willow and sheets of meadow diversified by curves of every color. Groves of oak at the head of the valley and dark crowds of pine and fir are gathered as if waiting turns to pass up the canyon of the Illilouette.

The arches overhead from this standpoint reach out fifteen to twenty feet beyond the perpendicular, and from the top of these arches oozing streams descend in large globular drops direct as plummets and wide apart, now and then hitting a grass tuft projecting from the mossy wall.

Underneath the diamond jets and white pearl tricklings with their gentle murmurs is the deep resonant roar of massive floods.

A delicate rainbow dust softens and brightens the black rocks, and bathes and freshens the ferns and grass tufts and the glossy evergreen foliage of the live-oaks within reach.... The pine trees overhanging the chasm at the very brink, fringing the rugged adamantine brows, are all shining and shimmering with mist. So also are a few small trees clinging to the face of the cliff, along narrow rifts and tablets.

May, 1871.

Clouds. Of all the Yosemite visitors, clouds are the most imposing. A few are valley-born, but most come from the adjacent mountains. They seem to wedge and scramble among the roughest crags and swim and glide and drift down the canyons, over smooth brows and sheer cliffs with peculiarly impressive gestures, and when a storm is breaking up, the light effects are marvelously effective and the grand show seems to be held in a walled room.

Undated.

Falls without Definite Channels. I have seen young birds and animals not quite at home in the bushes and boulders, trying to find their way through. So the young Sierra rivers seem not yet at home in their canyon channels, though rushing recklessly onward like a drove of excited bison, white cascade tails and manes streaming in the wind, meeting obstacles unforeseen, turning right and left, leaping barrier rocks, now outspread in thin plumes, scattered, divided; now closely compressed, in abounding energy moving onward resistless, singing grander songs than they know, unconsciously beautiful amid all their wild uproar, displaying beauty enough for a thousand staid well-dressed falls in quiet valley and glen. Nevertheless, in the most ungovernable displays of wild energy there never is wanting an inner spirit of repose, like that visible in the midst of the roaring turmoil and dash of ocean storms when with grim deliberation the mighty waves heave themselves against cliffs, break into bloom, and, falling back, gather again as well knowing what they are about.

How glorious are the ebbs and flows of the mountain streams with their thousand fountains spreading near and far, suddenly let loose by the sun, streams of music following everywhere, rising, falling, night and day, in drouth and flood, summer and winter, linking lake to lake, meadow to meadow, rocky, smooth, and rough, on their flowery adventurous ways!

May 8, 1871.

Snowstorm. The warm weather last month made everybody and everything lively. All the plants were budding and blossoming. Butterflies waved their petal wings over the newborn flowers, with myriads of lesser wings making all the air hum. Birds were building their nests, squirrels rustled among the dry leaves. The streams began their spring anthems, and even the rocks seemed to thrill and tingle in newness of warm life. But today storm-clouds appeared above the brows of the walls, moving with imposing majesty hither and thither, taking their places and ranging themselves in order like soldiers in battle array. Sharp winds cutting like lances came rushing through open spaces, and rugged detached flaws roared and rumbled down angular canyons on the sides of the valley, paralyzing and putting away insect life at a breath, while leaves and petals and budding branches were torn off and scattered broadcast among the rocks. Then the clouds, blending into one dark mass, shed their snow, heaping, freezing, drifting, burying plants and insects, quick and dead alike. The loaded branches of the trees bent low about the trunks, some of the willows and alders along the river-bank were bent to the ground in wide bossy arches, and the hushed silence of winter reigned supreme. One could not but pity the chilled multitudes overwhelmed like travelers in alpine avalanches. . . .

The snow soon melted, the bent branches sprang back to their places refreshed. The green Carices and patches of white violets are again visible over the shining meadows. The Libocedrus colors are brighter than before, and the golden pollen is being sown by the winds far and near, and new plumes are poising in the bland air of as fine a green and faultless a curve as ever were made.

The Douglas spruce is decked with yellow sprays. The pines have shed off the brown scales of their buds and are pushing forth sheaves of needles. The oaks with brighter array of young leaves seem covered with purple mist, growing fast. The ouzel leads the birds in song. All the lost life seems back by a miracle, as if instead of a crushing snowstorm only a refreshing shower had fallen. Nature loves man, beetles, and birds with the same love. With her storms of snow, hail, volcanic fire, and lightning, she seems to scatter firebrands, arrows, and death among her creatures, and so she does, but they are scattered as the stars are scattered in the heavens, each in its place, singing together in faithful harmony. Indeed every atom entering into the structure of a storm is measured and fitted to its place, has Nature's square and plummet laid upon it, and is inspired with unchangeable love.

Undated.

Nature, while urging to utmost efforts, leading us with work, presenting cause beyond cause in endless chains, lost in infinite distances, yet cheers us like a mother with tender prattle words of love, ministering to all our friendlessness and weariness.

Undated.

The astronomer looks high, the geologist low. Who looks between on the surface of the earth? The farmer, I suppose, but too often he sees only grain, and of that only the mere bread-bushel-and-price side of it.

Undated.

Sarcodes, notwithstanding its blowing crimson color and proud statuesque beauty of form, is singularly cold and silent; neither fragrance nor sympathy comes from its showy beauty. How far from the blessed Cassiope and lily, whose delicate bells seem to ring in tones of purest love, appealing to the finest and best that is in us!

Undated.

Light. I know not a single word fine enough for Light. Its currents pour, but it is a heavy material word not applicable to holy, beamless, bodiless, inaudible floods of Light.

Undated.

The Frost Crystals. During clear, cool nights in autumn, fine crops of crystals grow on the grass, and trees, and rocks, and even on sheets of snow when there is any. The effect after perfectly still nights when the sun rises is glorious. The size and beauty of the crystals, the sunlight sifting through them in irised color, make a show far surpassing what one could imagine about it.

Undated.

Some falls are not set back in a special home recess, while others have the rock fitted to them like the shell to a mollusc, or the case to an instrument or a jewel....

Undated.

Vernal and Nevada Falls in Winter. The Vernal, always staid and collected in her movements, gives scarce any sign of enthusiasm. Her broad summer sheet of white, vernal-tinted above and abundantly irised below, is now an irregular ribbon frayed on the edges, breaking farther down into a pattering shower of beads. The wall can be seen through its thinner places near the bottom, and on either side rows of drops course leisurely down behind each other like raindrops on a window-pane. To so small and so gentle a measure have its passionate springtime thunder-floods fallen.

The Nevada Fall is a series of lace-strips now uniting, now separating, pursuing their paths down the face of the grand precipice with an air of dignified leisure, scarce so much as hinting the big, booming, drumming notes of summer that go sounding forth over the listening landscapes, while crowding and hurrying its waters over the tremendous verge.

3. *Explorations in the Great Tuolumne Canyon* [1]

In September, 1871, I began a careful exploration of all the basins whose waters pass through the Yosemite Valley. Having health so good I knew nothing about it, unmeasured time, and perfect independence, I rejoiced over the rich wilderness that lay before me; lessons on mountain sculpture I might learn, the waterfalls I should find, the flowers, the forests, the animals, the glorious scenery night and day from mountain-tops and ridges and canyons; the numberless streams I should have to follow to their hidden sources; the nameless mountains to climb, the glacial rivers to trace in hieroglyphics of sculptured rocks, forests, lakes, and meadows, all with so much to show and so much to tell.

---

[1] This account of John Muir's early 'Explorations in the Great Tuolumne Canyon' is a portion of his article published under that title in the *Overland Monthly*, August, 1873. The part used follows the marginal revisions made by Mr. Muir.

This was my 'method of study': I drifted about from rock to rock, from stream to stream, from grove to grove. Where night found me, there I camped. When I discovered a new plant, I sat down beside it for a minute or a day, to make its acquaintance and try to hear what it had to say. When I came to moraines, or ice-scratches upon the rocks, I traced them, learning what I could of the glacier that made them. I asked the boulders I met whence they came and whither they were going. I followed to their fountains the various soils upon which the forests and meadows are planted; and when I discovered a mountain or rock of marked form and structure, I climbed about it, comparing it with its neighbors, marking its relations to the forces that had acted upon it, glaciers, streams, avalanches, etc., in seeking to account for its form, finish, position, and general characters. It is astonishing how high and far we can climb in mountains that we love, and how little we require food and clothing. Weary at times, with only the birds and squirrels to compare notes with, I rested beneath the spicy pines, among the needles and burrs, or upon the plushy sod of a glacier meadow, touching my cheek to its gentians and daisies. No evil consequence from 'waste of time,' concerning which good people who accomplish nothing make such a sermonizing, has befallen me.

Early one afternoon, after exploring the picturesque domes and ridges of the west rim of Yosemite Creek Basin, I reached the northmost tributary of the creek, a beautiful cascading stream with flowery banks which I followed to its head, wading across spongy patches of meadow, and climbing over fallen logs and heaps of boulders, to the top of the divide. This portion of the Merced and Tuolumne Divide is a smooth, sedgy tableland, holding a shallow lake. I made my camp in a grove of Williamson spruce [1] near the margin of the lake, and then proceeded to explore the plateau in a northeasterly direction. I had not gone far before I came in sight of a stately group of headlands, arching gracefully on the south, with here and there a feathery pine tree

[1] An old name for the mountain hemlock (*Tsuga mertensiana*). See Sargent's *Silva of North America*, vol. XII. — F. H. A.

on their sides, but vertical and bare on the north, drawn up side by side in exact order, rigidly curved, like high-mettled cavalry horses ready for a charge. From the base of their precipitous fronts there extends a large, shallow mountain-bowl, in the bottom of which ten smaller bowls have been scooped, each forming the basin of a bright lakelet, abundantly fringed with spruce trees, and bordered close to the water with yellow sedge. Looking northward from the edge of this nest of lakes, I observed several gaps that seemed to sink suddenly, suggesting the existence of a deep gorge running at right angles to their courses, and I began to guess that I was near the rim of the Great Tuolumne Canyon. Then, looking back at the wild headlands, and down at the ten lakes, and northward through the suggestive gaps on the rim of the plateau, scarce able to decide where to go first, veering for some minutes like a confused compass-needle, I at length settled on a steady course, along the top of a ridge, from near the edge of the lake-bowl in a direction a little east of north, until suddenly halted by a sheer precipice over four thousand feet in depth.

This stupendous cliff forms a portion of the south wall of the Great Tuolumne Canyon, about halfway between the head and foot. Until I had reached this brink, I could see only narrow strips and wedges of landscape through gaps in the trees; but now the view was bounded only by the sky. Never had I beheld a nobler congregation of mountains. A thousand pictures composed that one mountain countenance, glowing with the life of the sun. I crept along the rugged edge of the wall until I found a place where I could sit down to look and think at rest. The Tuolumne River shimmered and spangled below, showing two or three miles of its length, curving past sheer precipices, and meandering through groves and small oval meadows. Its voice I distinctly heard, and it gave no tidings of heavy falls; but cascade tones and those of foaming rapids were in it, fused into harmony as smooth as the wind-music of the pines.

The opposite wall of the canyon, mainly made up of the ends of ridges shorn off by the great Tuolumne Glacier that once flowed over and past them, presents a series of elaborately sculptured precipices, like those of Yosemite Valley. Yet awful as is the scenery of this magnificent canyon, it offers no violent contrasts in general views; for the mountains beyond rise gradually higher in corresponding grandeur, and tributary canyons come in from the ice-fountains of the summits that are every way worthy of the main trunk. Many a spiry peak rises in sharp relief against the sky; in front are domes innumerable, and broad, whale-backed ridges, darkly fringed about their bases with pines, through openings in which I could here and there discern the green of meadows and the flashes of bright eye-lakes. There was no stretching away of any part of this landscape into dimness, nor possible division of it into back, and middle, and foreground. All its mountains appeared equally near, like the features of one face — on which the sun was gazing kindly, ripening and mellowing it like autumn fruit.

The forces that shaped the mountains — grinding out canyons and lake-basins, sharpening peaks and crests, bringing domes into relief, from the enclosing rocks, carving their plain flanks into their present forms—may still be seen at work at many points in the High Sierra. From where I was seated, on the brink of the mighty wall, I had extensive views of the channels of five large tributary glaciers that came in from the summits towards the northeast, every one of which had eroded its channel down to the bottom of the main canyon. I could also trace portions of the courses of smaller tributaries, whose canyons terminated a thousand feet above the bottom of the trunk canyon. So fully are the works of these vanished glaciers recorded upon the clean, unblurred pages of the mountains that it is difficult to assure ourselves that centuries have elapsed since they vanished. As I gazed, notwithstanding the kindly sunshine, the waving of grass, and the humming of flies, the stupendous canyon, with its far-reaching branches, seemed to fill again with creeping ice, winding in sublime curves around massive mountain brows, its white surface sprinkled with gray boulders, and traversed with many a yawning crevasse, the wide basins of the summits heaped with fountain-snow, glowing white in the thin sunshine, or blue in the shadows cast from the peaks.

The last days of this glacial winter are not yet past; we live in 'creation's dawn.' The morning stars still sing together, and the world, though made, is still being made and becoming more beautiful every day.

When the sun was nearing the horizon, I, looking once more at the shining river, determined to reach it if possible. To right and left, as far as I could see, the descent seemed impossible; but from another jutting headland about a mile to the westward I had a commanding view of a small side canyon on my left, running down at a steep angle, which I judged might possibly be practicable all the way. Then I hastened back among the latest of the evening shadows to my camp in the spruce trees, resolved to make an attempt to penetrate the heart of the Great Canyon next day. I awoke early, breakfasted, and waited for the dawn. The thin air was frosty, but, knowing that I should be warm in climbing, I tightened my belt, and set out in my shirt-sleeves, limb-loose and ready. By the time I reached the mouth of the narrow way I had chosen, the sun had touched the peaks with beamless light. Exhilarated by the divine wildness that imbued mountain and sky, I could not help shouting as I bounded down the topmost curves of the canyon, there covered with a dense plush of Carex, easy and pleasant to the tread.

After accomplishing a descent of four or five hundred feet, I came to a small mirror lake set here on the face of the canyon upon a kind of shelf. This side canyon was formed by a small glacier, tributary to the main Tuolumne Glacier, which, in its descent, met here a very hard seamless bar of granite that extended across its course, while the less resisting granite in front and back was eroded, of course, faster, thus forming a basin for the waters of the canyon stream. The bar, or dam, is beautifully moulded and polished, giving evidence of the great pressure exerted by even a very small glacier whose channel is steeply inclined. Below the lake, both the sides and bottom of the canyon became rougher, and I was compelled to scramble down and around a number of small precipices, fifty or a hundred feet high, that crossed the canyon like steps of a gigantic stairway.

At the foot of the stairs I found extensive willow-tangles, growing upon rough slopes of sharp-angled rocks, through which the stream mumbles and gropes its way, most of the time out of sight. These tangles would be too dense to walk among, even if they grew upon smooth ground, and too tall and flexible to walk on top of. Crinkled and loosely felted as they are by the pressure of deep snow for half the year, they form as impenetrable jungles as I ever encountered in the swamps of Florida. In descending, one may tumble and slide and crash over them in some way, but to ascend some of them with their longer branches, dry and sharp, presented against you like bayonets, is very nearly impossible. In the midst of these tangles, and along their margins, small garden-like meadows occur where the stream has made a level deposit of soil. They are planted with luxuriant Carices, whose arching leaves cover the ground. Out of these rise splendid larkspurs six to eight feet high, columbines, lilies, and a few Polygonums and Erigerons. In these moist garden-patches, so thoroughly hidden, the bears like to wallow like hogs. I found many places that morning where the bent and squeezed sedges showed that I had disturbed them, and knew I was likely at any moment to come upon a cross mother with her cubs, unless I made a good deal of warning noise. Below the region of bear-gardens and willow-tangles, the canyon becomes narrow and smooth, the smoothness being due to the action of snow-avalanches that sweep down from the mountains above and pour through this steep and narrow portion like torrents of water.

I had now accomplished a descent of nearly twenty-five hundred feet, and there remained about two thousand feet more before I reached the river. As I descended this smooth portion, I found that its bottom became more and more steeply inclined, and I halted to scan it closely, hoping to discover some way of avoiding it altogether, by passing around on either of the sides. But this I quickly decided to be impossible, the sides being apparently as bare and seamless as the bottom. I then began to creep down the smooth incline, depending mostly upon my hands, wetting them with my tongue, and striking them flatly upon the rock to make them stick by atmospheric pressure. In this way I very nearly reached a point where a seam comes down from the side to the bottom in an easy slope, which would enable me to climb out to a portion of the main wall that I knew must be available from the live-oak bushes growing on it. But after cautiously measuring the steepness — scrutinizing it again and again, and trying my wet hands upon it — I was compelled to retrace my devious slides and leaps, making a vertical rise of about five hundred feet, in order that I might reach a point where I could climb out to the main canyon wall, my only hope of reaching the bottom that day being by picking my way down its face, clinging to the bushes in the seams. I knew from my observations of the previous day that this portion of the canyon was crossed by well-developed planes of cleavage that prevented the formation of smooth vertical precipices of more than a few hundred feet in height and the same in width. These may usually be passed without much difficulty.

After two or three hours more of hard scrambling, I at length stood among cool shadows on the river-bank, in the heart of the great unexplored canyon, having made a descent of about forty-five hundred feet, the bottom of this portion of the canyon above the level of the sea being forty-six hundred feet. The river is here fully two hundred yards wide [1] (about twice the size of the Merced at Yosemite), and well timbered with Libocedrus and pine. A beautiful reach stretches away from where I sat resting, its border trees leaning towards each other, making a long arched lane, down which the joyous waters sang in foaming rapids. Stepping out of the river grove to a small sandy flat, I obtained a general view of the canyon walls, rising to a height of from four thousand to five thousand feet, composed of rocks of every form of which yosemites are made. About a mile up the canyon,

[1] This statement is questioned. Perhaps 'two hundred feet' was meant.

on the south side, there is a most imposing rock, nearly related in form to the Yosemite Half Dome. This side canyon, by which I descended, looked like an insignificant notch or groove in the main wall, though not less than seven hundred to eight hundred feet deep in most places. It is one of the many small glacier canyons that are always found upon the south sides of trunk canyons that trend east and west.

The continuity of the north walls of such trunk canyons is also broken by side canyons, but those of the north side are usually much larger, and have a more steady and determined direction, being related to canyons that reach back to high glacier-fountains, while many of those of the south side are strictly local. The history of their formation is easily read: they were eroded by the action of small glaciers that lingered in the shade of the wall, long years after the exposed sun-beaten north wall was dry and bare. These little south-side canyons are apt to be cut off high above the bottom of the trunk canyon, because the glaciers that made them were swept round and carried away by the main trunk glacier, at heights determined by the respective forces of their currents. This should always be taken into consideration when we are weighing the probabilities of being able to reach the bottom of a trunk canyon by these tributaries.

Immediately opposite the point I descended there are 'royal arches' like those of Yosemite, formed by the breaking-up and removal of a portion of the concentric layers of a dome. All of the so-called 'royal arches' of this region are produced in the same way.

About a mile farther down the canyon, I came to the mouth of a tributary that comes in from the north. The glacier to which it belongs must have been of great size, for it eroded its channel down to a level with the bottom of the main canyon. The rocks both of this tributary and of the main canyon present traces of all kinds of ice-action — moraines, polished and striated surfaces, and rocks of special forms. Just at the point where this large tributary enters the trunk canyon, there is a corresponding increase in size and change in direction of the latter. Indeed, after making a few corrections that are obviously required, for planes of cleavage, differences of hardness, etc., in the rocks concerned, the direction, size, and form of any main canyon below a tributary are always resultants of the forces of the glaciers that once occupied them, and this signifies that *glaciers make their own channels*. In front of this great tributary the canyon is about half a mile wide, and nobly gardened with groves and meadows. The level and luxuriant groves almost always found at the mouths of large tributaries are very distinct in appearance and history from the strips and patches of forest that adorn the walls of canyons. The soil upon which the former grow is re-formed moraine matter, collected, mixed, and spread out in the lake-basins that are formed by the pressure of the incoming tributary. The trees are closely grouped, social and trim; while those of the walls are roughish, and scattered like storm-beaten mountaineers. Some of the lake-basin groves are breezy from the way the winds are compelled to tumble and flow, but most enjoy perpetual calm at the bottom of pits of air.

I pushed on down the canyon a couple of miles farther, passing over level floors buried in shady greenwood and over hot sandy flats covered with the common Pteris, the sturdiest of ferns, that bears with equal patience the hot sun of Florida and the heavy snows of the Sierra. Along the river-bank is an abundance of azaleas and brier roses growing in thickets, and in open spots a profusion of golden Compositæ. Tall grasses brushed my shoulders, and yet taller lilies and columbines rang their bells above my head. Nor was there any lack of familiar birds, bees, and butterflies. Myriads of sunny wings stirred the air. The Steller jay, garrulous and important, flitted from pine to pine; squirrels were gathering nuts; woodpeckers hammering dead limbs; water-ouzels sang divinely on foam-fringed boulders among the rapids; and the robin redbreast of the orchards was in the open groves. Here was no field, nor camp, nor ruinous cabin, nor hacked trees, nor downtrodden flowers, to disenchant the Godful solitude, nor any trace of lawless forces, no word of chaos or desolation among these mighty cliffs and domes; every rock is as elaborately and thoughtfully carved and finished as a crystal or shell.

I followed the river three miles. In this distance it makes a vertical descent of about three hundred feet in moderate deliberate rapids. I would fain have lingered here for months, living with the bears on wild cherries and berries. I thought of trying their board for a few days; but as I was in my shirt-sleeves and without bread, I began my retreat. Let those who become breathless in ascending a flight of stairs think of climbing to a bedroom four thousand or five thousand feet above the basement. Exhilarated and buoyant by what I had reveled in, I pushed up the first three thousand feet almost without stopping to take breath, making only momentary halts to look at striated surfaces, or to watch the varying appearances of peaks and domes as they presented themselves at different points; but towards the summit I became tired, and the last thousand feet seemed long indeed, although I took many short rests, turning again and again to see the setting sun blessing the mountains. I reached the top of the wall at sundown, then I had only to skim along a smooth horizontal mile to camp, and make my fire and tea. My meals were easily made, for they were all alike and simple, only a cupful of tea and bread. But mountain weariness and mountain hunger — how few know what these are!

No sane man in the hands of Nature can doubt the doubleness of his life. Soul and body receive separate nourishment and separate exercise, and speedily reach a stage of development wherein each is easily known apart from the other. Living artificially, we seldom see much of our real selves. Our torpid souls are hopelessly entangled with our torpid bodies, and not only is there a confused mingling of our own souls with our own bodies, but we hardly possess a separate existence from our neighbors.

The life of a mountaineer seems to be particularly favorable to the development of soul-life, as well as limb-life, each receiving abundance of exercise and abundance of food. We little suspect the capacity that even our flesh has for knowledge.

Oftentimes in climbing canyon walls I come to polished slopes that seem to be too steep to venture on. After scrutinizing them, carefully noting every dint and scratch that might give hope for a foothold, I have decided they were unsafe. Yet my limbs, as if possessing a separate sense, would be of a different opinion, and cross the condemned slopes against the remonstrances of the will. My legs sometimes transport me to camp, in the darkness, over cliffs and through bogs and forests that seem inaccessible to civilized legs in the daylight. In like manner the soul sets forth at times upon rambles of its own. Our bodies, though meanwhile out of sight and forgotten, blend into the rest of nature, blind to the boundaries of individuals. But it is after both the body and soul of a mountaineer have worked hard, and enjoyed hard, that they are most palpably separate. Our weary limbs, lying at rest on the pine needles, make no attempt to follow after or sympathize with the nimble spirit that, apparently glad of the opportunity, wanders alone down gorges, along beetling cliffs, or away among the peaks and glaciers of the farthest landscapes, or into realms that eye hath not seen, nor ear heard; and when at length we are ready to return home to our flesh-and-bone tabernacle, we scarcely for a moment or two know in what direction to seek for it. I have often been unable to make my muscles move at such times, as if the nerves concerned were broken, a state of things which, right or wrong, would probably be explained by want of food and extreme bodily exhaustion.

Few have anything like an adequate conception of the abundance, strength, and tender loveliness of the plants that inhabit these so-called frightful gorges. Had I been able, in descending this one small side canyon, to 'pluck up by the spurs' one of each of the mountain pines that I met, together with one of each of the other cone-bearing trees, my big resiny bouquet would have consisted of first, the short straggling *Pinus albicaulis*, then *P. contorta*, *P. ponderosa*, *P. monticola*, and *P. lambertiana*, *Abies magnifica* and *A. concolor*, the lovely ladylike mountain hemlock *Tsuga mertensiana* gilded by the sun, the noble Douglas spruce, the burly brown-barked *Juniperus occidentalis*, and the warm yellow-green *Libocedrus decurrens.* Had we gathered the shrubs, we should have had two species of maple, four willows, two dog-woods, two honeysuckles, three manzanitas, one kalmia, one mountain ash, one amelanchier, one vaccinium, one ledum, two ceanothus, one bryanthus, one cassiope, one azalea, two spiræas, one rose, two raspberries, one rhamnus, three ribes, and a few others.

This little canyon is a botanical garden, with dwarf arctic willows two inches high at one end, the largest trees, bush Compositæ, and wandy half-tropical grasses at the other; the two ends only half a day apart, yet among its miniature bogs, prairies, and heathy moorlands the botanist may find representatives of about as many climates as he would in traveling from Greenland to Florida.

I passed back over the Yosemite Divide next morning for more bread, glad that I had seen so much, and that so much more remained to be seen, saying to myself: These forty miles of canyon-wall rocks are stone books gloriously illustrated, and in pursuing my studies I must return to them again and again, as to a library where many a secret in mountain-building may be explained.

4. *Trip to Mount Clark and the Illilouette Basin,*
Thence up the Merced Canyon to Lake Nevada,[1]
Returning by Way of Clouds' Rest

Grand North Womb of
Mount Clark Glacier.
October 7 (?), 1871.

In streams of ice, of water, of minerals, of plants, of animals, the tendency is to unification. We at once find ourselves among eternities, infinitudes, and scarce know whether to be happy in the sublime simplicity of radical causes and origins or whether to be sorry on losing the beautiful fragments which we thought

[1] 'Lake Nevada,' upon the authority of William E. Colby, was undoubtedly one of John Muir's names for Lake Washburn. He also called it 'Shadow Lake.'

perfect and primary absolute units; but as we study and mingle with nature more, the pain caused by the melting of all beauties into one First Beauty disappears, because, after their first baptismal submergence in fountain God, they go again washed and clean into their individualisms, more clearly defined than ever, unified yet separate.

From the summit of Mount Clark I have seen a glorious sunset. It had been cloudy all day until towards evening; then the sun broke through the cloud gloom, irradiating the edges with a purple-and-golden flood of light, which fell on the mountain, making it glow in richest creamy yellow — intensely fine and spiritual in tone.

The summit of Mount Clark is sharp and most surprisingly curved.... It owes its curves and sharpness wholly to glaciers, and is difficult of ascent, but it is possible to go up on the east side. *Pinus flexilis*[1] groves climb nearly to the summit, and the slopes are bright with crystals and flowers, Erigeron and Spraguea and Cassiope, Compositæ, etc. Also the rocks are lichened.... The glacier which issued from between it and Gray Mountain has left a fine lake. The north sides of mountains in this group are strongly marked with steeper, newer walls and with full lakes....

The north and south lines are splendidly marked here and give grand and easy lessons in mountain structure. East of Clark is a young meadowless lake, newborn offspring of a small remnant glacier not quite dead, showing moraines distinctly.

I saw a hawk — black and white in equal big blotchy parts. Wings white beneath with some spots of black in the center. I observed him once before on the summit of Hoffman.

Silky white-leaved Eriogonum, beautiful in death, umbels of flowers on gravel in perfect beauty of color, dead. How apparent are the love and tenderness of God in the keeping of those

[1] *Pinus albicaulis.* According to Dr. Jepson, there is no *P. flexilis* in the Yosemite.

dear, delicate plant-children of His in places we are wrongly taught to call wild, desolate, deserted! God's love covers His world like a garment of light.

Eriogonum plants — pure fountains of life — are set in rock chips, and as the gravel accumulates about them, they lengthen the thick root cords that bind them to earth, rising higher and higher like seaweeds upon a wave.

Red Peak. *October* 8 (?).

Red Peak has five separate summits — slaty, zigzagged like a fence. The whole group was once united with the main summits of the Clark Range, having been cut off by glaciers. A little north of the Minarets, forty lakes are in view. A splendid tributary lies between Red Peak and Black Peak (Merced Peak), full of magnificent lakes, wavy bottom, broad, surrounded by peaks.

Birds fly about my head here.

Camp by glacial lake in
Upper Illilouette Basin
between Red and Mer-
ced Peaks. Night.

My day's journey ends in a full run home at dark. I find the lake, and know it by its bays and mossy rock islands. I feel the soft sedge and downy shore mosses, also the carpet bushes, and know them to be Phyllodoce. I halt and gaze at mountain shadows and stars down in the deep glass-sheet. I loiter to the other side to camp, saying, 'How full of beauty and love is this newest and fairest of God's creations!' ... Union of trees here, yellow willows with crooked black stems. Gray granite wall with cascade on south side. Sedges brown and green, dead flowers, pale glumes of grasses. White logs thrusting themselves out from all that color like bleached bones.... A singing bird. Seeds falling from trees. One butterfly.... A rill talking plainly.

Dawn.

Morning light rayless, beamless, unbodied of all its purple and gold. No outgushing of solar glory pouring in torrents among mountain peaks, baptizing them; but each pervaded with the soul of light, boundless, tideless, newborn from the sun ere it has received a hint of good or bad from our star.... The trees, the mountains are not near or far; they are made one, unseparate, unclothed, open to the Divine Soul, dissolved in the mysterious incomparable Spirit of holy Light!...

Mountain peaks stand around as if assembled by their Maker. They do not hold grandly to the sky. They do not brood with thought. They have no language, no gestures. They seem to wait only in the very special presence of the great Soul!

Thoughts upon Finding a Dead Yosemite Bear. Toiling in the treadmills of life we hide from the lessons of Nature. We gaze morbidly through civilized fog upon our beautiful world clad with seamless beauty, and see ferocious beasts and wastes and deserts. But savage deserts and beasts and storms are expressions of God's power inseparably companioned by love. Civilized man chokes his soul as the heathen Chinese their feet. We deprecate bears.

But grandly they blend with their native mountains. They roam the sandy slopes on lily meads, through polished glacier canyons, among the solemn firs and brown sequoia, manzanita, and chaparral, living upon red berries and gooseberries, little caring for rain or snow.... Magnificent bears of the Sierra are worthy of their magnificent homes. They are not companions of men, but children of God, and His charity is broad enough for bears. They are the objects of His tender keeping....

There are no square-edged inflexible lines in nature. We seek to establish a narrow lane between ourselves and the feathery zeros we dare to call angels, but ask a partition barrier of infinite width to show the rest of creation its proper place....

Bears are made of the same dust as we, and breathe the same winds and drink of the same waters. A bear's days are warmed by the same sun, his dwellings are overdomed by the same blue sky, and his life turns and ebbs with heart-pulsings like ours, and was poured from the same First Fountain. And whether he at last goes to our stingy heaven or no, he has terrestrial immortality. His life not long, not short, knows no beginning, no ending. To him life unstinted, unplanned, is above the accidents of time, and his years, markless and boundless, equal Eternity.

God bless Yosemite bears!

Rivers of Ice. As broad wind-rivers now flow around and above every dome and peak, so did those rivers of ice, the bottom current diverted now here, now there, in compliance with every slippery canyon, but the main flow sweeping grandly over all like a deep upper wind.

> Lake Nevada, My Heaven.
> *October,* 1871. Night.

A Storm. Black mirror lake seeming to have a surface polish. Pulse of waters to show life. No wind, hush of rapids. Pure spirit white of curving shore. Bushes and trees, straight and tall on mountain-side rising higher and higher; black trunks, gray branches with mixing of snow in their leaves and shadows. Rocks grayish, changing to deep white on upper brows. What black, arctic shadows, what caves of blackness densed and refined beneath those drooping, rounded, silvered rock brows!

Near-by a clump of tall pines at bend of lake shuts off all the distant mountain, leaving nothing but the clear, present, living, soul-awakening purity of heaven. Pine and willows calm; even the aspen quiet. A great dead log in the forest like a ghost — white on top, black with shadow beneath, that seems to lift it above the ground, and its whiteness glows almost like fire. . . . Heavens! what reflection! Every line, every shadow in fine neutral tint, clear, intensely pure as in full day, yet in this spirit, rayless, beamless light.

The glacier-polish of rounded brows brighter than any mirror, like windows of a house shining with light from the throne of God — to the very top a pure vision in terrestrial beauty. . . . It is as if the lake, mountain, trees had souls, formed one soul, which had died and gone before the throne of God, the great First Soul, and by direct creative act of God had all earthly purity deepened, refined, brightness brightened, spirituality spiritualized, countenance, gestures made wholly Godful!

Not a cloud-memory in the sky. Not a ripple-memory on the lake, as if so complete in immortality that the very lake pulse were no longer needed, as if only the spiritual part of landscape life were left. I spring to my feet crying: 'Heavens and earth! Rock is not light, not heavy, not transparent, not opaque, but every pore gushes, glows like a thought with immortal life!'... Then a veil of mist, thin like a breath, through which the stars shine scarce dimmed, issues from a brow above us, and trails slantingly over the top and out of sight....

Suddenly I hear the solemn wind-voices in the pines above me. All reflection is ruffled from the lake. It regains its voice of water lapping and tinkling on the sand. Stars leap from its waves and ripples. Lake Nevada has come back to earth with its own natural beauty, shored and mountained in terrestrial grandeur, wearing the jeweled garments of a winter storm!

                                                                   Morning.

The lake, bronzed, gleams through white mist. A thrush[1] dives, wading breast deep, down one side of log, up the other. He dives five times, flies to a branch, sings to say how happy. Summer, spring, all the treasured joy of a balmy day of June is in that song!

                                                               Another night.

I walk down from the canyon to the lake, and around the lake to gather wood. Warm lappings of Scotland memory. Distant mountains dim in storm. Lake, mountain, sky, all one black at last.... The drip, drip of water on my bed of cedar fragments. Slant snowflakes in light of fire.... Must sleep and wake....

                                                           Another morning.

The shadows are still smooth, close, and black; the rocks soft purple and white. Hard protruding masses, broken with splendid

---

[1] See footnote on page 87.

curves or rounded into brows, form a grand shelter for shadows. The snow-rim of the lake curving about the bays is still spotless white. The whole living lake with its seamless basin is a mountain cup perfect and pure, as if, like one unbroken snow crystal, it had come softly from the sky, not worn by ages of glacier grinding, and sculptured by storm, torrent, earthquake, and avalanche. Nevertheless, it is no longer Nevada in heaven, but Nevada on earth, wearing the garments of minted snow with terrestrial grace, a creature with whom mortals can blend. Nevada Lake died in storm, went to heaven, and, ere the dawn of a sun-measured day, came back to her mountain home.

I feel like a horse brought out to run. I know every cross-canyon before I come to it. On the south side I shall have to cross two large tributaries. On the north, if I can get below the tributary from the summit, and from Cathedral, I shall reach the moraine by climbing a chaparraled slope....

I saw a grove of *P. contorta*, waved all together like a field of wheat.... And I saw a rare bird in the deep grove of Dead Lake bottom. He had a black stripe each side of his head extending back from his bill. His breast was orange and speckledy like that of a way-cup.[1] Wasps' nests are abundant there, and bears to eat them....

Most civilized folks cry morbidness, lunacy upon all that will not weigh on Fairbanks's scales or measure to that seconds rod of English brass. But we know that much that is most real will not counterpoise cast-iron, or dent our human flesh.

Civilization makes desolation in the purest, most open newborn fields with their harvests of terrestrial beauty....

---

[1] 'Way-cup,' or 'Wake-up,' is one of the many names given the flicker, or golden-winged woodpecker. He is so named from one of his several cries — 'way-cup, way-cup, way-cup.'

I will brood above the Merced Mountains like a cloud until all the ice-rivers of this mighty system are fully restored, each in its channel, harmonious as a song.

Do you say send us something about ice and lakes? ... Bide a wee, I am not idle; you will hear some day around the South Dome, not the crash of worlds, but the crush of swedging, crevassing glaciers. ... Heaven knows that John Baptist was not more eager to get all his fellow sinners into the Jordan than I to baptize all of mine in the beauty of God's mountains.

> Returning to Yosemite Valley
> via Lower Clouds' Rest.

I sail softly through the canyons ... like a wind full of thistledown, or a winter's mist. Mount Watkins looms large in mass as South Dome, and is very like the south side of South Dome on its unbroken face....

I started on this walk with essences and crumbs.... My pack seems loose and light — unsubstantial as a squirrel's tail.

5. *Sierra Fragments*

> Yosemite. 1872.

Winter. In winter the sun seems to rise in the west and set in the east. While the east is yet dim, many degrees of the eastern sky being hidden by the lofty Half Dome and Glacier Point, Eagle Cliff suddenly glows with morning light, while all the valley beside is in shade. The light does not seem to lie only on the surface, but seems fairly to saturate and make the whole cliff glow to its center as if the light came from within. Eagle Cliff is, therefore, the winter Orient of Yosemite. From it as a source, apparently, the light gradually creeps eastward down the mountainside, rosily touching the highest pines, arousing all the groves about Indian Canyon, thence stealing outward over the meadows to the river-banks. About four o'clock in the afternoon, the valley grows dim like a half-lighted room. You may see no cloud nor well-marked shadows, but cold thin twilight gathers, nevertheless. You judge it must be sundown, and look west down the valley for the sunset, but it is not visible, nor can you see any sky on which a sunset might be. Then, turning to look up the valley to the eastward, you discover your missing sunset on Half Dome.

1872.

Birds. Our winter birds cheerily sweeten these shadowy days with their faithful, hopeful song. They are not many, but a happier set never sang in snow. First and best we have the water thrush,[1] a little dusky, dainty bird that sings deliciously all winter. No matter how frosty or stormy, go to the riverside and you will hear him. No icy, chittering cheeping, but rich, whole-souled enthusiastic music that, despite the cutting wind about your ears, will make you fancy you are in a flowery grove. He sings the best songs of the brown thrasher and the bluebird and robin, and he has borrowed from the streams to make the most delightful melody. This one bird makes a summer any time of year.

March, 1872.

Spring winds are sweet with the scented buds and flowers they have passed over. In spring everything feels the joy of fresh life. As soon as the snow melts, the hilltops are alive with myriads of tiny purple and yellow flowers. Even the rocks, as well as the plants and animals, thrill and vibrate to the vital sunbeams, and every bower on hill and glen is a bridal bower.

Storm succeeds storm, heaping snow on snow. From the canyon-walls and long white slopes of the mountains avalanches descend in glory, laying bare the very roots of mountains and domes.

Frogs are already singing in the meadow, and a fine hearty song they sing, not at all abashed or frightened by the thunder of the falls.

[1] Evidently the dipper, or water-ouzel, is meant. See the chapter about this bird in *The Mountains of California.*—F. H. A.

On the warm north side of the valley young Carex sprouts are an inch high and the male aments of the alder are about ripe. The Libocedrus is shedding its pollen, and some of the willows are putting out their catkins, and flies and a multitude of swelling buds are telling the coming of spring.

We always have plenty of hollow-voiced owls for echoes.

Glacial Records. Nothing goes unrecorded. Every word of leaf and snowflake and particle of dew, shimmering, fluttering, falling, as well as earthquake and avalanche, is written down in Nature's book, though human eye cannot detect the handwriting of any but the heaviest. Every event is both written and spoken. The wing marks the sky as well as making stir in sounding words, and the winds all feel it and know it and tell it. Glaciers make the deepest mark of any eroding agent, and write their histories in inerasable lines. And as we can in some measure read and recall the forms and songs of dried-up streams by walking in their channels, so we can read the history of glaciers by tracing their channels centuries after they have vanished. And here the difficulty is not so much on account of dimness as of magnitude; the characters are so large it is difficult to see them from top to bottom in one view.

Glaciers, avalanches, and torrents are the pens with which Nature produces written characters most like our own, and every canyon of the Sierra displays examples of this writing.

Much notice has been taken of the writing on the wall of the Persian king's palace, but there is a writing on every wall, and though, like palimpsests, these pages are written line upon line and crossed again and again, none of these old palimpsests is ever wholly obliterated, and no other effacement or obscurement is made save by the writing of other scriptures over those that have gone before.

Winds and streams following the pathways of the vanished glaciers are fine preachers and interpreters of their ancient grandeur, and never cease to proclaim it night or day.

On Cathedral Meadows.
Altitude 9820 feet.
Night of *August* 17, 1872.

In full moon, all the horizon is lettered and lifed. I want immortality to read this terrestrial language. This good and tough mountain-climbing flesh is not my final home, and I'll creep out of it and fly free and grow!

Tuolumne Divide.
August 21, 1872.

Grass, a species of Agrostis, with tall, unbranched, strong stem and panicle of purple flowers, arches and waves above the low velvet sod like tropic bamboos.

The cutting of perpendicular walls is difficult in basins and canyons, but easy on ridge-ends.... Glaciers move in tides. So do mountains, so do all things. In Yosemite Basin the heavy tide crushed Capitanwards....

The melting of snow, working evenly, spreads moraines. Snows soothe down and finish the work of ice, as dews follow and carry on the work of rivers.

There are no harsh, hard dividing lines in nature. Glaciers blend with the snow and the snow blends with the thin invisible breath of the sky. So there are no stiff, frigid, stony partition walls betwixt us and heaven. There are blendings as immeasurable and untraceable as the edges of melting clouds. Eye hath not seen, nor ear heard, etc., is applicable here, for earth is partly heaven, and heaven earth.

October, 1872.

Loneliness. There perhaps are souls that never weary, that go always unhalting and glad, tuneful and songful as mountain water. Not so, weary, hungry me. In all God's mountain mansions, I find no human sympathy, and I hunger.

Ten days ago I came down from the ice to get a supply of the Two Breads, but, alack, I found only one!

Yosemite Sculpture. While the snow-flowers for Yosemite glaciers were growing in the depths of the sky, the stones for Yosemite temple-walls were growing in the crystalline depths of the mountains. In the fullness of time came the glaciers, the offspring of the gentle snow, and opened the glorious temple to the sun. The young rocks glowed like silver, waterfalls began their anthems, waving pines flocked to their appointed places in the groves, the warm air blossomed with insects, and the mission of the ice was accomplished.

The brow of El Capitan and the crest of Eagle Cliff, and the cones of Starr King and of South Dome — these, like islands of silver catching the thin fountain light of the cold sunshine, swam on the ice-ocean as it swelled and heaved.

Plants and Humans. Some plants readily take on the forms and habits of society, but generally speaking soon return to primitive simplicity, and I, too, like a weed of cultivation feel a constant tendency to return to primitive wildness.

Well, perhaps I may yet become a proper cultivated plant, cease my wild wanderings, and form a so-called pillar or something in society, but if so, I must, like a revived Methodist, learn to love what I hate and to hate what I most intensely and devoutly love.

Yosemite Autumn River Pools. The traveler who has witnessed the grand displays of waterfalls and river floods in Yosemite Valley during the springtime, when the snow is melting on the far mountains, cannot readily conceive the sleep of water in tranquil Indian summer. Then the river forms a series of pools united by trickling whispering currents that steal over the smooth brown pebbles with scarce an audible murmur. The several pools are as sharply individualized as lakes. Their extreme beauty would scarce be suspected, for though almost currentless they are yet bright and cool and pure, because of the coldness of the nights and the willow shade on their banks.

Their shores curve in and out in bay and promontory, giving

the appearance of lakes. The banks are overhung with wild rose and azalea and sedge and grass, and above these in beautiful combinations and in the glory of autumn colors are willows and alders, dogwood and balm of Gilead, a full blaze of yellow sunshine above, cool shadows beneath, with only flecks of sunlight on the bottom. The strained light filtering through painted leaves, as through colored windows, produces a dreamy enchanted atmosphere like that of some old cathedral. The banks of moss and liverwort and tinted ferns that overlean the placid water are reflected with charming effect, colored light from above falling on colored leaves, sifting through and falling on the brown pebbles of the bottom of the pool, then partly reflected back into the bright foliage. The surface of the pool is stirred gently in some spots by bands of water beetles, or the swimming strokes of spiders and dipping of dragonflies. Now and then a trout shoots from shelter to shelter beneath fallen logs.

The noonday enchantment of these pools is entirely unlike any other effect, and must be experienced to be known. No wind stirs. The falls too are quiet, slipping down the grand precipices in lacework with scarce a sound, silently as sunbeams. The whole valley floor is a finely blended mosaic of greens and purples, yellows and reds. Everything is passive in appearance, even the unflinching rocks seem strangely soft in beauty of lines with their strength hidden and held in abeyance.

December 20, 1872.

The Valley is tranquil and sunful
And Winter delayeth his coming.
The river sleeps currentless in deep mirror pools,
The falls scarce whisper.
The brown meadows bask,
The domes bathe dreamily in deep azure sky,
And all the day is Light.[1]

[1] This was first written in straight prose in a Journal; later, on a scrap of paper, it was given the above line arrangement by Muir himself.

6. *Mountain Thoughts*

1872 (?).

The Sierra. Mountains holy as Sinai. No mountains I know of are so alluring. None so hospitable, kindly, tenderly inspiring. It seems strange that everybody does not come at their call. They are given, like the Gospel, without money and without price. ''Tis heaven alone that is given away.'

Here is calm so deep, grasses cease waving.... Wonderful how completely everything in wild nature fits into us, as if truly part and parent of us. The sun shines not on us but in us. The rivers flow not past, but through us, thrilling, tingling, vibrating every fiber and cell of the substance of our bodies, making them glide and sing. The trees wave and the flowers bloom in our bodies as well as our souls, and every bird song, wind song, and tremendous storm song of the rocks in the heart of the mountains is our song, our very own, and sings our love.

The Song of God, sounding on forever. So pure and sure and universal is the harmony, it matters not where we are, where we strike in on the wild lowland plains. We care not to go to the mountains, and on the mountains we care not to go to the plains. But as soon as we are absorbed in the harmony, plain, mountain, calm, storm, lilies and sequoias, forests and meads are only different strands of many-colored Light — are one in the sunbeam!

What wonders lie in every mountain day!... Crystals of snow, plash of small raindrops, hum of small insects, booming beetles, the jolly rattle of grasshoppers, chirping crickets, the screaming of hawks, jays, and Clark crows, the 'coo-r-r-r' of cranes, the honking of geese, partridges drumming, trumpeting swans, frogs croaking, the whirring rattle of snakes, the awful enthusiasm of booming falls, the roar of cataracts, the crash and roll of thunder, earthquake shocks, the whisper of rills soothing to slumber, the piping of marmots, the bark of squirrels, the laugh of a wolf, the snorting of deer, the explosive roaring of bears, the squeak of mice, the cry of the loon — loneliest, wildest of sounds. . . .

A fine place for feasting if only one be poor enough. One is speedily absorbed into the spiritual values of things. The body vanishes and the freed soul goes abroad. . . .

Only in the roar of storms do these mighty solitudes find voice at all commensurate with their grandeur. . . . The pines at the approach of storms show eager expectancy, bowing, swishing, tossing their branches with eager gestures, roaring like lions about to be fed, standing bent and round-shouldered like sentinels exposed. . . .

Sickness, pain, death — yet who could guess their existence in this fresh, abounding, overflowing life, this universal beauty?

Race living on race, killers killed, yet how little we see of this slaughter! How neatly, secretly, decently is this killing done! I never saw one drop of blood, one red stain on all this wilderness. Even death is in harmony here. Only in shambles and the downy beds of homes is death terrible. Perhaps there is more pleasure than pain in natural death, or even violent death. Livingstone declared that the crushing of his arm by a lion was rather pleasurable than otherwise. . . .

Bloody Canyon. Nature's darlings are cared for and caressed even here, and protected by a thousand miracles in the very home and brooding-places of storms.

Faint are the marks of any kind of life, and at first you cannot see them or feel them at all. But here is the blessed water-ouzel pleading, fluttering about amid the spray, and blending his sweet, small, human songs with those of the streams he loves so well. And many other birds who build their nests here, and the flowers with few leaves that bloom on the rocks as if fallen like snow from the sky.

And here the grasshopper jumps and springs his rattle, as if to say, 'Who is afraid?'

And the bumblebee singing every summer the songs sung a thousand years ago.

A flock of wild sheep move aloft on the crags of the walls, not lost and cast away, but seeming to say in fullness of strength and ease: 'Here we are fled, and here is our home and safe hiding-place.'

One thinks of the redmen with flesh colored like the rocks, and sinews tough as the granite, who for thousands of years have dragged in files through these silent depths, clad in dull skins and grass, with mountain flowers stuck in their black hair and their wild animal eyes sparkling bright as the lakes.

Only the unimaginative can fail to feel the enchantment of these mountains.

Nothing is more wonderful than to find smooth harmony in this lofty cragged region where at first sight all seems so rough. From any of the high standpoints a thousand peaks, pinnacles, spires are seen thrust into the sky and so sheer and bare as to be inaccessible to wild sheep, accessible only to the eagle. Any one by itself harsh, rugged, crumbling, yet in connection with others seems like a line of writing along the sky; it melts into melody, one leading into another, keeping rhythm in time.

The cleanness of the ground suggests Nature taking pains like a housewife, the rock pavements seem as if carefully swept and dusted and polished every day. No wonder one feels a magic exhilaration when these pavements are touched, when the manifold currents of life that flow through the pores of the rock are considered, that keep every crystal particle in rhythmic motion dancing.

Tissiacks seldom have lofty domes to give grace to their strength. They are mostly stout, thickset mountains with spread bases for strength, because they have been born of two great streams and overflowed and much eroded. Glaciers eat their own offspring.

Books. I have a low opinion of books; they are but piles of stones set up to show coming travelers where other minds have been, or at best signal smokes to call attention. Cadmus and all the other inventors of letters receive a thousand-fold more credit than they deserve. No amount of word-making will ever make a single soul to *know* these mountains. As well seek to warm the naked and frostbitten by lectures on caloric and pictures of flame. One day's exposure to mountains is better than cartloads of books. See how willingly Nature poses herself upon photographers' plates. No earthy chemicals are so sensitive as those of the human soul. All that is required is exposure, and purity of material. 'The pure in heart shall see God!' . . .

Water Music. When in making our way through a forest we hear the loud boom of a waterfall, we know that the stream is descending a precipice. If a heavy rumble and roar, then we know it is passing over a craggy incline. But not only are the existence and size of these larger characters of its channel proclaimed, but all the others. Go to the fountain-canyons of the Merced. Some portions of its channel will appear smooth, others rough, here a slope, there a vertical wall, here a sandy meadow, there a lake-bowl, and the young river speaks and sings all the smaller characters of the smooth slope and downy hush of meadow as faithfully as it sings the great precipices and rapid inclines, so that anyone who has learned the language of running water will see its character in the dark.

Beside the grand history of the glaciers and their own, the mountain streams sing the history of every avalanche or earthquake and of snow, all easily recognized by the human ear, and every word evoked by the falling leaf and drinking deer, beside a thousand other facts so small and spoken by the stream in so low a voice the human ear cannot hear them. Thus every event is written and spoken. The wing scars the sky, making a path inevitably as the deer in snow, and the winds all know it and tell it though we hear it not.

We all know and wonder at the writing on the wall of the Persian king's palace, forgetting that there is a writing on every wall. Every glacier makes its mark, writes its history, and as we can read and recall the music of streams by going along their dry channels, so we can read the history of ice, the creature who above all others makes the heaviest mark, the largest track, and here the difficulty is not because of dimness, but because of magnitude, like an alphabet written too large. As tributary after tributary enters a trunk the channel is always increased to a corresponding degree. Winds and streams of water are the best interpreters and historians and preachers of ice now dead. Yosemite winds, Yosemite waters, glorious proclaimers, apostles of the combing ice, never cease to preach it night and day!

In snowstorms, flakes flicker and die in the lee of bare domes, or bear steadily aslant to their place on the white bosom of the sea.

Yosemite meadows in spring are covered with mirrors, small shallow lakelets that reflect with marvelous distinctness the overlooking mountains and every tree and bush and fine sculptured marking. These lakelets are memories of the great glacier lake of Yosemite, which, while the Tenaya Canyon above South Dome and the canyons of Illilouette and Nevada were yet blocked with ice, stretched its green waters from wall to wall. Its sublime rocks were reflected as now, but themselves, not yet dimmed with the rust of rains and snows, gleamed bright as the water. Gradually the lake was filled with the washed pebbles of its feeding streams, its waves were driven farther and yet farther from the walls, until at length only these mirrors are left to illustrate and enliven its history. Just as a man gradually fills to motionless rigidity by deposits from his life-supplying veins until like a sandy sediment-filled lake he rests in old age with only the main central river of life flowing through his body, yet with many a memory of the ardent days of youth.

We rode along the warm rim of Yosemite, to our left the magnificent battlements of the Hoffman and, on beyond, Cathedral Peak and its group of spires. Thick silence broods the Valley mansion. From the walls the splendid shafts of the pine, clustered and tall, rise stiffly as if to support the near low starry sky.

Some people miss flesh as a drunkard misses his dram. This depraved appetite stands greatly in the way of free days on the mountains, for meat of any kind is hard to carry, and makes a repulsive mess when jammed in a pack.... So also the butter-and-milk habit has seized most people; bread without butter or coffee without milk is an awful calamity, as if everything before being put in our mouth must first be held under a cow. I know from long experience that all these things are unnecessary. One may take a little simple clean bread and have nothing to do on these fine excursions but enjoy oneself. *Vide* Thoreau. It seems ridiculous that a man, especially when in the midst of his best pleasures, should have to go beneath a cow like a calf three times a day — never weaned.

Indian Summer. Calm, thoughtful peace and hushed rest — the pause before passing into winter. The birds gathering to go, and the animals, warned by the night frosts, relining their nests with dry grass and leaves and thistledown.

The sun glowing red; the mountains silvery gray, purplish, one mass without detail, infinitely soft.... The far-spreading meadows brown, yellow, and red. The river in the foreground silver between bosky willow-banks, green and orange and lemon yellow, every leaf and spray reflected in the mirror-water, its beauty doubled....

In the yellow mist the rough angles melt on the rocks. Forms, lines, tints, reflections, sounds, all are softened, and although the dying time, it is also the color time, the time when faith in the steadfastness of Nature is surest.... The seeds all have next summer in them, some of them thousands of summers, as the sequoia and cedar. In holiday array all go calmly down into the white winter rejoicing, plainly hopeful, faithful... everything taking what comes, and looking forward to the future, as if piously saying, 'Thy will be done in earth as in heaven!'

Spring. Quick-growing bloom days when sap flows fast like the swelling streams. Rising from the dead, the work of the year is pushed on with enthusiasm as if never done before, as if all God's glory depended upon it; inspiring every plant, bird, and stream to sing with youth's exuberance, painting flower petals, making leaf patterns, weaving a fresh roof — all symbols of eternal love.

Nature's literature is written in mountain-ranges along the sky, rising to heaven in triumphant songs in long ridge and dome and clustering peaks.

Mountains. When we dwell with mountains, see them face to face, every day, they seem as creatures with a sort of life — friends subject to moods, now talking, now taciturn, with whom we converse as man to man. They wear many spiritual robes, at times an aureole, something like the glory of the old painters put around the heads of saints. Especially is this seen on lone mountains, like Shasta, or on great domes standing single and apart.

Gain health from lusty, heroic exercise, from free, firm-nerved adventures without anxiety in them, with rhythmic leg motion in runs over boulders requiring quick decision for every step. Fording streams, tingling with flesh brushes as we slide down white slopes thatched with close snow-pressed chaparral, half swimming or flying or slipping — all these make good counterirritants. Then enjoy the utter peace and solemnity of the trees and stars. Find many a plant and bird living sequestered in hollows and dells — little chambers in the hills. Feel a mysterious presence in a thousand coy hiding things.

Go free as the wind, living as true to Nature as those gray and buff people of the sequoias and the pines.

Wind. How far it has come, and how far it has to go! How many faces it has fanned, singing, skimming the levels of the sea; floating, sustaining the wide-winged gulls and albatrosses; searching the intricacies of the woods, taking up and carrying their fragrances to every living creature. Now stooping low, visiting the humblest flower, trying the temper of every leaf, tuning them, fondling and caressing them, stirring them in lusty exercise, carrying pollen from tree to tree, filling lakes with white lily spangles, chanting among pines, playing on every needle, on every mountain spire, on all the landscape as on a harp.

Go east, young man, go east! Californians have only to go east a few miles to be happy. Toilers on the heat plains, toilers in the cities by the sea, whose lives are well-nigh choked by the weeds of care that have grown up and run to seed about them — leave all and go east and you cannot escape a cure for all care. Earth hath no sorrows that earth cannot heal, or heaven cannot heal, for the earth as seen in the clean wilds of the mountains is about as divine as anything the heart of man can conceive!

Chapter III : 1873

'The mountains are calling me, and I must go.'

JOHN MUIR, not to be diverted from his quest of scientific evidence of glacial action, brought his work to a climax in the year 1873, in a series of the 'longest and hardest trips' he had ever taken in the mountains. This was also his year of most prolific journal-writing. Beginning with the substantial 'Sunnyside Observations' made during the early storm-bound months, as soon as the season permitted, he set out to make and record his ultimate explorations of the summit peaks and glaciers, including an ascent to the Minarets and Ritter, and to the 'topmost stone' of the Matterhorn.

Returning to the 'bread-line' after a five-weeks sojourn to eat and sleep 'deep and fast' for a few days and nights, he set off in mid-September on an excursion into the Kings River region. As he tramped down the range, he paused at the head of the young San Joaquin River to name a 'wide-winged mountain' after his friend Emerson. From there he climbed over the divide to follow up the Kings River tributaries, ascending Mount Tyndall and nameless other peaks; then, passing southward and eastward, he climaxed the journey by being the first man to scale Mount Whitney from the east side. After this feat he made a 'simple saunter' northward to Lake Tahoe.

Summoned to the cities by the growing demand for his writings, John Muir in December reluctantly turned his face toward 'the wastes of Oakland,' where he lived for ten months in the home of his friends Mr. and Mrs. J. B. McChesney, harvesting his discoveries and experiences into a series of magazine articles.

- - - - -

I. *Sunnyside Observations, with Notes Written on Floor of Yosemite Valley*

Yosemite. *January* 1, 1873.

The December just completed was three-fourths sunshine, one-fourth tranquil, continuous rain, and this evening finds Yosemite filled with low heavy foundering clouds that are still giving rain as if their fountains, long open, were inexhaustible.

January 2.

A day of subdued cloud-filtered light with a few gleams of free sunshine. The falls were grandly frosted, and for a few minutes in the morning were gloriously irised. This evening more rain.

Water-ouzels are busily engaged in diving in the shining river. Each one alights in the water with a dainty, slanting glint. I see first a stroke of wings, then a head held under the water in the rapids, making the current flow over it in a smooth, white, crystalline shell like a bell glass.

January 3.

Rain in the morning, the valley filled with clouds gradually clearing. The point east of Yosemite Falls looms grandly up into the sweet clear blue or is overarched with a downy dissolving cloud of the same texture and color as that which muffles it below. The rock islands thus made in the clouds, rising and subsiding like volcanic islands in the sea, are extremely beautiful with their close white snow and clumps of dark pine and sculptured forms well developed by snow on projecting shelves and tablets. Tissiack in the afternoon wears gorgeously illumined clouds which she is ever changing, now drawing her dazzling robes close about her throat, now dropping them back on her shoulders in noble curves. The white of the snow on her head is close-pressed. White clouds show grandly as they hover caressingly above. The skirts are bare of snow along the bottom, occupying as they do a warmer climate than the head, and are fringed with dark fir trees.... The Tissiack is never so impressive as when decked in her first snows. She is then so palpably adorned and so glorious in her robes and jewels that she is acknowledged Queen of all the valley rocks. Were all the other features of the valley removed, she would still, all alone, form a shrine for the worship of the world. In the sunset her crown glows in warm sunshine showing dimly through cloud mufflings.... Richly colored red and gray clouds block the lower end of the valley from wall to wall, stretching from Cathedral Rocks to Capitan, clouds fringed with burning yellow light above and now and then a long tattered streamer.... Thus light coming up the valley from these heavy luminous clouds makes a rich blend with that from the great glowing dome Tissiack.

January 5.

Clear, calm, sunful. A sharp earthquake shock at 7.30 A.M. Rotary motion tremored the river.... A boulder from the second of the Three Brothers fell today.

Insects are sporting in the free mild air. Grass tips are growing.

Domes laden with rosy-tinted snow rise above the shadow-filled valley at the sun setting, like islands on a dark sea. Clouds' Rest is intensely white, not having yet lost any by avalanche.

January 6.

A fine gentle day with irised waters and delicate fibrous clouds. Yellow at sunset, the Domes are warmly mantled with light over their furry snow.

Picked up metamorphic pebbles on the river-bank which came from Mount Lyell or the Merced group. Found another pair of ouzels, and a kingfisher.

Instead of narrowing my attention to bookmaking out of material I have already eaten and drunken, I would rather stand in what all the world would call an idle manner, literally gaping with all the mouths of soul and body, demanding nothing, fearing nothing, but hoping and enjoying enormously. So-called sentimental, transcendental dreaming seems the only sensible and substantial business that one can engage in.

January 7.

A big yellow butterfly is flapping steadily up the valley as if going on some definite errand — a fine illustration of our January sunshine. A mild day; the falls are irised and gently swayed and edged above with rockets. The Domes are clothed upon with Light as with a garment.

January 8.

A delightful day of sunshine of that warm as well as bright kind that calls butterflies from their coffins and mottles the air with humming wings.

The meadow willows are full of the intentions of spring. Their buds are swelling, and some that I examined showed the infant catkins close wrapped in silk down ready for the spring resurrection. Grass, too, is growing green in the warm sand slopes of Indian Canyon, and the manzanita flower racemes are near an inch in length. The rapids of the river spangle with ineffable brilliancy, flashing forth the forms of every pebble of their beds. ... Long cobweb tresses stream from the bushes to tall mint stalks and float free in air in astonishing abundance.

Winter is not a whit less tender and lovable than summer.

Found charcoal in river-bank section of shallow lacustrine deposit. Also a small cove with a few weed roots dangling like gauze in front, mica scales and washed sands strewing the floor. A few delicate strands spun by spiders stretch across the floor, and from the steamy exhalations rising to the call of the sun minute dew beads are strung from end to end. These are now irised. Compared to them in delicacy and size the ordinary morning dewdrops on grass blades and twigs are heavy as thunderbolts. These irised beads are of all the forms of water the most delicate and beautiful I have ever seen....

January 9.

A gentle bonnie day spent in the Illilouette Canyon. Sunshine somewhat dim as if material was being gathered for a storm. Birds now in the valley are a jay, two woodpeckers, some pigeons, a bluebird, a wren, a titmouse, two or three small sparrowy and linnety birds, and a few ducks. Buds of black currant and azalea are swelling.

Sunset.

Tissiack is bathed in yellow light with a veil of clouds stretched across a little below the vertical portion, with gaps in the clouds showing sunny patches and gauzy flecks revealing the yellow light beneath. Down the valley direct light is streaming through the lower clouds. The wind is blowing the tops of the clouds aslant, making them appear like huge overleaning rock-crags red and yellow.

How eloquently every pebble in the margins of the river is expressed in a shimmer of living silver!... Tissiack, now that the sun is set over the valley wall, is itself like a very sun — the whole immense summit glowing red hot.

My Sunnyside Camp, on
North Wall, Yosemite Valley.
January 10.

Rain, three-eighths of an inch. Windless all day. Fine weaving of clouds going on, most abundant about the warm lips of Indian Canyon....

January 13.

An irregular garment of cloud covered the Cathedral Rocks, extending across the valley in front like a veil formed of angels, but also drooping back of the rocks. This garment extended from near the bottom of the rocks to the top, and high above the upper portion was one downy-edged massive cornice of glowing silver, of a texture refined beyond thought and burning with a whiteness like that of a blast furnace. On the projecting bosses of the lower portions of the robe the sun burned with equal whiteness, but in thinner portions. Back in the shadows the dark unlighted rocks appeared holding here and there a handful of pines that seemed to be gazing from windows of fire.

Around these gauzy portions of cloud there were shown all shades of light, from silver fanned to whitest glow down through every tone of gray to the blackness of the shadowed rocks and pine trees. Now the summit of the fir-covered mountain beyond the Cathedral Rocks would appear in patches lifted high in the soft air as if borne aloft by fondling clouds, their massive blackness and vividness contrasting with the white hairy clouds that carried them, their outlines wavering and altering as if seen through the depths of a clear swelling lake. Now one of the three Cathedral Rocks would shine dim through the melting fibers, then another, and then all became visible at the base, their gray feet feathered with black firs. . . . Now the tops would rise to view, one by one or all together, with an endless abundance of glowing white folds. . . . Several times the lowest outstanding rock rose like a lighthouse from a foamy ocean, with long wreaths of whitened spray. Altogether a most palpably and infallibly glorious creation of clouds, sunshine, forests, and huge granite rocks. These are called dreary days of mist and rain, but they are beautiful and joyful as sun-storms when only floods of light are falling.

Up the valley Starr King is overshadowed by a dim round down-curling mass of cloud like smoke coming from the cone of the summit as from a crater. South Dome has been magnificently turbaned the greater portion of the afternoon with white downy tissue. Some of these clouds have a baseless, unsolid appearance, but most seem to live calmly and strongly on invisible foundations. Repose is spoken in every gesture, and Cathedral Rocks have all the fervor and white light of cascades and falls, without any of their dazzling, throbbing thunder.

Frosty morning. Clear sunshine all day. Falls well bordered
with frozen spray cracked off by the sun. It fell on the rocks
below with a heavy echoing cannonade till noon. Seen from
Black's, the Upper Fall about 11 A.M. was irised from the base to
a height of three or four hundred feet with glowing sun-beaten
ice.

The forest trees, seen from the meadows, are showing forth
their winter beauties without stint. The pines, half their needles
silvered in the light, wearing a great deal of warm yellow with
their green, rise in single spires, or in groups and massive bands,
ever fresh, ever more and more beautiful to the eye that loves
them. Ranging in front of the pine groves and hiding their bases
are the oaks, leafless and stricken through and through with light,
yet giving a most delicate effect of color and also of form. The
color in the bright sunshine is dark gray and purple, with noble
trunks and arched branches clearly cut in outline black. In
front of these rise, here and there along streams, rows of poplars
white as milk. Then appear the more delicate yellow and purple
of willows and brown sheets of last year's Pteris, and yellow of
the meadow itself.

A bright, calm, crystal day full of peace.... The sunny air is
tingling with infinite wing-beats of newborn insect people.... No
lack of pigeons, or squirrels, or banks of fragrant Aplopappus
and ferns with club moss, *Cheilanthes gracilis*, and Gymnogramme, all tingling at the roots. Also asters, pentstemons,
Eriodictyon. Laurel flowers are open....

Winter Landscape. The Lyell group of mountains show their
peaks striking sharply into the dark sky with streaks of foamlike
waves. Irised clouds hover above them, and here and there a
cloud caught on a peak trails horizontally like a banner. The
shadows of the peaks lie clearly outlined on the ample sheets of
snow below that, like drapery, conform to the rugged anatomy

of the landscape. Beneath the snow-fields one sees the forests dark by contrast.

In front of all this grand picture of mountain and forest are laid the massive, snowless gray walls of the valley, swooping majestically to the bottom and planting their feet firmly among pine groves and meadows. The river glows like a mirror, and in every rapid the sun is sowing spangles. The meadows are lovely in color — yellow with sedge, and brown with patches of fern. The sun glowing on smooth bosses, and groves of erect taper pines with their shadows, produce light and shade among the rocks.

The winds sing on among the pines and firs. Jays scream lustily. The snow is melting into music. There is scarce any frost at night. Little wrens and mouse-like chickadees appear in considerable numbers.

Clouds' Rest is nearly solid white. Tissiack is white on her crown, and white-streaked and dusted in her many folds of rock clothing. Starr King has a dense black forest of fir sweeping up his slopes from the north. The top of his cone is white on the north, gray on the south.

Shadows from Glacier Point as a center sweep up the valley, shearing off the glow from river and meadow.

All the fields of God, whether reposing in the garments of winter or of summer, sing of gentleness and love.

January 16 (?).

It is now springtime on this north side of the valley, though winter rules on the south. The ant-lions are lying in wait in their warm dry sand-cups, rock ferns are being unrolled, club-moss carpets are covered with fresh-growing points, laurel flowers are nearly open, honeysuckle vines are rosetted with young leaves, and some small grass-plots near the brow of the Lower Fall are giving off summer fragrance.

There is no mystery but the mystery of harmony, no inexplicable caprice, no anomalous or equivocal expression on all the

grandly inscribed mountains, although all causes that lie
within reach and are readable to our limited vision are only
proximate and lead on indefinitely into the impenetrable mystery of infinity.

The features of the Sierra landscapes grew as grows the grass.
Crystallized beneath the heavy folds of overlying rocks, perhaps
a mile thick or more, there in the darkness they lay awaiting
development, like winter-bound gardens waiting for summer.

The Glacial Period may almost be said to separate the knowable
from the unknowable in geological history, so great is the difference between the blurred, foreshortened glimpses we may have
into pre-glacial history compared with the vivid clearness of the
post-glacial page.

Drifting about among flowers and sunshine, I am like a butterfly or bee, though not half so busy or with so sure an aim. But
in the midst of these methodless rovings I seek to spell out
by close inspection things not well understood. Still, in the work
of grave science I make but little progress.

If, in after years, I should do better in the way of exact re-
search, then these lawless wanderings will not be without value
as suggestive beginnings. But if I should be fated to walk no
more with Nature, be compelled to leave all I most devoutly love
in the wilderness, return to civilization and be twisted into the
characterless cable of society, then these sweet, free, cumberless
rovings will be as chinks and slits on life's horizon, through
which I may obtain glimpses of the treasures that lie in God's
wilds beyond my reach.

Shadows. Sitting in my Sunnyside Camp, with the valley
outspread before me like a map, it is interesting to watch the
movements of the lights and shadows on the floor. The main
masses of shade mark the boundaries and locations of the residual
glaciers that lay in the shadows protected by them long after

the main trunk glacier of the valley was melted. The residual glaciers, of course, formed terminal moraines, and it is upon these terminal moraines that the principal groves are planted. This explains the relationship they bear to the shadows, the shadows controlling the glaciers, the glaciers controlling the position of the moraines, and the moraines governing the position of the groves.... Some lake-basins carved from the hardest granite owe their existence to the same cause by the control shade exercised over the eroding tool. These mighty results and many others have followed so unlikely and feeble and imponderable an agent.

Comparing the north and south walls of the valley, they are seen to be mostly the same as to structure of rock, but unlike in sculpture. The north wall is comparatively plain and massive in style. Its canyons are on a grand scale, and all open down the valley in the direction of the flow of the tributary glaciers that entered the main trunk. The one exception is the canyon west of Upper Yosemite Fall. The south side, where the shadows dwell, is deeply sculptured; and the greater and more constant the shadow belonging to any portion of the wall, the more deeply is that portion sculptured, and the side canyons are mostly small and their courses variable in direction of trend, some opening down the valley, some up, some zigzag, since only the residual glaciers were much concerned in making them, with the exception of Pohono Glacier.

This after sculpture is on the greatest scale between Cathedral and Sentinel Rocks, glaciers of small size having lingered there furnishing soil for the forest now growing beneath them. Not being much rounded or crushed, the moraine material of those small glaciers resembles avalanche detritus, and is, indeed, covered by earthquake talus in many places. Thus the large grove that fills this portion of the valley is accounted for.

The next largest mass of timber in the valley is near the head, growing upon moraine material deposited by a glacier that lingered in the Illilouette Canyon and was connected with another that lingered beneath the shadow of Glacier Point. The fine curve of the grove sweeps from the west side of the face of Glacier Point Mountain into the valley, nearly across to Royal Arches and around to the South Dome. This was the largest of the residual glaciers — two miles long.

Another had its source on the east side of the Sentinel. The grove at the base of the Sentinel is on the moraine of this glacier, mostly covered lately by avalanche boulders.

Another lingered half a mile above the Sentinel, giving material for the grove that occurs in front of this portion of the wall. The upper part of this last moraine is covered by an earthquake avalanche.

Some of the rocky slopes which have furnished ground for the principal groves are almost wholly the results of earthquake shocks, only a small portion scarce appreciable having been derived from ordinary weathering. Many of the groves are growing upon soils that have no direct relation to moraines, such as that one which runs out into the meadow in front of the Yosemite Fall. This one is growing upon a rocky slope that was formed by Yosemite Creek during some great flood, the material having been derived from an earthquake talus, the result of an avalanche that was shaken from the cliff on the west side of the falls.

Another fine grove, showing no dependence on shadows, extends from the foot of Indian Canyon well out toward the middle of the valley. This grove is like that in front of Yosemite Creek and is growing on a delta-shaped flood deposit, of moraine material mostly.

And all around the walls on both sides of the valley, the width of the forest border is seen to depend upon the amount of wash of floods, excepting the moraine groves, which, as we have seen, depend on shadows.

All the high ground in the bottom of the valley is rocky and has been formed by floods, earthquakes, and glaciers, and all, without exception, is more or less clothed with forest trees, such as firs, pines, cedars, and a few oaks.

River Bends. It will now be easy to account for the meandering course of the river through the valley. It is jostled from side to side by the rocky slopes upon which the groves are growing. The first great bend at the head of the valley is evidently measured by the moraine slope of the residual Illilouette and Glacier Point Glacier. Then the wash from Indian Canyon shoves the river back against the south wall. There it is shoved north again by the wash from the Sentinel Rocks. Then south again by the wash from the canyon between El Capitan and Three Brothers, with minor bends between these main bends due to minor taluses, moraines, and flood washes.

Thus it appears that everything here is marching to music, and the harmonies are all so simple and young they are easily apprehended by those who will keep still and listen and look: however far these harmonies may extend beyond our powers, they are simple enough on the surface. . . .

Since I came to Sunnyside this time, the snow mantle has become thin and broken on the north side of the valley, wasting like a cloud. But a strip still remains unbroken on the south side beneath the main shadows of the walls.

Yosemite Fall. Suppose that in the midst of its headlong descent, with all its whirling fairy fabric of spray and rushing comet masses, the fall were suddenly frozen solid and carried bodily out into the middle of the valley, where we could go around it and see it on all sides. In the sunshine, what a show it would make — a colossal white pillar . . . lavishly and intricately adorned with airy flowing drapery exquisitely sculptured as if chiseled out of whitest marble! . . .

January 17.

Bland, warm. Fly people abundant in merry dancing hosts, keeping Nature busy.

The finest grove of live-oaks is here on the east side of the Sunnyside garden, and just beyond it a grove of Libocedrus

with undergrowth of smaller oaks, laurel, Rhamnus, etc., together with pentstemons, mints, bahias, rock cresses, ferns,
mosses, honeysuckles, Gnaphalium, and Aplopappus. Here is
the finest scenery of Sunnyside itself, with glorious views of the
valley.

A beautiful fringe of Adiantum grows on a seam near the side
of the fall whose delicate fronds float and waver in response to
every movement of the moist breath of the water.

January 19.

Long cobweb streamers adorn the dead ferns and goldenrods,
flying loose in the air to a great height. Insects are daily increasing in numbers and vigor — a glad, wonderful multitude.

January 20.

The sunny, balmy air is mottled with insects, buds are swelling,
and all sounds and sights more than half belong to summer.
Water-ouzels are songful.

January 22.

This cluster of January days so balmy and so bright is a
winter summer. Butterflies know no better than to come forth
in search of flowers.

January 23.

The few fibrous brushes of white cloud are too transparent to
materially impoverish the sunshine.... 'Cracks' would not be
of any controlling power in the formation of yosemites, because
when the bottom of an ice-stream is forced into a narrow, tortuous lane, it is very nearly passive, being deprived of power by
isolation from the main current. Many instances of this death
of ice may be found in canyons above Yosemite. In general,
the narrower a canyon-bottom is, the less it has been pressed and
polished and enlarged. Only water-streams are greatly influenced
in action by fissures.

January 25.

A half-bright abundance of white and gray and black and irised clouds ... wandering hither and thither far out over the summits of the range, and around the valley-walls. Tissiack is veiled.

January 26.

The falls are glorified with iris. On the black rocks east of the Lower Fall the changing winds spread a fabric of spray in colors fine and pure as ever tint the lips of shells. Thus the savage black rock is known by the company it keeps and the clothes it wears to be unsavage, pure, divine. And in all the blackness of the world, moral and material, there is a mantle, more or less variable and visible, of beauty. Strange that the falls should shoot their comets and booming thunderbolts through so gorgeous a storm of color without halting or swerving in gait or speed. Once I saw the fall halted by a wind upon which it poured and bulged flat out as if falling upon a floor.[1]

January 27.

Insect wings are abundant as blades of grass. The ouzel sings with increased vigor as the stream-music becomes louder.

January 28–29.

Not a cloud in the bright and windless sky.

Yosemite Valley Floor.
January 30.

Snow all day — eight inches. About 3 P.M., a storm swept up the valley with great force and steadiness. Flakes seemed to go on without thought of alighting. The meadows have become a boundless plain — not a rock is to be seen. I observed an ouzel in the ripples of the river, dipping and feeding and singing, giving no heed to the storm.

[1] See Muir's *Mountains of California*, vol. II, pp. 176–77.

January 31.

This morning one foot of snow. The day snowily inclined, with few narrow minutes of sun. At 10.30 P.M., snowing fast. Sunshine weeks and snowshine weeks placed together.

February 1.

Two feet of snow fell last evening. Still snowing all day and this evening. Calm, the air full of snow as if coming from inexhaustible fountains. The snow is damp at the bottom of the valley; therefore it is clogged and aggregated on all kinds of foliage, and branches, and old stumps and rocks. It lies in largest masses on the flat fronded branches of firs and the mounded close foliage of the live-oaks, and it bends and welds together the tassels of the pines.... The fall is booming grandly, but is seldom seen on account of continual snow.

The ouzel [1] is on his favorite feeding ground. He dives nineteen times in forty seconds. He heeds not the roar of avalanches, the heavy masses of snow from banks and trees, and the constant upspringing of pines. He would not cease singing or feeding for an earthquake. Waters of rapids when they flow under a muffling snow-bank are full of tones identical with those flowing under the feathers of an ouzel. Jay sings as if a piece of melting ice were in his throat. An eagle perches on a dead Libocedrus, allowing the snow to collect on his shoulders. Woodpeckers are busy pecking at the undersides of oak limbs, and on knots, passing the time, saying little beyond a few complimentary nods on meeting.... The sunny delta of Indian Canyon is a favorite abode of birds.

February 2.

Still snowing this morning, steadily and fast. Rocks capped grandly, entablatured in three marking days of storm. Pines gloriously loaded and massed, grand caves of darkness with bits of brown trunk corniced. Jubilee of snow descending ceaselessly from the fields of light.

[1] See Muir's *Mountains of California*, vol. II, pp. 12-15.

When a boy in bonnie Scotland I was more exhilarated with the coming down of snow than with the springtime fragrance of the hawthorn hedges or ... the glorious sky music of the lark and mavis.

February 5. 10 P.M.

Sky clear, moon shining over Eagle Peak, casting fine shadows on the meadows. Calm. Trees splendidly laden, mist rising from the river.

February 5.

The storm past, every feature of the valley is shining in new beauty. Every rock and tree and stump, leafy or leafless, is blooming in glorious snow-flowers. No mountain rock of the walls wears snow robes with so queenly a mien as Tissiack. She looks alive, sublime in every feature. The massed sublimity of snow jewels laid on every pine and bush and carved tablet and niche brings them forward. Snow avalanches from rocks continue all day, with noise and appearance of waterfalls bursting into existence and fading suddenly, like meteors. Also the constant thuds from heavy-laden pines,[1] of snow avalanches or cascades, disturbed by a wind breath or sun gleam. The topmost accumulations on tall pines are frequently set free, and as they pour through the other lower branches jostle their white fur also, and thus frequently the whole snow-laden spire is enveloped in a cascade of dusty broken snow. A sudden breeze on the breaking up of a storm will dislodge vast quantities of heaped-up snow, from oak and pine groves at once, filling the air as with a cloud. Many unsound branches are thus removed and Nature's orchards pruned. The live-oak collects a smoother crown of snow and wears it longer than any other tree. The black oak has a beautiful appearance — black trunks and limbs laden with intensely white unshaded snow.... Pines have dark caverns beneath the whorls of snow-bent aggregated branches; so also has the Libocedrus, but these are not so dark. Fine bland, cloud-

[1] See Muir's *Our National Parks,* pp. 272–73.

less shine all day. Tree and rock shadows on smooth bossy snow-sheets. . . . This evening, the valley is filled with pale, half-transparent frozen vapor. . . . The banks of the river are nobly rounded and shaded. I place cans for birds in feeding-places. . . .

February 6.

A day of open, effortless, gentle sunshine which poured slant-ingly into the deep snow of the meadows and set free much that was variously lodged on the north wall, keeping up a continuous cannonade. Mild evening. Enchanting moonlight.

February 7.

Smooth equal sunshine. Squirrels beginning to venture out. Snow delicately colored rose towards evening. Booming of snow avalanches still heard. Poor flies creeping on snow; many small wings folded never to move again.

February 9.

Five and a half inches of snow. Clear at 10 P.M. Trees grandly snowed. Rocks muffled.

February 10.

Today a fine bland block of tranquil light. Booming of ice from edges of the fall. Moon centered in a halo of noble dimensions.

February 11.

Sublime clouds around valley-walls at a height of two thousand feet, in huge embankments, frequently fired atop with sunshine. Falls coming out of clouds. . . . Blendings of cloud and rock, fine beyond thought.

February 13.

Snow last evening — four inches — sufficient to robe and blossom the rocks and trees. Magnificent cloud effects. Huge banks on Cathedral Rocks. Mild and gentle wintriness.

February 14.

A fine sunful, ripe day. At night, Venus, glowing gloriously like a moon, casts distinct shadows from the rocks and trees.

February 15.

Slight rain, then snow at night. Veiling of the walls, peak after peak disappearing in gray, trees gradually dimming to indistinct forms. . . .

February 16.

One or two inches of snow fell last evening. This forenoon, mostly sunful, preluded a gradual coming on of clouds marshaled about noon in most imposing abundance and order all along the upper half of the walls. Their bosses and fringes were now aglow in the white fire of the sun, now gray in shadow with delicate tones of purple. About 1 P.M., I was roused from a book I was reading by a most earnest and powerful wind-rush. Looking out from the window, I saw that the valley was filled with a mighty river of air, muddy and opaque from the plentiful dust of broken snow-flowers, which, as they stormed swiftly and horizontally past up the valley, afforded plain proof of the velocity of the wind. Never before have I beheld so sudden and loud a rush of storm. Anxious to experience as much as possible, I ran out to the meadow. The pines swayed and waved and sang in sublime manner. . . . A group of nestlings could not show more glad eagerness at the approach of their parents with food than did these pine groves at the coming of the snow. One has not seen a pine tree in its grandest mood who has not partaken with it of the banquet of winter storms. Perhaps to clear-visioned angels a pine tree feasting tranquilly on summer sunbeams is more interesting than when, lion-like, it is consuming the grand storms of winter snows and winds, but to mortals the sublimities of sun-storms are too spiritual to be appreciated.

The air when one looked into the sky was streaked, marbled with dark sooty veins, apparently by the unequal distribution of the snow.

The storms of winter which so exalt and glorify mountains
strike terror into the souls of those who are unacquainted with
them, or who have only seen the lights of cities, but to anyone
who is in actual contact with the wilderness, these storms are
only emphatic words of Nature's love. Every purely natural
object is a conductor of divinity, and we have but to expose
ourselves in a clean condition to any of these conductors, to be
fed and nourished by them. Only in this way can we procure
our daily spirit bread. Only thus may we be filled with the Holy
Ghost.

February 20.

Loud thundering of ice from the borders of Yosemite Fall
continued all day. I witnessed a grand snow avalanche back of
Leidig's winter house, exactly like a fall of water. When it
arrived at a horizontal bench, it filled it as a stream does a basin
and then flowed on.

February 22.

Our outside mountain doors are well locked, and who will
be able to disturb us? All this grandeur of sunlight and storms
was made for ourselves. A water-ouzel sings most joyously —
many of his notes are pure laughing. He cares not for my noisy
axe which I am plying within twenty yards.

February 23.

The pines are stripped of their snow-flowers and stand motionless in hushed sunlight. . . .

The vegetation of the Sierra does not slant up against the
cold frosty sky of the summits to end in a sharp colorless edge
of lichens. There are ten flowering plants of large size that go
above all of the pinched blinking dwarfs which almost justify
Darwin's ungodly word 'struggle,' and burst into bloom of
purple and yellow as rich and abundant as ever responded to the
thick creamy sun-gold of the plain.

February 24.

Rain during the night changed to snow at daylight, accompanied by a strong wind which carried the snow in fine dust horizontally through the trees and over the meadows. Fair and bright with grand cloud effects at noon.... An imposing snow avalanche fell from the narrow gorge east of Yosemite Fall, lasting more than a minute. It was at first a homogeneous roaring mass of dazzling brightness, then most of the back-streaming, enveloping snow-dust disappeared and the body of the avalanche was seen as distinct in structure and motion as a waterfall, thinning until the gray rock showed through, then leisurely closing.

February 27.

A water-ouzel, looking like a little clod of mud, is singing gloriously on the edge of the river, a mellow laugh in his song. The wind blows cold and sighs stormily among the pines. The sky is black. The falls tell of a storm to come. The ouzel reads all lovingly.

February 28.

A few hailstones and snow crystals fall, but it is mostly sunshine. The snow and ice of Yosemite Fall are constantly booming. The cone at the foot is over three hundred feet high.

March 1.

A fine snow cascade east of Yosemite Fall at 2.40 P.M. It paused momentarily in a forest of dark live-oak, then reappeared. ... Universal and immovable repose characterize all of the deeds of God. Repose is as visible in the so-called ragings of storms and crash and roar of avalanches as in the sleep of mountains in sun-calm.

March 5.

The wind is swaying and stripping the falls to shreds. Dim all day.... Snow. Ouzel sings angelically.

March 7.

The snow was hard in the morning, making a pavement which I tested in a rollicking walk.

Sunnyside Bench.
March 10. Before sunrise.

Owing to the great depth and sheerness of its walls, and to the westerly trend of the valley, the difference between the climates of the north and south sides of the Yosemite, both in summer and winter, is far greater than would at first sight appear, greater indeed than between countries on the same level hundreds of miles apart in a north-and-south direction. For not only is most of the south wall in constant shadow all winter, but it is covered with snow wherever snow can lie. Meanwhile, the north wall is bathed in sunshine every clear day both winter and summer, and this sunshine beats down vertically or nearly so, raising the temperature of the rocks far above that of the air. Moreover, the walls of many a recess radiate and reflect the heat from side to side, reverberating until an almost tropical climate is produced. The granite on the north side is, of course, also much warmer in summer, and much of that heat is stored up. Therefore the winter snow that falls on these rocks is speedily melted off without the aid of winter sunshine. Flowers bloom there every month of the year, and butterflies are hatched out and may be seen any day, except when storms are in progress and for a few days after they have ceased. All the birds know this difference of climate well and resort to the warmest nooks of the north side to spend the winter; so do the people who winter in the valley, though they are less generally and completely compliant with the weather, some of them not being able to afford both a winter and summer residence. In the Swiss alpine valleys, the same annual migration of half a mile or so is made.

This morning I rolled some bread and tea in a pair of blankets with some sugar and a tin cup and set off for my favorite Sunnyside Camp on the first bench of the north wall, east of the head of the lower Yosemite Fall, about five hundred feet above the level of the valley. It is a charming spot with abundance of water close at hand, a wild vineyard, fernery, and flower garden with picturesque groves of live-oak, Libocedrus, and pines, and views up and down the valley, while the interesting gorge between the upper and lower falls is near enough for sauntering to at one's leisure, and beside these advantages, it has an easy way out of the valley and up to the higher forests, by way of Indian Canyon. Birds, too, to keep one company, are here, and views of morning and evening lights and now and then noble storm scenes, with many of Nature's Yosemite extras thrown in from time to time. A' that and a' that and far more are the advantages of my Sunnyside Camp even in winter.

There is no frost this morning on the north side of the valley, and no snow is left on the walls, while the south wall has about half its surface covered, and the floor of the valley over the meadows presents a sheet of smooth unbroken white, which yesterday morning was frozen hard enough to bear my weight in walking over it. Here and there a few brown willow-tips appear above the snow. And in groves along the edges of the débris slopes are patches of bare ground, also beneath the tall pillars of the yellow pine where the snow has been shed off by the outsweeping branches.

A Small Cascade near Camp. As I climbed the bench of Sunnyside, I came to this small, slipping, whispering cascade coming down the smooth granite brow above the terrace in a thin sheet of lace. In front of it on the bench terrace there is a fine garden of Woodwardia and wild vines wherein most of its waters are absorbed. I dipped my face into it and drank, as it rippled over my head and shoulders in sparkling spray. Living water — who knows how much of life these mountain streams carry! I shouted and rubbed my palms against its tiny ripples, devout as a worshiper of the Ganges or Nile. Then I stripped and bathed in it, exhilarated as never before by any bath in any lake or stream in all the Sierra.

Sunrise.

The edges of Mount Starr King begin to glow silvery white.
Then one by one the burnished bosses of Glacier Point catch
the light and flash it off in wider masses; now it moves down
to the green bush-clad débris at the base. First Eagle Point and
the forest spires beyond Cathedral Rocks catch large masses of
sunlight. The sunshine on Eagle Point is shaped like a wedge
reaching downward as the sun rises higher. Now it touches the
tips of the pines, and in a few moments they are aglow. Pines
never are more impressive than while receiving their sunshine
breakfast. From tree to tree in slow economy the sunshine creeps,
in light-wedges widening and lengthening, touching the valley
floor and thrusting a long lance-point across the meadow toward
the base of Sentinel Rock.

Now the grove of black oak, leafless but already tinged with
purple from swelling of the buds, catches the light, then one pine
in the upper meadow, and another in the opposite grove. Soon
the wedge of Glacier Point meets that from Eagle Point, and
gradually these widen until in half an hour the valley has a light
belt three hundred to four hundred yards wide. Each black
trunk of the oaks casts a well-defined blue shadow on the snow.

No part of the fall is visible, only an uprising, boiling mass of
irised spray above the rocks. Looking towards the sun the glossy
leaves of the live-oak sparkle and spangle as if wet with dew.

Now comes the sun over the shoulder of Tissiack, and speedily
all the bottom of the valley is lighted as far as Lamon's, or to a
line drawn from the foot of Washington Column to Glacier
Point.

At 7.22 A.M. the wedge of light reaches from the base of the
north wall over the meadow to the river, and fifteen minutes
later, as far as to the rocky point a fourth of a mile farther up the
valley, where it meets the light from Glacier Point. A few minutes later, all the groves of Indian Canyon catch the glow and
immediately the birds that dwell there begin to stir. The jays
chatter and chuckle in low, tranquil tones as well as scream.

Bird song and chatter increase as the sunshine grows in warmth, and now from the groves of Indian Canyon comes a fine hearty chorus of morning praise.

The falls, like the birds, sing louder as the sunshine is felt, melting the snow and icicles along the banks above the walls. No part of the Upper or Lower Yosemite Fall is visible from my camp, save the upboiling spray from the foot of the latter. The spray, however, is the best of it, rising in grand convolving masses of brightly irised dust. None of it is white, but it seems to be wholly made up of finely beaten rainbows.

9 A.M.

The most noble of all avalanches glides from Glacier Point down into the forest with long back-flowing hair, noble mien and voice.

I came up to my camp by the path of a snow avalanche that the trail follows through laurel groves a little to the west of Indian Canyon. I noticed some sturdy, tough live-oaks that the rock and snow avalanche had cut off. One nearly three feet in diameter had been cut sheer in two by a single stone. Nature does some heavy pruning in her Yosemite groves, and at times roughly weeds her gardens. . . .

The scouring action of these avalanches on their beds is of a peculiar kind owing to the half loose, half firm way the graving and grinding grit and stones are held in the snow-current, and the mixed action of the clean snow itself and the wiping leaves and branches. Some of these channels are more than a mile long and are occupied every winter — a few minutes per year. Besides the Glacier Point avalanche, and the one just described, there is an avalanche channel on each side of the Sentinel Rock, and another between El Capitan and the Three Brothers, making five, corresponding to the number of the main waterfalls. There are also many others smaller and less marked.

A Snowstorm. When a snowstorm is coming on, the clouds descend and clasp the mountains from summit to base. Then follows usually an interval of brooding stillness. Small flakes and single crystals in glinting zigzags at length appear, falling with inimitable gentleness from the gray sky. As the storm progresses, the thickening flakes darken the air, and soon the rush and muffled boom of avalanches are heard, but we try in vain to catch a glimpse of them until rifts occur in the clouds and the storm ceases. Then, standing in the middle of the valley, we may witness the descent of half a dozen or more within the space of a few minutes or hours, according to the abundance and condition of the snow.

The boom and rush and outbounding energy of a great snow avalanche [1] far surpasses the conception of those who have never seen one. When the mass first slips on the upper fountain slopes, a dull rush and rumble is heard which increases with steady heavy deliberation, seeming to come nearer and nearer. Presently the white flood is seen leaping wildly out over some precipitous portion of its channel with ever-increasing loudness of roar and boom, decked with long back-trailing streamers worn off by friction in rushing through the air like the spray whorls and banners of a waterfall. Now it appears in an open spot, now shoots back of live-oak clumps and chaparral patches, leaping from bench to bench, spreading and narrowing, throwing out fringes of rockets... airily draped with convolving, eddying gossamer spray. These cataracts of snow, however unlike those of water in duration, are like them not only in form but in voice and gesture. In the snow-falls we detect the same variety of tones, from the loudest low hollow thunder-boom to the small voices in highest key, but there are none of the keen kiss-and-clash sounds so common in some portions of waterfalls where separate bolts strike full or glance on beveled ledges. But we see the pearly whiteness with lovely gray tones in half-shadows, the arching leaps over precipices, the narrowing in gorges, the expansion into lacelike sheets upon smooth inclines, and the final dashing into up-whirling clouds of spray.

[1] See Muir's *Our National Parks*, pp. 273-74.

March 11 (?). 9.15 A.M.

The ice-cone [1] at the foot of the Upper Fall is black in mouth, and ragged and broken like a crater. ... Within the crater there is a grand laboring of convolving, silvery-gray spray. The spray is heavy, and, when belched and whirled over the lip and cast up a hundred feet or more into the air, speedily comes down and follows the slope of the hill.

A wild play of light illumines the mouth of the cone, in the midst of the spray and water of ever-varying forms and densities. At first, this would be regarded as a type of wildest uproar and disorder, like a maelstrom, but it should be interpreted by the calm circle of light which environs it. Every dark and terrible abyss in nature is lighted with a like circle of Love. ...

The spray travels to a grove of pines, making the wet trees brightly, warmly green. Stricken with billows of irised foam they wave nobly. ...

Today I crossed a slope of avalanche snow at the foot of a brook. I found a place to stand where if one had stood a few days ago he would have lost twenty lives if he had had that many. God scatters firebrands, arrows, and death among the fairest and dearest of his mountains, but they are scattered as stars, orderly. ...

Last night I dreamed I stood with a friend on the edge of a precipice shaken by an earthquake. The rocks started to fall. I said: 'Let us die calmly. This is a noble death.' But it settled and we escaped. ...

On edge of gorge between
Upper and Lower Yosemite Falls.

It is a bright sunful day, two or three o'clock in the afternoon. You are facing northward with spray driven about you, but you do not feel the spray — the sights and sounds and tremendous energy of the crowd of waters preventing all knowledge of yourself. You are standing on the edge of a black deep gorge at the

[1] See Muir's *Mountains of California*, vol. II, pp. 170–74.

bottom of which you see the intensely white water, rushing with speed and sounds that inspire you at once with terror and admiration.

On the left hand you see the Little Diamond Cascade bestowing its tribute direct to the heavy flood at your feet. About seventy yards up the gorge you see the Lace Cascade adding its precious waters to the grand show. Just above the Lace Cascade you see two falls on the main stream near the head of the gorge, and yet farther up on the right side are three falls descending side by side from the ice-cone to the channel below. Beyond the ice-cone rises the plain sheer wall of gray granite, streaked here and there with dark stains. The sun is shining on its face, and sixteen hundred feet down the middle of the precipice floats the grand Upper Yosemite Fall. A widening column of water that falls so far through the air is no longer like water at all. It is more like the finest substance of gossamer cloud, but with displays of energy so tremendous and with such mighty voice we gaze upon it oblivious of all else.

The other falls near, with their marvelous forms and voices, are in the meantime hardly recognized before this king of falls. It takes all the eye like a blazing comet or meteor when one is star-gazing. Out of the blue sky into the black crater, the vast torrent pours irised foam that rises and falls, filling the air and transfiguring everything about it in spiritual light — gray cliffs, black wet rocks, white ice-hill, trees, brush, fringes, boulders, and the surging, roaring torrents that, escaping after tremendous churning, proclaim the triumph of Peace and eternal invincible Harmony. The combined voices of the many waters beneath and about you prevent all save the heavier of the tones of the grand fall from being heard. Look at the comet-shaped masses shooting out from the ledge two hundred feet below the brow of the fall. They launch into the free air with an assurance and grace of gesture that is altogether God-like.

The finest effects are to be seen about a thousand feet below the top. Lower, the general effect is more cloudlike, though even at the bottom there is no discord or confusion, while the rainbow dust makes all·divine, adding radiant beauty and peace to glorious power.

March 12.

Awoke in the night at 2 A.M. The huge mass of Yosemite Point, the lofty brow to the east of the fall, stood over me like a spirit. Its magnificent sculptures were revealed in the light and shade of the full moon, and above the brow, like a crown, was a silvery-gray cloud rayed on the edges like an aurora.

Tissiack, with broad shoulders snow-mantled, was flooded with the moon-storm of light, and she seemed to crouch while receiving her silvery baptism. The forests along her base were black and opaque, but the snow-sheeted meadows were white as the moon herself.

When I awoke again at sunrise, my live-oak canopy was glistening with the first sunbeams, whiter than usual from filtering through a filmy cloud. Heavier clouds began to form, with bosses atop that burned yellow and red. Cloud mountains took form over the brow of Tissiack. Soon the color vanished from them, and the heavy rocklike cumuli gave place to fleecy sheets of darker texture, which, sailing overhead, made eclipses at the rate of one every minute, now a pulse of shadow, now a pulse of sunshine. . . .

Blessed weather, all sunshine and moonshine — both the very essence of Light! Instead of pouring into the valley trough to be absorbed in groves and grassy meadows and gray walls as a river is absorbed in sand, or only to be flashed about on glossy leaves of oak and pine, the Light breaks in wide-sheeted spangles on a thousand wet mirror-rocks, and is beaten to a foam of reflected radiance on the snow that covers the meadows.

My camp is glorious in sounds, for not only is the Lower Fall near, but the Upper also. And all the falls and cascades between are blended into a massy roar like the sea in storm on a rock-bound coast, and marked by thunderous explosions of air caught

against ledge and pavement, with innumerable under- and over-
tones carried past on the varying currents of the wind. I seem to
be in the heart of the great Yosemite organ, and the sounds and
songs flow past in surging cascades like water, interrupted now
and then by storm-winds, but kept well together in the main, not
dissipated like spray or smoke.

Though I am thus in a kind of instrument, a fountain of music,
I cannot rest much, for masses of wind from the fall occasionally
come tumbling down on me with so sudden and heavy a pressure,
and are so suddenly removed, that I am left vibrating on the
elastic branches of my bed, while all about me is motionless.
Now and then a small airberg drops plump in my direction. . . .
These air masses are of a wide variety of form and size and temperature; almost all are rugged in outline and angular—more like
rocks than air would be thought capable of becoming.

Fern Ledge. On a bench leading out from camp to a point
above the Lower Fall is a beautiful fringe of ferns, kept fresh
with spray and pulsing in unison with every movement of the
fall, registering each sound and motion could we but read the
record. The maidenhair fern of feathery lightness is particularly
compliant to the fainter impulses from the waters, fairly floating
its fronds on soft wavelets of sound, moving each division separately at times, fingering the music delicately as if playing on
invisible keys. On the same shelf are bushes of white and purple
Spiræa, laurel, lilies, and mosses.

> Looking up from the foot of the
> first fall at the bottom of the
> gorge,[1] between Upper and Lower
> Yosemite Falls. 10.30 A.M.

Terrible energy, roar and surge, flapping, dashing, storm-
sustained exultant power!   Glorious maelstrom of irised foam
taking forms and movements of flame — a halo of beauty encompassing the wild uproar night and day. . . .

[1] See Muir's *Mountains of California*, vol. II, pp. 158–59.

Far up above the wall of the gorge two white threads issue from a grove. A little later having become a small cascade, after various feats of sliding and tumbling, they at length arrive at the brow fifty feet above the base of this wild fall.... The very noise might frighten the tiny stream aside, but without hesitancy it divides into three strips of lacework and pours straight down into the heart of the wildness. When two-thirds the way down, it encounters the buffetings of the irised flames, and the smallest one on the left is lapped and blown to spray and carried back over the brow to fall upon a plot of grass and lilies, and water a thicket of oak chaparral. The other two strips make out to break across the flaming rainbow down to the foaming fall.

This first fall is about twenty-five feet high. At its foot, the water enters a pool and is calmed, then passes immediately into another pool, from which it goes nobly over the wrinkled brow of the great Lower Fall, its surface adorned with diamonds.

Next above is a large fall of a hundred feet. The face of the rock over which it pours is not vertical, but, owing to the impetus already received from the steepness of its upper channel, it is enabled to bound forth with all the freedom of motion which a strictly vertical fall would enjoy.

A short distance above this is a third fall, occupying the deepest and most inaccessible portion of the gorge. A view of the head of this fall may be had from the east edge of the gorge immediately above it, and a view of the lower part may be obtained from a point one hundred and fifty yards farther down the edge on the same side, by holding on to a live-oak branch and leaping out over the rounded brow.

Between the base of the Upper Fall and the head of the Lower are six main falls, separated by a few deep rest basins, and snowy rapids and three tributary falls descending on the west side.... Thus if both the Upper and Lower Yosemite Falls were a-wanting, this black gorge, a third of a mile long, bright with this white out-blooming of nine falls and cascades, lovingly companied and adorned with ferns and fringing oaks and all the glories of

Light, would still be one of the most wondrous works of God, and of itself well worthy a visit to the Sierra. . . .

Though the dark hall of these glorious waters is never flushed with the purple of morning or evening, it is warmed and cheered by the white glow of noonday, and filled with rainbow foam as with common air.

Laurel sheds fragrance from above. Live-oaks — those fearless mountaineers — hold fast to angular seams and lean over with their bright mirror-leaves and fringing sprays. Lilies also look down.

These rocks have not been stained by the foot of man. . . . Here is an ecstasy of water and wild melody. . . . One bird, the ouzel, loves this gorge and flies through it, merrily stopping to sing, not on the water's edge, but down on a boulder half buried in foam.[1]

All terrestrial things are essentially celestial, just as water beaten into foam, crystalized in ice, or warmed into vapor, or steam, dew, or rain, is still water.

Light is dashed upon the earth and changed in appearance just as water is dashed into spray. All human love is in like manner Divine Love.

Wild Geese. The last two nights I heard wild geese on their way across the range, I suppose, to the Mono and Walker Lakes. They were traveling by moonlight.

A Rainstorm. After vibrating between cloud-shadows and balmy summery sunshine, dark clouds begin to gather today, and soon scatter rain on the grand landscape, falling this side the valley on sun-warmed rocks, making a fragrant steam from odors of leaves and flowers. I make haste back to camp to prepare for a wet day. The morning clouds, usually so silvery or rosy-golden, as if spread out for beauty and only as decorations for the sunshine, now become sober black and, losing all distinction of texture, darken all the valley.

[1] See Muir's *Mountains of California*, vol. II, pp. 158–59.

Bird song in the rain. . . . Delicious balmy, flower-opening rain, bathing and laving the great granite brows as gently as if refreshing a summer garden. Drip, drip down through the live-oak groves and honeysuckle tangles of Sunnyside. A fragrant steam foretells the coming of a thousand flowers. Winter-hidden beetles and chrysalids and worms respond drowsily to the call of the warm rain's reviving touch. Fern coils grow in every cell; all the life of Sunnyside, rejoicing, tingles with the gift of the mountain clouds.

Vapor from the sea; rain, snow, and ice on the summits; glaciers and rivers — these form a wheel that grinds the mountains thin and sharp, sculptures deeply the flanks, and furrows them into ridge and canyon, and crushes the rocks into soils on which the forests and meadows and gardens and fruitful vine and tree and grain are growing.

After the first hearty dash of rain, there is a lull of perfect stillness while the clouds crawl slowly in a squirming manner as if getting ready for steady work. At length the calm is broken by lightning that seems to strike midway between Glacier Point and the Royal Arches. This first stroke is followed by two others with long rolling billows of sound that beat on the massive walls in oft-repeated echoes. Then down comes the gentle rain, falling steadily all day. Fog-masses formed by the mixing of colder with warmer air creep here and there, while the outlines of the great headlands of the walls up and down the valley are softened and loom immensely higher in the gray misty atmosphere.

At night, the bottom of the valley is crammed full of black, heavy clouds with ragged prongs that look like the roots of upturned pines, much harder, more barrier-like than the rocks in this light. Later the valley is filled with silver-gray mists to a depth of about one hundred and fifty feet, and out of the level mist-bed tower the dark spires of the pines, as if rising out of a lake, while the oaks and smaller trees are submerged. Occasionally the mists vanish, leaving the snow-clad meadows and their border groves bare for a few minutes; then, coming down the Tenaya and Illilouette Canyons with imposing bluff-like fronts, they unite and roll over them again as the glaciers once did.

March 13.

Sunlight touched the top of the wall at 7.30 this morning, and in thirty minutes reached down to the upper comets.

The solid heads of the comet-like water masses of the Upper Fall [1] are close, dense white in color, like pressed snow from the friction of the air, the portion worn off forming the tail, which is drawn back like combed silk, faint grayish films of shadow appearing between the white lustrous threads, while the finer outer spray, whirling in back currents, is pearl-gray throughout.

Between the main comet masses are large cave-like shadows which descend with the water, often maintaining their forms unchanged to any appreciable extent throughout a descent of a thousand feet.

The shadows cast on the precipice along the side of the fall are jet black, and, being constantly in motion, are very striking in appearance when seen from the upper bench — the tips of the comet shadows distinct and chasing each other down the face of the cliff, mimicking or rather repeating every movement and gesture of the fall.

Tree Shadows. The shadows of the valley pines on the snow with their dense foliage are simple pillars of blackness, perfectly solid; but the leafless oak shadows are delicately outspread, every branch reproduced with marvelous fineness of detail, looking flat like a pressed fern, or like a black fossil fern of the coal laid on a white surface.

Live-oak. Gray limbs interlocked, outer sprays drooping low, swaying, pendulous as weeping willow, upper sprays erect,

[1] See Muir's *Mountains of California*, vol. II, pp. 151–52.

bristling, eagerly drinking the sun. . . . A splendid mountaineer!

The finest live-oak in the valley overhangs Tenaya Falls. It is about eight feet in diameter and its trunk is as irregular as a pile of boulders, while its long pendant branches reach over and dip into the spray, greatly enhancing the beauty of both falls and tree. The soil this ground monarch is growing on consists of boulders, rough and angular, six to ten feet in thickness, with a sprinkling of humus in the joints.

From such noble dimensions this species, in climbing out of the valley by the canyons of the sunny north side, is gradually dwarfed, until at an elevation of thirty-five hundred feet above the floor it assumes the form of a slender shrub a few feet in height, growing in dense chaparral patches, forming beds on which one may sleep softly like a Highlander on heather.

A Cliff Stream. A charming little cliff stream springs from the base of the precipice near the fall, and in descending to yet lower levels keeps close to the wall, into which it has not yet had time to wear itself a channel. Therefore on the smooth face of the cliff its unconfined waters are spread in a wide thin sheet, divided here and there by a slight unevenness of the rock. Then, meeting a horizontal bench, it is again united, forming a tiny network of lisping, purling, tinkling cascades in contrast with the thunder-toned fall at its side. Here it refreshes a mat of club-mosses, there a clump of ferns or grasses that have found a standing-spot on the cliff, or it slips proudly through a little garden of fern and lily, known only to itself and the birds and winds.

A Cliff Garden. Upon a narrow terrace twenty feet above the Sunnyside bench is one of the dearest of cliff gardens. It can be reached by climbing a seam in the cliff a few yards east of the cascade. The fissure terminates about six feet above the garden, from which point one may easily jump down into it, and climb out again by holding on to grass tufts. In this garden grow the soaproot lily, Brodiæa, larkspurs, pentstemons, purple and yellow

Mimulus, echeverias, and a few tufts of ferns, Pellæa and Gymnogramme, grasses and sedges and mosses in delightful bosses, and here and there a tall mint.

The cascade is here divided, one division flowing through the garden on the east side, the other on the west. A portion of the east-side branch glides into a grassy and shrubby tangle at the entrance of the garden, and after bathing the roots of the larger plants and the mosses that cover them, it runs out over dome-covered rocks in front and falls in a shower of drops which the sun colors like beads.

Small as this dainty wall-garden is, it contains two pools, both of them close to the wall; the bottom of one is gravelly, the other is muddy at the east end, but is clear at the west where the stream enters it.

March 14.

Irised Waters. On the very brink of the wall about forty yards to the east of the Upper Fall in some fissured crumbling rocks is a standpoint from which a fine view is obtained of the great fall, from within two hundred feet of the top to the bottom — perhaps the best view looking from above. A more impressive but nerve-trying view may be had from the rounded brow of the precipice close alongside the torrent as it leaps out into the air at the beginning of its descent. The view from here also extends from a point about two hundred feet below the head, where the fall strikes a ledge and bounds out in characteristic comet-shaped masses all the way to the bottom, where from the foreshortening effect the whole tremendous flood seems not to strike with rock-breaking power, but to sink and settle with infinite softness, like down in a still room. . . .

At twelve o'clock this spring day, the whole column is magnificently colored, the iris bow in front being broken up and mingled with the rushing comets until all the fall seems stained with rainbow tints and poured forth from some color fountain, leaving no white water at all visible from top to bottom. The solid masses seem to push the irised spray ahead of them, which streams back, enclosing the head of each comet until all the mighty flood is glorified. ...

The Upper Yosemite Fall,[1] before passing over the brink at the head, comes forward calmly and gracefully, swaying this way and that. About a hundred yards from the brink it makes a fine cascade of twenty-five to thirty feet descent, at the foot of which its waters are calmed in a deep round pool. After resting here it makes another cascade and enters another pool somewhat smaller than the first. From this second pool the stream glides swiftly down a hollow curve in a sheet of superb lace until it reaches the tremendous verge, where the lace structure is broken up and dashed into a coarse mealy mass in passing over some rough corrugations of the brow of the precipice. Thus prepared, it goes to its grand descent covered with a crisp mantle of glassy crystalline beads and foam.

Sunset.

From Sunnyside bench the trees in the bottom of the valley — Libocedrus and yellow pines — are beautiful in the mellow sunlight, the spires proudly erect, rising out of the snow-carpet, the foliage singularly warm and yellow, the tips free and as well defined as church spires though massed below. A few tall dead masts stand peeled from top to bottom, and one that has not been long dead still retains its leaves, which are rich brown in color.

Beyond the lighted portion of the valley the groves in shadow are strikingly somber by contrast, apparently as black as they look on a moonlit night. The meadow snow in shadow is cold blue.

Voices in the Wilderness. We seem to imagine that since Herod beheaded John the Baptist, there is no longer any voice crying in the wilderness. But no one in the wilderness can possibly make such a mistake, for every one of these flowers is such a voice. No wilderness in the world is so desolate as to be without divine min-

[1] See Muir's *Mountains of California*, vol. ii, p. 143.

isters. God's love covers all the earth as the sky covers it, and also fills it in every pore. And this love has voices heard by all who have ears to hear.

Everything breaks into song just as snow-banks in the spring burst forth in loud rejoicing streams. Yosemite Creek is at once one of the most sublime and sweetest-voiced evangels of the Sierra wilderness.

The evening sky is dark, the walls of the valley nearly black.... White clouds, ragged and loosely fringed, come stealing over the dark forest of firs and along the south wall from Cathedral Rocks to the Sentinel. The tops of the rocks are dipped in mist, but they still show dimly through it. Now a gap in the clouds reveals an island-like section of the forest that seems to float wholly separate from the earth.

Small flakes of cloud wander hither, thither over the groves, brushing the tips of the tallest pines and mazing lingeringly as if seeking something they have lost.... The falls sing stormily, for the wind is blowing hard and telling something unusual. This wind comes not down from the snowy summits of the range, but from the sea, distilling through the redwoods, threading ferny gulches and flower-stained bosses of the blue Coast Mountains, flowing over the San Joaquin plains now in golden bloom, breathing up rocky canyons, through sweet rosiny pine woods, pressing and breaking against these huge granite cliffs like sea-waves on a rock-bound coast, with the tones and fragrances of a thousand thousand flowers.

And to every rocky dome and brow of Yosemite and to every garden and grove of my Sunnyside is granted a portion of this precious wind in as special a manner as any prayed-for blessing bestowed upon saints.

Long lances of moonlight thrust through the clouds fall upon the meadows sharp and stiff like icicles....

Night shadows of the oak groves of Sunnyside are somber in day, black in the moonlight. Ebon bars are cast on the ground from branchless smooth boles so opaque and sharply defined that in my night walks I frequently step over them carefully as over fire-blackened logs that lie in my way. Or I have gone around the ends of the larger shadows as being too big to climb over. Sunnyside with its wealth of oak and mistletoe would furnish a fine retreat for Druids. However savage and superstitious, they would look noble, I fancy, beneath these gnarled limbs and drooping foliage. And what a grand place for a Druid temple! Granite blocks of every size for Stonehenge altars!

March 15.

A slight sprinkling of warm rain. Crag-like clouds stand round about the walls as if eager to be employed. The Cathedral grove is a favorite lounging place for clouds before and after storm. When set a-crawling and squirming by a sudden gush of sunshine, they glow whiter than snow.

El Capitan is thinly veiled from top to bottom by delicate gauze. Starr King is crowned with cloud, at times wrapping the whole cone warmly in plush folds, at other times becoming a level bar pushed down, leaving the top bare and sunlit. But what rock is so caressed and fondled by clouds as Tissiack?

The pine groves and the meadows of the level bottom of the valley do not cover unsightly heaps of fragmentary rubbish as a beautiful enameled lid sometimes covers a boxful of unsightly scraps. But a smoothly ice-planed floor of granite underlies the pebbles and crystalline sands spread out for the use of meadows and groves. The meadows become every year smaller as the surface is raised by the deposits of floods, while the area covered by trees is constantly increasing. In a few centuries all bottom would be forested were fire and the axe and trampling cattle kept out of it.

What is 'higher,' what is 'lower' in Nature? We speak of higher forms, higher types, etc., in the fields of scientific inquiry. Now all of the individual 'things' or 'beings' into which the world is wrought are sparks of the Divine Soul variously clothed upon with flesh, leaves, or that harder tissue called rock, water, etc.

Now we observe that, in cold mountain altitudes, Spirit is but thinly and plainly clothed. As we descend down their many sides to the valleys, the clothing of all plants and beasts and of the forms of rock becomes more abundant and complicated. When a portion of Spirit clothes itself with a sheet of lichen tissue, colored simply red or yellow, or gray or black, we say that is a low form of life. Yet is it more or less radically Divine than another portion of Spirit that has gathered garments of leaf and fairy flower and adorned them with all the colors of Light, though we say that the latter creature is of a higher form of life? All of these varied forms, high and low, are simply portions of God radiated from Him as a sun, and made terrestrial by the clothes they wear, and by the modifications of a corresponding kind in the God essence itself. The more extensively terrestrial a being becomes, the higher it ranks among its fellows, and the most terrestrial being is the one that contains all the others, that has, indeed, flowed through all the others and borne away parts of them, building them into itself. Such a being is man, who has flowed down through other forms of being and absorbed and assimilated portions of them into himself, thus becoming a microcosm most richly Divine because most richly terrestrial, just as a river becomes rich by flowing on and on through varied climes and rocks, through many mountains and vales, constantly appropriating portions to itself, rising higher in the scale of rivers as it grows rich in the absorption of the soils and smaller streams. . . .

Human Love. To ask me whether I could endure to live without friends is absurd. It is easy enough to live out of material sight of friends, but to live without human love is impossible. Quench love, and what is left of a man's life but the folding of a few jointed bones and square inches of flesh? Who would call that life?

March 16.

Snow-Banners.[1] A pale blue sky without a single cloud. A strong storm-wind roaring like the ocean among the deep lanes and bays of the valley. Sunnyside is protected, yet many a surging wind-wave breaks among her groves, shaking and worrying the live-oaks as if the winds, like bulldogs, had fastened upon them and would not let go while a leaf or branch remained. Such storms account for many a strange curve and kink in the grain of trees, so wide-topped and leafy and offering such leverage to wind-dogs. . . .

I set out at noon. All the way up the mountain the storm-notes were so loud about me that the voice of the falls was almost drowned by them. Noble wind-torrents were swooping down the Tenaya and Nevada Canyons, filling the valley, boiling and eddying in side canyons, glancing around the outstanding headlands and angles and sweeping upward over the rim of the valley as the greater glacier did when it was in its prime.

When I had reached an elevation of about eight hundred feet above Sunnyside, where the summits of the peaks of the Merced group come in sight over the shoulders of the South Dome and Mount Starr King, and the curves that sweep between them, I was surprised to see them adorned as I had never seen them before. From the wave-like crest of Mount Clark and the sharp pyramidal peak of Gray Mountain, there streamed magnificent banners of drifting snow showing in dazzling splendor on the black blue of the sky. These banners were at first curved upward from the narrow point of attachment, then continued in long-drawn-out horizontal silky-looking sheets, for a length of at least three thousand feet, judging by the known height of the mountains and their distances from each other. Relieved against so dark a sky, the effect was indescribably grand. The banner from Mount Clark was broad and variable, the breadth increasing with the distance from the point of attachment. That from Gray Mountain was the finest, the attachment delicate and clear, and

[1] For an account of snow-banners seen on what may have been another occasion, see Muir's *Mountains of California*, vol. I, pp. 47-54.

the widening of the swaying sheet regular. From the small serrated peaks between, smaller snow-flags were flying, and from
one of the Red Mountain summits also.

The tremendous currents of the north wind were sweeping up
the curves of these mountains just as the glaciers they once nourished swept down. A steady supply of wind-ground, wind-
driven, mealy, frosty snow was just being spouted up over the
peaks and then carried southward in those lustrous wavering
banners, the texture of which seemed as substantial as banners
of the finest silk, and their waving motions were as plainly visible as if they were on a flagstaff above one's head. There was
no confusion, no dusty opaque mass of drift anywhere to dim the
effect; all the snow seemed concentrated on these finely formed
and proportioned banners, and their movements were graceful
and noble beyond those of any flag or streamer I ever beheld.

Undated.

Fall Music. In the warm sunny season of May and June, when
the snow is melting fast, then the falls play their loudest music.
Heavy tones like those of a great organ and muffled thunderclaps
and gasping, drumming sounds occur at variable intervals and
are readily heard, under favorable circumstances, a distance of
four or five miles. The Upper Fall possesses far the richest as
well as the most powerful voice of all the valley falls. Its tones
vary from the sharp hiss and rustle of the wind among the glossy-
leaved live-oaks, and soft, sifted hush tones of the pines, to the
loudest rush and roar of storm-winds and avalanches among the
crags of the summit peaks. All these and many more are in such
abundance as we might expect when we know the richness of
the field whence these waters were gathered. . . .

Undated.

The Lost Arrow, or rather the isolated spire on the west side
of it, is to me one of the most wonderful rocks of the valley. For
about one hundred feet of its length it is entirely free and separate from the wall. This separation has been effected by the dis-

integration and falling away of a narrow vein of less resisting rock which united it to the main wall on the north. The action of a small stream has evidently aided atmospheric disintegration in separating it. Seen from a ledge of the valley wall, a little to the west of the arrow's point, it is very striking. Its cross-sections show it to be nearly a prism. There is a similar prism on Mount Watkins.

2. Trip to the Upper Canyons of the Middle and North [1] Forks of the San Joaquin River, Including an Ascent to the Minarets South of Mount Ritter [2] and to the Summit of the 'Matterhorn.'

August 13 (?), 1873.

I set out from the foot of Red Mountain,[3] August 11. Camped the first evening in the bottom of the West Fork double canyon. Then crossed the Merced Divide the evening of the twelfth, weary and heavy-laden, and camped at the bottom of one of the San Joaquin canyons between two lakes, on a narrow isthmus of rock upon which grew one living tree, a clumpy and well-developed *Pinus albicaulis*. A large dead one lay by it which I used for fuel. This canyon is broad and basiny, containing at least ten lakes, separated by bosses of bright gleaming rock, and rich forests of *Pinus albicaulis, Pinus contorta* and spruce,[4] and fine meadows, garden-like both in shape and flowers. I found here a new plant, Chænactis. And a new butterfly; also a hungry wood rat. He stole my spectacles and barometer during the night, from beneath a heavy stone. I recovered the barometer, the case much gnawed, but not the spectacles. Also the same animal may have stolen the lid of my teapot. At least here it disappeared.

[1] This stream called the North Fork of the San Joaquin River is really the North Fork of the Middle Fork of the San Joaquin. The true North Fork of the San Joaquin has its rise south of the Mariposa Grove.

[2] For the account of Muir's ascent of Mount Ritter in 1872, see his *Mountains of California*, vol. 1, pp. 61–85.

[3] Red Mountain (Peak) belongs to the Clark group, and is one of the fountain summits of the Illilouette Basin.

[4] *Tsuga mertensiana.* — W. L. J.

This morning, I climbed out of this into another yet grander canyon of the Joaquin, with noble side canyons adorned by all kinds of gardens and groves. The rock is slaty and granitoid, much veined and mixed and seamed like marble where cut smooth by the ice. I saw where a snow avalanche in the west-side canyon had broken off about as many pine trees as it had uprooted. Also it had slid over some young ones and left a few standing marked in such a way as to show the avalanche depth at that place to have been about twenty feet. Broken pines, two and a half feet in diameter, and sound, were broken off close to the ground. One was broken five or six feet from the ground, which grew again.

Noble mountains rise close about me, and far southward are the highest of the range with glaciers on their sides. A stream sings lonely.

Storm, thunder, hail on water, grasses wincing. Rush of new-made cascades.

August 14 (?).

Bright after rain. The sun is stealing down the opposite wall, touching every boss and cove in turn, now dyeing to the bottom every rock and meadow. The mountains are wet, washed, reeking. The birds are late; they do not care to come from their snug nooks in leafy cedars and pine, and dry places in the rocks. The butterflies are late, too; their wings are damp. How fresh every mountain, every flower! The sun is not an hour high, yet the gentians are open. . . . No sign of injury from hail — only here and there a petal lip hanging low. The cascade by my side sings with a new voice. . . . Raindrops and hailstones as well as glaciers and snow-banks are heard in it now. The huge canyon like one life seems born again.

An hour later: A hard climb up the canyon-side among wet bushes. I am now near the top of the cascade. A white sun-stained cloud glows in the notch at the top of the mountain —

the sun just over it. . . . The sun-silvered stream seems the cloud itself coming down, leaping glad from boulder to boulder, often high enough to break into smoke-like spray. About the base is a fringe of willows and tall purple spikes of Epilobium. Trees grow on the wall above on both sides.

All the animals were apparently ready to own me as a friend and brother today. At noon I slung down my heavy bundle, which was galling my shoulder, and sat down to cool and take breath beneath a Williamson spruce. After resting a moment I began to sketch a view up the canyon, and soon became wholly absorbed. After I had been working ten or fifteen minutes, holding my page steady with my left hand, a chipmunk came whisking beneath my upcurved arm and sat, evidently astonished, upon the middle of my sketch, then whistled and sprang off to a tree within five or six feet of me, and stood looking over at me, the very picture of puzzlement.

Last night I slept under a grand old, experienced juniper that had been nearly overthrown by a snow avalanche from the mountain dome above, but after coming down on his elbow, maintained his ground and grew and prospered, though never erect again. A grand tangle of branches and all kinds of recesses were about his roots. I cleared away stones and built up a wall to keep from rolling over a precipice of two hundred feet. An arm of the tree also held me. Just below my tree was a smaller one with protecting branches. On the very brink of the wall beneath is a deer's bed, a most complete concealment and resting-place and lookout. An hour or so before dark, I heard a sharp, loud snorting and, looking down into the valley spread maplike, I soon discovered a doe. She bounded up the mountain-side where bushes and a gentle slope made it possible, often stopping to look back at me. Meanwhile I sat perfectly motionless. My brown shirt was like the junipers, and evidently she lost sight of me. After long and most perfect caution and deliberation she turned and moved slowly towards me, stopping constantly to look and smell. Her movements as she ran down the mountain-side, over logs, over rocks, over all kinds of brush, were most admirable, never straining although making enormous leaps. After coming within sixty or seventy yards of where I sat, she sniffed eagerly like a dog, turning up her head and trying different directions, then, catching my scent, she bounded off suddenly, snorting loud and quick until out of sight round the edge of a small fir grove. Soon she came on again with the same caution, again caught my scent, and again bounded away. This she repeated four or five times, after which I ceased to watch her. I think her fawns were hidden away in the chaparral. It is always to such wild and unfrequented canyons that they go for the purpose of rearing their young. I have often seen their beds while forcing my way over and under the most impenetrable jungles of Ceanothus, manzanita, and chinquapin. This morning I saw two fawns groping their way down the mountain, called by their mother.

While I sat watching the deer as she bounded and snorted, a squirrel, evidently excited by the noise, came out and climbed a big boulder beneath me, and looked on at her performances as patiently and attentively as myself. Still nearer to me a fussy chipmunk, too heedless for such affairs, busied himself about a dainty supper which he obtained in a thicket of shad-bushes whose fruit was ripe.

Just before night a bird that was new to me came into the juniper as if at home. He was about the size of a canary, ash-colored, his wings dark and striped with yellow bars.

When I first came in sight of the various glacial fountains of this region, I said there must be a very deep and slaty-walled canyon at their meeting-place — a kind of slate Yosemite — and I feared it might be difficult to cross in going to the Minarets. I was right, for certainly this Joaquin Canyon is the most remarkable in many ways of all I have entered. An astonishing number of separate meadows, rich gardens, and groves are contained in the canyon, and it is a composite-slate Yosemite Valley with huge black slate rocks that overlean, and views reaching to the snowy summits. I knew not what most to gaze at — the huge black precipices above me, the beds of white violets and ferns about me, or the grand hemlock and fir groves with the young Joaquin (thirty feet wide, two feet deep, with current of three miles per hour) flowing through them. 'Surely,' said I, as I snuggled myself away among the roots of my juniper, 'this has been a big feast day. Plants, animals, birds, rocks, gardens, magnificent clouds, thunderstorms, rain, hail — all, all have blessed me!'

August 15 (?).

During the night a rascal wood rat (Neotoma) tried to steal my hatchet.[1] ... (Strange what frightening capabilities small creatures have in the night. The Englishmen at Lyell were afraid.) Before I went to bed I examined the various rooms in the big cedar, but discovered no nest of mouse or squirrel. However, in the night I was awakened from sound sleep by something about or on or under my head. After the performances of the light-pawed or -jawed rat that stole my teapot lid and spectacles and barometer, I kept my barometer in my bosom and made a pillow of my provisions, but did not think the hatchet in any danger. Yet, after repeatedly waking and watching, I caught sight of a large wood rat trying to drag it from beneath the tree by a buckskin string with which the handle was tied. I thumped with it on the tree trunk to frighten him, but in a minute he would return, chattering his teeth in a way provokingly like a rattlesnake.

On starting down the canyon-side this morning, I was surprised to find the largest of the four Sierra gentians standing about carelessly in the bush as if quite at home. ... Found it very abundant all the way to the bottom, and all over the bottom, and up the other side, in all kinds of soil and company and climates. Although one of the commonest plants, I had seen it but once before this side the Sierra.

Also, an hour before sunset, I discovered a noble anemone (*Anemone occidentalis*, not before found south of Lassen Peak),

[1] See Muir's *Our National Parks*, pp. 220–22.

gone to seed, and silky heads of Akenia an inch and a half in diameter. The flower must have been at least two inches. Growing
in among rough slate avalanche blocks on the east side of the canyon at an elevation of eighty-four hundred feet, just above the
entrance of the big lower Ritter Cascade, I found this, my first
anemone, in California. The sight sent me bounding to a certain
hillside in Wisconsin where the *Anemone nuttalliana* came in
clouds in spring, and a dozen species of goldenrods and asters
gathered and added gold to gold, and purple to purple in autumn.

August 16 (?).

Last night I camped between a close pair of lakes. Discovered
a new Composita plant. Glorious and yet more glorious groves of
Williamson spruce [1] — I might say I never saw spruce groves
before! Spruce cones at this stage of growth may have their
scales removed, leaving the bracts on the axis. Scales are beautifully marked and show plainly they are altered leaves.

Had a glorious thunderstorm. Thunder avalanches descended
the mountain-sides, crossing from peak to peak with suddenness
and splintering angular notes as if the sky was of cast iron and
had suddenly exploded to fragments. How the pines waved and
sang their rain-prayers! How every needle thrilled with joy!
How every grass panicle along the meadow-edges and every spiky
broad-leaved Carex fluttered and waved its flag to every lightning
throb! And when all had received their share, how deeply they
rested, and how fully they evidenced their regeneration!

Saw a boulder fall a thousand feet down the mountain-side this
afternoon, loosened by rain.

Watching the raindrops plashing and breaking upon the polished surfaces of the rocks near my camp, I could not but admire
their wonderful strength to bear all this direct hammering together with that of the hail, and all the corroding fogs and dews,

[1] See footnote, page 69.

and long-lying snows for thousands of years. Yet their polish is
as brilliant as the tranquil surface of a sunny crystal lake.

Two days before, when I crossed the Merced Divide, I felt I was
going to a strange land, yet not so, only turning over a new page;
and a blessed page it was, engraved with ten lakes and forests
and sculptured rocks uncountable. . . .
I never saw such a meeting of lowland and alpine plants of all
kinds — trees, sedges, grasses, ferns — as in this Joaquin composite-slate Yosemite. Perhaps it is because streams descending
directly from the summit mountains on both sides, bring down
all higher seeds, making a kind of natural botanical garden. It
has at least one Wisconsin darling — Anemone. No canyon is
dangerous that holds a single anemone — a sufficient redemption.

This day I found *Primula suffrutescens* gone to seed (altitude
ninety-five hundred feet). Had glorious thunder and rain close
up at the foot of Ritter and the Minarets! Oh, how they rose
above their resting stature when the rain bathed them and the
lightning played among their black spires! What tempest waves
of air dashed and surged among these rocks of the great air ocean!
After the storm I heard a grouse calling her young on the meadow-
side close by. I was sheltered from rain in a clump of Williamson
spruce. Will ascend Minarets tomorrow. . . .

The higher we go in the mountains, the milkier becomes the
Milky Way.

Ascending the Minarets
August 17 (?).

Set out early for the glaciers of the Minarets, and whatsoever
else they had to give me. The morning was bright and bracing.
I walked fast, for I feared noon rains. Besides, the Minarets have
been waited for a long time. . . . I saw six woodchucks — high
livers all of them, at an altitude of ten thousand, seven hundred
feet — on a sloping moraine meadow. They were fine burly
sufficient-looking fellows, in every way equal to the situation.

Water-ouzels in a lake diving.... This day I found *Primula suffrutescens* abundant nearly everywhere on my way, up to eleven thousand feet, one place in flower.

On the very summit of one of the passes of the Minarets, standing in the stiff breeze at an altitude of eleven thousand six hundred feet, I found two currant bushes new to me in species, and two half-benumbed bumblebees upon them, and a fly or two. The big Gentiana I found yet higher. Saw one with its corolla split down as if a hailstone had fallen into it.

I never before saw rocks so painted with yellow and red lichens as are the sides of the passes among the Minarets.

My camp last night was in the uppermost grove of a canyon running nearly south from one of the many deep recesses of Mount Ritter. I knew that the glacial canyon that drained away the ice from the many chambers of the Minarets ran parallel to this, and was separated from it by a huge wall impassable opposite my camp, but I hoped to be able to cross higher up. Was soon on top and crossed without difficulty. The Minarets were now fairly in my grasp. I had been crossing canyons for five days. This was certainly the last. Their appearance from here was impressively sublime because of their great height, narrow bases, linear arrangement, and dark color. They are the most elaborately carved on the edges of any slate summits I have seen. Four lakes lie like open eyes below the ample clouds of névé that send them water. These névé slopes are large, and wonderfully adapted in form and situation for picturesque effects among the black angular slate slabs and peaks. I observed the lines of the greatest declivity upon several of these slopes, no doubt due to the slipping and avalanching of dry winter snow upon their steeply inclined surfaces.... Wind streaks snow, but not thus....

I thought of ascending the highest Minaret, which is the one farthest south, but, after scanning it narrowly, discovered it was inaccessible.

There is one small glacier on the west side near the south which I set out to examine. On the way I had to ascend or cross many névés and old moraines and rock bosses, domish in form, with which the bottom of the wide canyon is filled. At the foot of a former moraine of this west glacier is a small lake not one hundred yards long, but grandly framed with a sheer wall of névé twenty feet high. ...

Beautiful caves reached back from the water's edge; in some places granite walls overleaned and big blocks broke off from the main névé wall, and, with angles sharp as those of ice, leaned into the lake. Undermined by the water, the fissures filled with blue light, and water dripped and trickled all along the white walls. The sun was shining. I never saw so grand a setting for a glacier lake. The sharp peaks of Ritter seen over the snow shone with splendid effect.

The ascent from this rare lakelet to the foot of the moraine of the present glacier is about one hundred feet. This glacier has the appearance of being cut in two by a belt of avalanche rock reaching from the first of the cliffs above the glacier to its terminal moraine. This avalanche is quite recent, probably sent down by the Inyo earthquake of a year ago last twenty-sixth of March. Slate is able to stand earthquake action in much slimmer forms than granite. ... This is the first instance I have discovered of a large avalanche on a glacier.

I wished to reach the head of this glacier, and after ascending a short distance came to where the avalanche by which I was making the attempt was so loosened by the melting of snow and ice beneath that it constantly gave way in small slides that rattled past with some danger. I therefore tried the glacier itself, but the surface was bare of snow, and steps had to be cut. This soon became tedious, and I again attempted to thread my way cautiously among the ill-settled avalanche stones. ... At length I reached the top, examined the wall above the glacier, and thought it possibly might be scaled where avalanches descended. I worked cautiously up the face of the cliff above the ice, wild-looking and wilder every step. Soon I was halted by a dangerous step, passed it with knit lips, again halted, made little foothold, with my hatchet. Again stopped, made a horizontal movement and avoided this place. Again I was stopped by fine gravel covering the only possible foot- or hand-holds. This I cleared away with the hatchet, holding on with one hand the while. It is hard to turn back, especially when the danger is certain either way. But now the sky was black as the rocks, and I saw the rain falling on the peaks of the Merced group. Besides, the rocks in the shapes they took seemed to be determined to baffle me. I decided to attempt return. After most delicate caution in searching for and making steps, I reached the head of the glacier. Cut steps again, and began to work my way down. Now the first lightning-shot was fired among the crags overhead. The wind hissed and surged among the spires. Ritter was capped with cloud, yet loomed aloft higher. Oh, grandly the black spires came out on the pale edges of the storm-cloud heavy behind them!... More and more lightning, but as yet no rain fell here, although I saw it falling about Lyell. I ran down the avalanche excitedly, half enjoying, half fearing. Clad free as a Highlander, I cared not for a wetting, yet thought I should endeavor to reach camp at once. I started, but, on passing along the face of the grand spire, made a dash for the summit of a pass nearly in the middle of the group of Minarets. I reached the top by careful climbing. The pass is exceedingly narrow, some places only a few feet wide, made by water and snow acting on dissoluble seams in the slate.... Walls of the pass were vertical, and grand beyond all description. A glorious view broke from the top — another glacier, of which I could see a strip through a yet narrower gorge pass of the east side, volcanoes of Mono, a section of a fine lake, also a good portion of the Inyo Mountains. A splendid irised cloud sailed overhead, the black clouds disappeared, while the wind played wild music on the Minarets as on an instrument.

I began to descend the east side of the glacier, which reached to the top of the pass. I had to cut my steps very carefully in the hard, clear ice, as a slip of one step would ensure a glissade of half a mile at death speed. I wedged myself most of the way through the lane-like gorge between the ice and wall, all the time cutting steps. Came out at last after many a halt for consideration of the position of my steps, and to look back upward at the wild, wild cliffs, and wild, wild clouds rushing over them. It was a weird unhuman pathway, and as I now look back from my campfire, it seems strange I should have dared its perils.

August 19 (?).

On again reaching the main portion of the glacier I was delighted with its beauty, its instructive crevasses and ice cataracts, and with the noble landscape reaching far south and west. This glacier is a perfect mass of yawning crevasses on the right side, as if its bed had been suddenly heaved up in the center. It ought not to be set foot upon by solitary explorers, as many of the most dangerous of the crevasses are slightly snow-covered even this late in the season. . . .

I descended the left side after deliberating whether I should not attempt the right in order to look down into the wide gaping crevasses — some four or six feet deep — but on reaching the terminal moraine discovered that I had narrowly escaped some crevasses that reached nearly to my track.

I sketched, and then ran northwestward around the great wall that formed the left bank of the glacier. Here I discovered another glacier, not so wildly broken and shattered, but exceedingly beautiful in its fine swooping curves, its splendidly sculptured walls, and its moraines and lakes and general scenery. In form it is nearly triangular, much crevassed near the head, and finely barred with curved dark-stained lines at the foot. I attempted the ascent up the middle of the glacier to the head, but it was so bare and hard I was compelled to cut every step in the most careful manner. Wearied of this I cut my way over to the snow-covered portion on the right side, which is steeper in its curves, yet much safer. I thought its crevasses all too narrow to be dangerous, but on looking into one was startled to find myself on the brink of one wide enough to take me in. The snow-bridge was not more than six inches thick above it. The ice was clear below a depth of four feet, but I could not see down more than ten feet owing to the curve of the walls. All these marginal crevasses have convex sides uppermost. I walked carefully, sounding with my hatchet, and reached the top safely. From here, on looking down the opposite side of the main Minaret wall, I was glad to see I should be able to get over to the west side, and so home to camp. The Minaret Pass is eleven thousand three hundred feet, and the glacier about five hundred yards wide by three or four hundred long.... I reached camp early, unwet, unwearied, unhurt, after a rich day whose duties domineered me.

August 20 (?).

In descending the canyon-side of the Joaquin slate Yosemite, I could see the young river beneath flowing from grove to grove and through meadows with here and there a strip of white that marked a cascade, but, in trying to follow it up among the mountains at the head, I lost it entirely. In vain I tried to discover its course by some green forest or willow strip. It appeared to sink. Only dry, empty canyons were seen, excepting the long distant strip of white water which came from the great west glacier of Ritter. In following up the river two days later, which is a favorite pursuit with me, I discovered the cause of my difficulty. The west dip of the slate is so developed at the upper end of the canyon that the stream is huddled up and made to flow out of sight, as if on its edge.

Three miles above the mouth of the Minaret tributary on the main stream there is a beautiful fall, forty or fifty feet wide. It is approached through a fine tall grove, and has the sublime background of the mountains.

Today I saw a weasel, a handsome fellow, trying to catch a young robin.

Going along this morning past a huge old juniper trunk which had been torn from the mountain above by an avalanche, I saw a grouse [1] rise from the side of it, and fall as if dead in the grass at

[1] See Muir's *Our National Parks*, p. 238.

my feet. She allowed me to approach within five feet, then flapped and limped and made dying squawks. I thought she must have a nest in one of the many nooks of the big trunk whose brown bark was hanging about it in flakes favorable for concealment and nest-making. I wanted to find the eggs, but, on going close to the log, was surprised to see young grouse rise from my very toes, nearly as large as their mother.

The grouse cares long and well for her little ones. I have often noted her motherliness in her notes to her young, and in her deprecating cries for their sakes, when man appears. How far ahead of even the oldest grannies among tame hens, in true wisdom! . . .

The first grand cascades on the San Joaquin are five miles above the Minaret tributary and flow from one small lake-basin to another lake-basin, showing how ice rises in passing the lips of basins.

> Camp at head of main
> Joaquin Canyon.
> Altitude 9800 feet.

Here the river divides, one fork coming in from the north and the other from Ritter. The Ritter fork comes down the mountain-side here in a network of cascades, wonderfully woven, as are all slate cascades of great size near summits, when the slate has a cleavage well pronounced. The mountains rise in a circle, showing their grand dark bosses and delicate spires on the starry sky. Down the canyon a company of sturdy, long-limbed mountain pines show nobly. All the rest of the horizon is treeless, because moraineless. How fully are all the forms and languages of waters, winds, trees, flowers, birds, rocks, subordinated to the primary structure of the mountains ere they were ice-sculptured! When all was planned and ready, snow-flowers were dusted over them, forming a film of ice over the mountain plate. And so all this development — the photography of God.

Linnæus says Nature never leaps, which means that God never shouts or spouts or speaks incoherently. The rocks and sublime canyons, and waters and winds, and all life structures — animals and ouzels, meadows and groves, and all the silver stars — are words of God, and they flow smooth and ripe from his lips.

The branches of no pine have so long a downward swooping curve as those of *Pinus monticola*. The ends are tufted and conspicuous, the bark redder the higher you go. And it is nobler in form, the colder and balder the rocks and mountains about it.

Never saw so many large gentians as today (*Gentiana frigida*). I counted thirty-three flowers on a patch not much larger than my hand — some flowers one and a quarter inches in diameter.

Ursa Major nearly horizontal, and has gone to rest. So must I. I am in a small grove of mountain hemlock. Their feathery boughs are extended above my head like hands of gentle spirits. Good night to God and His stars and mountains.

August 21 (?).

Clouds, dense and black, come from the southwest, over the black crests and peaks of Ritter, the lowest torn and raked to shreds. The cold wind is tuned to the opaque, somber sky. Now a little rain, snow, and hail.

I wish to cross to the east side of Ritter today, visiting the main north glacier on my way. I have never crossed these summits, and fear to try on so dark a day.... At noon, the clouds breaking, I decide to make the attempt, climbing up the fissured and flowery edge of the rock near Ritter Cascade. At the top of the Cascade I find a grand amphitheater, where glaciers from Ritter and peaks to the north once congathered — now meadow and lake and red and black walls with wild undressed falls and rapids and a few hardy flowers. From the rim of this I have a glorious view of high fountain glaciers and névés, with pointed spears and tapering towers and spires innumerable. Here, too, I find a white stemless thistle, larger than the purple one of Mono.

Yet higher, following the wild streams and climbing around inaccessible gorges, at length I am in sight of the lofty top crest of Ritter and know I am exactly right. The clouds grow dark again and send hail, but I know the rest of my way too well to fear. I can push down to the tree-line if need be, in any kind of storm. A glacier which I had not before seen heaves in sight. It is on the north side of the highest of the many spires that run off from Ritter Peak as a center. It is one of the best I've seen in this wild region, occupying a most delicately curved basin and discharging into a lake.

I come in sight of the main North Ritter Glacier, its snout projecting from a narrow opening gap into a lake about three hundred yards in diameter. Its waters are intensely blue. The basin was excavated by the Ritter Glacier and those of Banner Peak to the north at the time of their greater extension. The glacier enters on the east side. On the west is a splendid frame of pure white névé, abounding in deep caves with arched openings. One of these has a span of forty feet, and a height of thirty. This frame of névé comes down to the water's edge and in places reaches out over the lake. Most of its face is made precipitous by sections breaking off into the lake after the manner of icebergs. These are snowbergs. The strength of the glacier itself is so nearly spent ere it reaches the lake, it does not break off in bergs, but melts with many a rill, giving a network of sweet music.

Wishing to reach the head of this glacier, I follow up the left lateral moraine that extends from the flank of the mountain to the lake until I reach the top. Then I try to cross the glacier itself where it is not so steep. But I find it bare, unwalkable ice. The least slip would send me down a slope of thirty-five degrees, a distance of three hundred yards, into the lake. I begin to cut steps, but the snout of the glacier is so hard, and every step has to be so perfect, that I make slow and wearisome progress with blankets on my back. I soon realize it will be dark ere I succeed in crossing, therefore I resolve to retrace my steps and attempt crossing at the foot of the glacier close to the lake where it is narrower, and if I find that too dangerous from crevasses, to go around the lake, which last alternative, from the roughness of its shore, would be no easy thing. I find the ice on the end of the glacier softer and cross, then push rapidly up the right moraine to where the glacier becomes much more easily accessible on account of the lowness of its slope and because its surface is roughened with ridges and rocks.

In going up between the edge of the glacier and the moraine I discover the main surface stream of the glacier has cut a large cave, which I enter and find is made of clear-veined ice — a very unhuman place, with strange water gurgles and tinkles.

The lower steep portion of the glacier is alive with swift-running rills. A larger stream begins its course at the head of the glacier, and runs in a westerly direction along near the right side. It has cut for itself a strangely curved and scalloped channel in the solid green ice, in which it glides with motions and tones and gestures I never before knew water to make.

The glacier receives its snow from the north side of Ritter. It has two motions, one nearly north, the other W. 10° N. In flowing north it is soon stopped by the steep wall of Banner Peak, and its course is then westward down the canyon. In grating along the right bank, pressed hard against it by the northward thrust, it grinds the stones of its moraine in a way that crushes off their angles, and gives them forms that I never before could account for. It also gives to the bank thus crushed a finish I have often observed without knowing its cause. It is similar to that of a gorge down which earthquakes have sent stone avalanches. I have been noting for some time the curved dirt lines on the snouts of glaciers. It is well seen here, apparent in sections made by the stream, and in some which I made with an axe, that this surface marking consists of the weathered ends, or edges rather, of the successive layers of which the whole glacier is made up by the seasons' snows. Wherever a line with greater quantity of dirt appears, a section of the clean ice beneath always shows that the greater dirtiness is correspondent with a seam or layer of harder, bluer ice. These layers of ice are curved more or less in a vertical direction — some places being nearly on edge. This bending of the layers proves a bending of the whole glacier — always most bent where most opposed in its motion. The motion of these small steeply inclined glaciers is not at all in the direction of their surfaces throughout all their thickness, but rather in the direction of their bottom curves. That is, when they occupy basins of this form, which most of them do. This upward motion of the lower ends of glaciers explains not only these vertical stratifications but the condition of their terminal moraines, which always have slopes exceedingly steep, as if added to and fed from the bottom upward with outer face convex. The motion of ice being greater in the middle of the glacier would make the surface of the terminal moraine convex. ....

This handsome glacier is not much crevassed, but has a large, conspicuous schrund, and is finely curved, both from slip of winter snow and from the swedging of its flow. Reckoning the long and much-crevassed glacier between the Minarets and Ritter on the east slope as belonging to Ritter, and also those of the large half separate crest on the north of Ritter, then the glaciers of Ritter number six together, having many large névés with feeble glacial motion. All of these glaciers are in picturesque basins, and are exceedingly beautiful in their edges, which comply in most graceful lateral curves to all of the salient and re-entering angles of the black basin rock, and in the vertical curves of the bottom.

When my observations on this most interesting glacier (the main North Ritter Glacier) were finished, it was near sunset, and I had to make haste down to the tree-line. Yet I lingered reveling in the grandeur of the landscape. I was on the summit of the pass, looking upon Ritter Lake with its snowy crags and banks, and many a wide glacial fountain beyond, rimmed with peaks, the wind making stern music among their thousand spires, the sky with grand openings in the huge black clouds — openings jagged, walled, and steep like the passes of the mountains beneath them. Eastward lay Islet Lake with its countless little rocky isles; to the left, the splendid architecture of Mammoth Mountain,[1] and in the distance range on range of mountains yet unnamed, with Mono plains and the magnificent lake and volcanic cones,

[1] Mr. Muir probably refers to the mass which includes Electra and Foerster Peaks on the Merced San Joaquin watershed.

sunlit and warm, between. To the eastward over the Great Basin swelled a range of alabaster cumuli, presenting a series of precipices deeply cleft with shadowy canyons, the whole fringed about their bases with a grand talus of the same alabaster material. Here and there occurred black masses with clearly defined edges like metamorphic slate in granite. Beneath these noble cloud mountains were horizontal bars and feathery touches of rose and crimson with clear sky between, of that exquisite spiritual kind that is connected in some way with our other life, and never fails, wherever we chance to be, to produce a hush of all cares and a longing, longing, longing. . . .

My feet soon awakened to their work — a mile of rock-leaping on slopes of water-washed moraines — and I soon came in sight of a wind-bent *Pinus albicaulis* that ensured fire for the night. I walked on soft cushions of Carex, and I saw here and there a daisy and a stone fringed with Bryanthus. I descended a long, icy névé, felt weary, and halted in the first dwarf-pine thicket I came to. The wind blew wildly all through the night. I hedged myself and fire roundabout with boughs to deaden the blows of the wind that came down in craggy avalanches. The outlet of a little glacier sends down its stream over a precipice near-by, making music that is heard in the pauses of the wild wind. . . .

A new bird came to look at me just at dark.

Near-by I found a skull and a pair of the big horns of a mountain sheep. I struck the skull between the horns a dozen blows with the hatchet without breaking through. Although very old, it is far stronger than bulls' heads. Hunters' stories that the sheep leap from precipices, alighting on their heads, may be true. Certainly such heads would not break so easily as legs. . . .

August 22 (?).

Morning clear and lovely. The stony isles of Islet Lake[1] are lighted as if the sun arose for them alone. The finely curved

[1] The full name is Thousand Islet Lake.

mountain-fronts to the north show forth their arches and pillars to fine advantage with the splendid arrangement of light and shade. The sun silvers the cascade as if getting its waters ready for the spangles of the lake. The glaciers are tinted with rose, but ever and anon comes a black, foggy cloud down-sweeping from the western summits, and applies its chill, somber folds to the surface in every part, filling also all the chambers and vaults of the black walls around them.

I moved over into a valley belonging to a cluster of glaciers nearer Lyell, ready to ascend Matterhorn,[1] and camped on the edge of a glacial lake in the protecting lee of a clump of *Pinus albicaulis* at an altitude of ten thousand four hundred and fifty feet.

August 23 (?).

Wildly came the wind all through the half-lighted night, clouds black and cold hanging about the summits. But will attempt the Matterhorn. Coffee, and away free-legged as any Highlander. Arrived at the summit after a stiff climb over névés and glaciers and loose, rocky taluses, but alack! the Matterhorn was yet miles away and fenced off from the shattered crest I was on, by a series of jagged, unscalable crests and glaciers that seemed steeper and glassier than any I had seen. After studying the situation like a chessboard, narrowly scanning each spiky wall and its glacier-guarded base, I made up my mind to the unhappy opinion that it would be wrong to incur so many dangers in seeking a way from this direction to the peak of Matterhorn. I concluded to spend the day with three glaciers to the left towards Ritter, and seek the Matterhorn again next day by ascending a canyon leading up from the north.

Yet, in order to make sure of the practicability of even that route, I scaled the peak next me on the left[2] to get a wider view of the jagged zigzag topography. On reaching the top (twelve thousand feet), I saw it was possible to descend to one of the

[1] Probably Rodgers Peak.
[2] Probably Mount Davis, altitude 12,308 feet.

glaciers which before seemed to threaten so much, and that at its top it was not only snow-covered but less steeply inclined, and that on its shattered, precipitous head wall there was a narrow slot, three or four feet wide, which I could reach from the head of the glacier, and possibly descend on the south side into what promised to be a canyon leading up to the highest névés of Matterhorn, towards which I could see a long, easy spur coming down from the summit.

It is hard to give up a brave mountain, like the Matterhorn, that you have counted on for years, and the upshot of this new view was that I began to scramble down towards the first glacier that lay beneath me, reached it, struck my axe into its snow, and found it in good condition — crisp, yet not too hard. There were some crevasses that threatened, and in some places the schrund yawned in what is called a cruel and infernal manner, but I escaped all these, passing the schrund by a snow-bridge, and reached the narrow gap (eleven thousand seven hundred feet altitude). There I found, to my delight, I could clamber down the south side, and that after I reached the edge of a little lake in which snowbergs were drifting, the rest of the way to the Matterhorn peak was nothing but simple scrambling over snow slopes, over the snout of the Matterhorn glacier, across moraines, down the faces of fissured precipices, up couloirs, threatened with avalanches of loose stones, on up higher, higher, peaks in crowds rising all around — Dana, Hoffman, Ritter, pinnacles of the Minaret group — and I could see the Merced group clouded with ice and snow, and glaciers near and far, and a score of lake-gems. With many a rest for breath and for gazing upon the sublimity of the ever-changing landscape, and without any fierce effort or very apparent danger, I reached the topmost stone.

I take bearings of a few peaks and glaciers, and give myself up to the glorious landscape until night draws nigh. Then down again. I discover a narrow avalanche spout or channel in the northeast shoulder which leads to Matterhorn Lake, a gem at the foot of the vertical north face of Matterhorn. This lake is one

of the highest in the range (altitude twelve thousand feet). . . .
The Matterhorn breaks away at the top in huge rough-edged
faulted precipices, and is approachable by the way I came up
from the canyon of the San Joaquin. This is easy to find, but
outlandish for civil travel. It is also approachable via the Rush
Creek Canyon, which was the way of my descent, but this is
hard to find.

The Matterhorn is built of black slate, blacker slate, and gray
granite, mingled and interfused and belted and conglomerated,
making a strange-appearing section on the north side. In some
places the granite penetrates the slate in crack-like seams. In
other places it appears to have got into large solid slate masses,
like plums into a pudding. In some places the slate and granite
seem to be about equal in quantity, and even mixed in the form
of boulders, but planes of cleavage run unbrokenly through both
slate and granite. In Matterhorn and adjacent mountains I
observed some instances of curved dome cleavage which is not
common at this elevation. The granite of the Matterhorn par-
takes of all the forms in its crumbling and disintegrated condition
that the neighboring slate does — all highly metamorphic.

The glacier of the north side that is nearly inactive has been
long dying, as evidenced by its terminal moraines reaching for
two hundred yards from the lake, draggled out or heaped irregularly according to periods of weakness or strength. Northwesterly
is another glacier, emptying into a lake, three hundred yards in
diameter — a rare gem in color and setting (altitude eleven
thousand five hundred feet). The water of Matterhorn Lake
flows into this, and the waters of both, after descending in cascades three hundred and fifty feet, enter a long, picturesque,
islanded lake, one thousand yards long. This last lake receives
the drainage of three glaciers, and its outlet is a fine stream
forming one of the feeders of the Middle Fork of San Joaquin.
The second of these three lakes is nobly surrounded by mountain-
peaks and glaciers and clouds of pure white névé. Its south shore
has many precipices of icy névé rising from the water's edge. Not
a plant grows there, but the north shore has sedges, and on one

place where a high wall-faced rock reflects the sun heat is a
garden containing ... gooseberries ... one goldenrod, a noble
thistle, and the bush Epilobium, fringes of which are in flower,
of as rich a purple as ever worn by high-bred plant of the tropics.
Besides there are nodding Carices in clumps, and Senecio, and
hairy lupines and Spraguea — all this at an altitude of eleven
thousand five hundred feet, with a circumference of snowy mountains looking at it, and a chill, icy lake clashing its frosty waves
at its foot.

The outlet of these three lakes, on its way down the mountain-
side to the big open valley that goes straight to the east or
Thousand-Mile branch of the San Joaquin, cuts itself a passage
beneath a narrow but deep névé. The entrance of this is imposing,
the roof splendidly sculptured and arched. The glad and bold
stream enters this gateway at a bound. In crossing the névé
I narrowly escaped falling into a crevasse twenty or thirty feet
deep and three feet wide.

Had a glorious walk home, reaching gardens yet richer and
richer. I reached camp on the lake-side, sifted full of mountain
air, rich, happy, and weary....

August 24 (?).

Wild, unweariable wind all the night and day. Crossed over
to the Tuolumne Valley.... The air on the divide is full of insects,
and when seen in the sunlight with the eyes protected by the
crest of the mountain they appear like transparent flecks of silver.
So also do the rocks and bushes. A stiff wind blows the insects,
driving them like chaff. Some with fluttering wings fly against
the wind.... I have observed the air full of silvered cobwebs
above the summit of Lyell. None of these insects or webs are at
all visible under other circumstances.

August 25 (?).

I camped for the night near the right Lyell Glacier on a steep
mountain-side at an elevation of eleven thousand two hundred
and fifty feet. It was cold and windy, and difficult getting wood,

and even more so getting water. I climbed up the mountain and brought down a hatful of snowballs, enough for coffee night and morning. Then I built up a place for a fire with rocks, to keep the wood from rolling down the slope, and also for a bed. It is a wild, loud, frosty night. Bread about gone. Home tomorrow or next day.

3. *Sierra Fragments*

1873 (?).

Alpine gardens of gentle evanescent beauty adorn the most enduring granite, closely embraced by strong rocky bosses.

Many alpine plants form tufts on rock fissures or seams, short stems massed into cushions, perennial roots perhaps a century old, fed by invisible ooze from some snow-fountain above it. Many ivesias and eriogonums offer fine examples of this sort, old perhaps as the trees.

Much of the soil of the Sierra gardens is composed of crystals, and the glow and gleam and glint of the embedded gems at different times of the day might well exhilarate the flowers that grow among them.

Beautiful, tender flowers grow upon the lava lips of Mono craters, pines ascend their ashy slopes, and it is just where the glaciers have crushed heaviest that the greatest quantity of beautiful life appears.

Lake Tenaya.

Some lakes in canyons have but one inflowing and outflowing stream coming and going in music at head and foot, while the lake itself, deep in its canyon bed, is silent, seldom stirred by any wind. Others are centers of song, serene yet tremulous, fed by a lacework of rills singing all the summer among the encompassing mountains.

Tenaya is a basin full of mountain water as pure and blue as the sky that covers it. Its shores are singularly solid and bare, giving the impression of a burnished bowl cut out of the hard porphyritic granite. Only at the east and west ends is there any soil — mostly thin meadow made by filling in. The north shore is particularly bald and bold, and displays some of the finest polished granite ice-work to be seen in the range. When the sun is shining after rain, the rocks reflect the light almost as brightly as the water, and it is not easy in some places to say where the one begins and the other ends. . . .

The lake is about a mile long and half as wide. Remarkably fine junipers stand on the north and northeast shore, eight to ten feet in diameter, and not much more in height. . . . Very little of the shore is sandy. The inlet is Cathedral Creek; another of some size comes in from the space between the north end of Cathedral Peak and the south end of the Hoffman Range. Another comes from the Hoffman slopes farther west. All these streams are fine dancers, and the outlet is the best of all, as it goes down its majestic granite stairway to the Yosemite.

A few months ago I came to this beautiful lake just as the setting sun was glinting its lowest and last rays on the smooth surface. . . . All its compassing rocks and mountains, glacier-moulded and polished from top to bottom, looked silvery gray. I gazed awhile at the glowing purple slowly creeping higher and suddenly fading. Then I sought a camping-ground on the enameled emerald meadow. After bread and bed work was done, wood collected for the campfire, and all small cares calmed for the day, I strolled to the shore in the moonlight to enjoy the tranquil beauty. A two-minutes walk on a velvet lawn brought me to the rim of sand that separates lake and meadow, and the scene from here was charming and impressive. The only sounds were the tinkling of small wavelets stirred by the night wind, and the hushing low and soothing tones of Cathedral Creek a mile away passing over its rocks in rapids. The moonlight rayed and spangled on the ruffled surface, and the whole lake bosom was silvery bright except where a rock or tree cast a shadow.

Here is death and resurrection every day, and just now Lake Tenaya seemed to have passed from the sunlight to another life —

another world. In the early evening, she had appeared as a surface painted by the sun. Now she seemed unfathomable, without connection with her rocky basin. The rocks too, reflected in that wondrous mirror, seem to have parted with their weight and to float in the calm light, decked with ebon shadow, all their broidery of grove and sculptured niche, brought out in a strangely ethereal floating, trembling, bodiless, spiritual way.

Tenaya Fall blooms among trees like a flower in the grass. This, I think, is the most picturesque fall in the valley. Unlike all the rest of the choir in their bare rocky sublimities, it is richly embowered in trees and flowering shrubs and ferns. Broad arching fronds of tall woodwardias are kept ever fresh in the beaten spray, and gentle, floating maidenhair and emerald mosses float and arch in sheltered caves and recesses, making this one of the most beautiful of all the Yosemite cataracts.

Yosemite.

Ouzel. The life of an ouzel seems an echo of the mountain streams. In the coldest weather he dives and breaks forth again in sweet summer strains of song. Wherever you go tracing the streams you will be cheered by his song. He has not the full-spoken passionate fire of the Scotch mavis, nor the gushing ecstasy of the bobolink, nor the conventional exactness of the brown thrasher's music, nor yet the strong, clean articulation of the mockingbird. His song is low, sweet, and fluty, expressing the very heart-peace of nature, steadfast as a star in its shining, through dwelling in the midst of the blare and glare of wild torrents.

How interesting it would be to keep close beside an ouzel all his life, and be present at his death-bed! Surely there would be no gloom, no pain. I fancy he would vanish like a flower, or a foam-bell at the foot of a waterfall.

We may conceive of other birds coming from other countries, as if Nature had prepared a new home for them, saying, 'Here, little ones, you will find shelter and food, seeds and berries from season to season, and leaves and mosses for nests.' But the ouzel seems an outgrowth of the streams themselves, derived from them like flowers from the ground, as if the pebbles around which the waters had sung for ages had at length been overgrown with feathers and flown away, preserving all the music that had passed over them to be given back to blossom again. . . .

4. *Tuolumne Days and Nights*

About 1872.

Snatch a pan of bread and run to the Tuolumne.

How blessed it would be to be banished to the Canyon! There are many witches, say the Indians of Yosemite. Blessed witches! Even banishment to the lonely tundras of Siberia would be a blessing to many.

In whatever mood the lover of wildness enters the Canyon, he speedily yields to the spell of the falling, singing river, and listens and looks with ever-growing enthusiasm until all the world beside is forgotten.

The Canyon has something grand and good for everybody. The hardy hunter will glory in the tangles and rock dens of the grizzly bear, the parks loved by the deer, the groves where the grouse find grateful homes. The poet will listen to the river as a poet of marvelous range, thundering, leaping, tinkling. The artist will see all combinations of rocks and waters, piled clouds in the sky, storms driving overhead and up and down the canyons as in a closed hall, and ethereal lights glowing on the rim and polished crystal windows of the Cathedral and Castle Rocks. To the lover of gentle beauty where are gardens finer, with their bosky dells, glades, gardens, and ferneries with humming bees and birds?

The Tuolumne Canyon is a street of the sublime Sierra City more than twenty miles long. Its rocks, one to five thousand feet high, present themselves on either side, like works of art elaborately sculptured, many places polished on which the sun glances as upon windows of glass. Here are all kinds of architecture, gables high and low, cornice and cusp, windows and arched entrances adorned with plants. Through the canyon flows a river clear as crystal, bordered with trees, Cassiope, fairest of shrubs, and sunny meadows here and there. Nature's best gardens are here in deepest repose, fountains of wild ever-playing water falling in every form — the endless song of Creation shaking the devout listener into newness of life. He who enters will hear a music which will never cease to vibrate in his life throughout all its blurring moil and toil, and to the beauties of this Tuolumne Street he will fain recur with fond memories, and all his material gains will rise in the balance against the riches garnered in her clear and rocky wilds.

Indian Summer in Tuolumne. Nature with eternal earnestness keeps her autumn holidays, giving gold to the maple leaf, crimson to Vaccinium, and scarlet to the leaves and berries of the mountain ash, mellowing her sunbeams, hushing her winds and streams, robing the harsh rocks in soft rosy light, gathering leaves and burrs into hollows, feeding her squirrels and birds, sparing no deed of loving beauty as if she were gardening this canyon above all others.

Ambrosial days of thick gold. Crimson yellow of clouds. Small isles of mist transfigured. Water moving in luxurious languor, thick-swelling rolling folds gliding in silence. . . .

Sunlight lying thick and rich, brooding calm in the groves of pine down the rugged ice-carved canyon. . . .

Bars of crimson at sunset, glorifying the edge of night.

Long after the valley lies quiet in shadow the domes still sing warmly with sunshine and echo it back to the meadows.

At night the great walls, unrelieved by any alternations of light and shadow, loom in unbroken masses to the stars.

The green pines far above wave and sing to the touch of the grand old winds.

Nowhere is God's love coined into more beautiful forms. Everywhere is timelessness and infinite leisure.

As a man in his books may be said to walk the world long after
he is in his grave, so the glaciers flow again in their works —
these stony books.

Glaciers. In walking the Tuolumne street, the most unquestion-
able evidence can bring us to see that all this fabric of granite
with forests and streams was born of glaciers. In the beginning,
the first flakes, settling softly or driven like meal-dust in frosty
winds, seem like down with no greater erosive power than feathers, but during the lapse of the plentiful seasons, the wind-driven
snow-feathers are densed and organized. No longer jostled by
storm-winds, they brood motionless like spirits outspread. Yet
mountains long predestined are being born, rock-ribs are being
crushed in the coils of the invincible snow-companies until the
mountains are dug and delved, here deep for this canyon and
these forests, there shallow for warm shrubs and flowers ... the
glacier seeming to sleep yet ever awake and working incessantly;
clumsy and unmanageable in appearance, yet as a tool in the
hand of Nature accomplishing work of meadow and lake and
garden and grove and glossy polished dome, delicate and fair in
architecture as that of a coral or a shell.

Lakes, too, shining like eyes round and oval in the fullness of
time, are transformed to meadow and planted with all that is
most precious in gardens of Nature's own, forming the fairest
memories of the age of ice.
In the recession of the ice-sheet, which was at once an instrument and a creation perfect in itself, the whole flank of the
mountains was not at once laid bare, like a piece of ocean bottom
rising above the waters in a single night. As the ages rolled on,
the icy shore crept backward and upward like a shadow, islet
after islet, dome and crest rising above the white waning sea....

Life for Every Death. There need be no lasting sorrow for the
death of any of Nature's creations, because for every death there
is always born a corresponding life. And what life shall follow the

death of the glacier, what creation shall come to that sea bottom on whose cold burnished rocks not a moss or dulse ever grew! In smooth hollows crystal lakes will live, to sandy beds sedges will come. Pines and firs will feather the moraines, advancing like an army and followed by the dearest flowers and happy animals, and instead of a robe of white ice will be a robe of yellow light upon the new Edens of the Sierra!

Life in the Canyons. How little noted — how little known! Only the few large animals, one species of deer, two or three bears, the panther, bobcat, fox, badger, coon, skunk, and wolf, but seldom even these are seen save by the skilled hunter, and not once in a score of years does even the hunter pass through these canyon-streets of the mountains where meat or skins cannot be carried out without greater cost than they are worth. It is no thoroughfare for mule or horse. Only Indians afoot cross the range by passes that lie at the head.

But how full of life for the adventurer pure and simple! But of those with courage enough to meet the toil and danger scarce any at all come here. Not love of beauty, but love of gold with its good and evil is the grand force that pulls and pushes people into such wilds as these.

The Tuolumne and Merced are twin rivers rising in the same cluster of fountain peaks, and both splendidly endowed with the gifts of song and beauty of form. On all these higher branches lakes abound, shining discs showing like fruit on a fertile tree in an orchard, while they bloom white in a thousand falls great and small.

One can get through the canyon well enough with patience and love and the strength that comes of it, though there are not wanting places that put even good mountaineers to their mettle.

We are brought into contact with the tuneful spirit of creation, silvery mists and flaming sunsets, when the wild young river is still singing the song of creation.

The music of winds and waters among the great rocks and trees is solemn and telling. Birds are here also, but not enough to fill the great spaces. Song sparrows, trilling, like singing children in a grand cathedral. But I have oftentimes wondered that the finer music of insect wings is unnoticed. The hawk screaming above the nest on the cliff has a startling distinctness; the clamorous cranes, herons, geese, crossing from side to side of the range, make strange sounds in the thin air three miles high.

It is in these garden dells and glades, in peaceful spots where the winds are quiet, holding their breath, and every lily is motionless on its stem, that one is wholly free to enjoy self-forgetting. Here is no care, no time, and one seems to float in the deep, balmy summertide after being thoroughly awakened and exhilarated by the dangers and enjoyments by some grand excursion into the thin deeps of the sky among the peaks.

Robins and ouzels sing with the river all through the canyon wilderness, and bees doze and feast on goldenrod and mint. A finished land, though new. Nature has taken time to tune the waters and bushes and myriad reeds of the trees, and to tend the Bryanthus bells, and to coil the ferns.

When the day is done, the bough bed made, bread eaten, the fire glows cheerily in the majestic depths of darkness, giving back the sunshine gathered from the sifting sunbeams of a hundred summers, gathered in cells like honey, illumining the flowers in the grove or the one tree, the arching grass-tuft, the daisy. How living they look — like fairy spirits clad in forms woven from sunbeams! What expression in their faces, and as we lie dressed, how profound the calm, and how profound the action! We hear the heart-beats of Nature, the tides of the life-sap in every cell, the swirling eddying currents. The river song is louder, filling all the canyon. The fall nearest, how much louder, more impressive the roar, and how much firmer the low tones and wider the range! How many birds and beasts are awake and waiting; how many feet and wings are active, eyes shining like stars! The bear and the deer and the mice, marmots, moles, and gophers — a curious sturdy race living always in the dark, to whom the daisy-starred sod is a sky. And as we lie, our face to the heavens, the stars how they shine! As we gaze, they call us into the far regions of thought, singing the song of Creation's dawn. From the bottom of a canyon where I lie alone, I hear that song the best, as the constellations swing into view over the rim of the rocks, the same stars the shepherds looked at on the plains, thousands of years ago. How many tubes are pointed at them even tonight. Mills of God, every one of them grinding out gusts of light, sending a blessing to each living creature in the sea, on the land, in every nook and corner, the height and depth of the round globe itself. On this shining spark in the firmament every crystal is throbbing, sleeping, yet waking — the quartz, mica, feldspar, tourmaline, hornblende, garnet. What a picture of celestial industry is beheld in the heavens! What a storm of harmonious motion, enduring forever, abating never! Worlds in motion are pulsed through space like the beating of our own hearts, like the myriad globules in the blood of plants and animals.... Everything in the wildness is revolving through life like the stars in their places, always within measured bounds though seemingly boundless.

A Cascade. Impetuous and majestic, in all its wild, snowy thundering, ever making tremendous displays of power and motion, and exuberant joy, beautiful and steady as sunshine, uncontrollable as an avalanche, earnest as fire, glad as the stars and as calm, the cascade is a fit voice for the tremendous wilderness. Lilies lean forward, bathing in the finest spray, ringing their bells in time with the wild thundering heart-beats. Think of camping beside it, seeing it white and undefined like a ghost in the dark, mixing it with our dreams....

5. *An Autumn Saunter* [1]

Casa Nevada below Nevada Fall.
September 29, 1873.

Mrs. A. G. Black, Yosemite.
John Muir, Oakland, Calif.
A. G. Black, Guide, Yosemite.

Made delightful trip around from Glacier Point across the valley of Illilouette and down through Little Yosemite to the main valley via Nevada and Vernal Falls.

Glacier Point commands a noble and instructing view of the Tenaya Canyon, down whose ample channel descended the great Tenaya Glacier, which played so important a part in the excavation of Yosemite Valley. The less regular, crooked, and dome-blocked channel of the South Lyell Glacier is also well seen from here, as well as those of Yosemite Creek, Hoffman, and Illilouette Glaciers. These five principal Yosemite glaciers, now feebly represented by the streams of the same name, are united in the upper end of the great valley as a focal point, concentrating their erosive energy and flowing down and out of the valley as one grand glacier.

From Sentinel Dome a still more comprehensive view of the channels and foundations of these old ice rivers is obtained. The knotty domes and ridge-waves of Yosemite Creek, the steep-descending groove-shaped valleys of the Hoffman, the bold, simple furrow of the Tenaya, the broad, irregular pathway of the Nevada or South Lyell, interrupted by domes as a water-stream is by boulders, and the wide flash-shaped basin of Illilouette, ridged with moraines and manifesting its noble ice-wombs in sublime simplicity among the bounding peaks of the so-called 'Merced group.'

[1] This little narrative, discovered in 1930 by C. A. Harwell, Yosemite Park Naturalist, was written by John Muir into the old Register of Casa Nevada, a tiny hotel once maintained by Mr. and Mrs. F. A. Snow on Table Rock just below Nevada Fall. The date, September 29, 1873, is one of the unsolved puzzles of Muir's time-reckoning, since according to his Journal of September, 1873, he was on that day exploring the southern tributaries of the San Joaquin River, and was that same night encamped at Big Meadows. It is possible that one or the other date was later affixed by him as approximate.

While we lingered upon Sentinel Dome a fine storm was observed in progress among the black, jagged summits of the Lyell group. Long bent tresses of rain descended from the base of a dense, bluish-black cloud in which some of the peaks were outlined dimly, while others were wholly obscured down to their shoulders. Above the dark rain-cloud a series of fine-grained, light-colored cirri were laid in exquisite combinations, and these again were surmounted by white, bossy sun-filled cumuli glowing upon the tranquil azure. We watched the motions of the storm as it swept leisurely northward, bathing the grateful mountains in its path, settling down upon each in turn with a fondling gentleness of gesture that is utterly indescribable. Nor were there wanting the majestic tones of the lightning in deep rumbling explosions, reverberating from peak to peak with greater and greater faintness.

After dark, as we looked back towards the head of Little Yosemite, a belt of cloud appeared drawn across from wall to wall that shimmered with lightning in every pore.

The autumn tints of the Rubus, maple, and wild cherry were most enchanting, the latter covering the banks of Illilouette with a mist of yellow. We reached Snow's weary with delight, and the Nevada sang us asleep.

6. *Down the Sierra to Mount Whitney* [1]

Camp at Clark's Station.
September 19, 1873.

It is almost impossible to conceive of a devastation more universal than is produced among the plants of the Sierra by sheep. Clark's Meadows is fast changing from wet Carex to a sandy flat with sloping sides. The grass is eaten close and trodden until it resembles a corral, although the toughness of the sod preserves the roots from destruction. But where the soil is not preserved by a strong elastic sod, it is cut up and beaten to loose

[1] Mr. Muir's companions on portions of his trip were Galen Clark, Yosemite pioneer, Dr. Albert Kellogg, botanist, and William Sims, an artist.

dust and every herbaceous plant is killed. Trees and bushes
escape, but they appear to stand in a desert very different from
the delicately planted forest floor which is gardened with flowers
arranged in open separated groups. Nine-tenths of the whole
surface of the Sierra has been swept by the scourge. It demands
legislative interference.

The basin of Big Creek, south of the Big Trees, is one of the
finest sugar pine forests of the range. But the fir of Slate Ridge
has some disease which is making many branches yellow. 'Not
so two years ago,' says Clark. . . .

> On divide between Chiquito
> Creek and South Fork Merced.

In descending the sandy slope of the divide, Clark got a
glimpse of a frightened bear beating a hasty flight from the
open hillside to the dense chaparral that filled the bottom of the
gulch. At noon, we lunched on the edge of a beautiful brown
and yellow meadow spread with many a bay projecting into a
forest of fir and pine (altitude seventy-one hundred feet).

From the divide we obtained a glorious view of all the heads
of the San Joaquin River excepting the north forks. Through the
haze and smoke we could see Mount Gabb and Mount Abbot,
and Red Slate Peak. Chiquito Buttes were on our left. These
are a cluster of domes, not well developed, which have been
laid bare by the great Joaquin Glacier.

> Camp on Chiquito Creek.
> *September* 20.

After a long, weary search through many miles of dry tamarack [1]
flats, we camped in the deep picturesque channel of Chiquito
Joaquin. This is the head of a fine forest basin, with rocks
marvelously colored. We saw deer. . . .

The wild cherry is abundant on the rough hillsides at an elevation of six thousand to eight thousand feet. The berries are
bitter, but beautiful, and droop gracefully.

---

[1] Tamarack pine, otherwise known as the lodgepole pine (*Pinus contorta*).

No pine in the forest compares with *Pinus contorta* in graceful-ness of habit and airy, feathery lightness. I have seen many specimens as graceful in motion and in form as the fairest grasses. It is badly named.

Nooned upon a delightful untrodden meadow, over which insects joyous and busy hummed in the sunshine....

Camp on Middle Fork,
San Joaquin, Boulder Valley.
September 22.

Made tedious progress, horses and mules tending constantly to roll down canyon.... The average slope of San Joaquin Canyon, Middle Fork, is twenty-three degrees for a height of seventy-five hundred feet.... Aldery gulches, cool water, crimson and yellow sand.

The glowing sunset beams fall in bars through the forest, touching with marvelous power the hushed waiting firs, yel-low ferns, and glistening grass stems that grow in sandy moraines, and the asters and goldenrods along our camp meadow....

Two large bucks with branched antlers like dead pine roots are splendidly framed in willows as they stand in grasses over their backs — noble inhabitants of this wild forest.

September 23.

In the morning we ascended four hundred feet to a magnificent basin containing groves of fir and *Pinus contorta* and yellow meads and lakes, with gray, rocky, sparsely feathered ridges around. Then a long ride over a succession of sandy flats, some well chaparraled, others remarkably open with bars of light and shade purely divine. No track but that of the deer....

At noon we come to a lava plateau. Granite is now breaking into a labyrinth of rocks of every conceivable form, with ponds and winding meads innumerable. We arrive at a large tributary of the South Fork about noon, and at Mono Creek at 3 P.M.... Gentians a meadowful, one foot high. Many roses and Linosyris.

We pass a lake lovely with ducks and rippling glassy dark

mountains nobly sculptured, sheer to the water. Pines and junipers stand picturesquely on the rock headlands.... Hum of bees, and dragonflies.... Altitude seventy-one hundred feet.

Camp. *September* 24.

In this poor rocky wilderness are wild roses with scarlet hips, yarrow and goldenrod, also mint, Monardella, and asters....

The direction of the striæ in this section is about N. 33° W. Mount Gabb and Mount Abbot in the distance....

Camp in South Fork, Joaquin Canyon.
September 25.

View very grand and universal. Ritter the noblest and most ornate of all. Red Slate Peak very noble — black, white, and red. Then Mount Gabb and Mount Abbot and an inseparable field of high peaks at the head of Joaquin. One is castle-shaped and gray. Another towers dark on the right side of the canyon as we look. Snowy peaks loom yet beyond at the head of Kings River Forest. It grows dusk ... shadows are stealing over the many meads and groves of Joaquin Canyon....

South Fork, San Joaquin.
September 26. Night.

I am camped at the head of this San Joaquin Yosemite this ninth day, in a small oval dry lake-basin, with a rocky wall fifty feet above the river. The basin here is filled with fallen trees, and fringed with willow. I set out this noon for the icy summits, leaving Clark, Billy, and Doctor Kellogg in camp on a moraine with the animals. I have a week's provisions.

My fire cannot be seen far, hid as it is in willows and rocks. No Indians will find me. I shall be too high for them tomorrow evening. I hear the river rushing. It is a fine, calm, starry, soothing night. Many goldenrods and gentians grow in the meads of this yosemite.

September 27. Dawn.

A calm Indian-summery morning; a hazy light wrapping all the high mountains, shadows creeping, barred with light, summits tipped with the first spirit light of dawn.

At San Joaquin Cascades.

The river forks here, the left valley being deeper and in every way larger, but far the greater portion of water comes from the right cascades in vertical short leaps.... Here granite meets slate — slate on the south side, granite on the north.... Mostly the granite and slate, however mixed, yet are separate and meet sharply, but in some places they seem to blend. However, in all cases every crack and seam and joint belonging to one is continued on through the other. Granite has essentially the same physical structure as slate....

This San Joaquin Canyon all the way up from the lowest to the highest yosemite is very rough. One is constantly compelled to ascend knobs and buttresses that rise sheer or steeply inclined from the water's edge.... The scenery from the first main fork is very grand. The walls are steep and close, fold on fold, rising to a height of three thousand to four thousand feet....

I saw a fine band of mountain sheep,[1] light gray in color, white on the backs of hips, with black tails. All horned, they seemed strong and moved with great deliberation, led by the largest. They crossed the river between the steps of the cascade where the channel was blocked and bridged with big boulders. On level ground they were close together, but in ascending the steep mountain wall were far apart in single file, zigzag. They were in full sight at first, only forty yards distant, and seemed to have been scattered; two or three were on my left on the opposite side of the cascades. Had they been tame sheep, they would have run off to destruction. This first lovely fellow studied the intentions of the flock, and met them by making leaps on glacial bosses that made me hold my breath. They first had to cross

[1] See Muir's *Mountains of California*, vol. II, pp. 46–52.

loose angular blocks of granite, which they did splendidly; then they leaped up the face of the mountain just where I thought they wouldn't, and perhaps couldn't, go. I could have scaled the same precipice, but not where they did. I could have followed in most places only by being barefooted.

Like the true mountaineers they are, they never seemed to hurry. They went up over the polished bosses in admirable order and with no noise, each showing a separate will, some lingering to look back at me, some hastening on to the front. Each individual crossed the river, unlike tame sheep. Tame sheep often jump at a rock-face and fall back, but here was not one false step. These are clean and elegant, the others dirty and awkward. These are guarded by the great Shepherd of us all, those by erring money-seekers. They passed in review before me; so also did deer — the sheep for the rocks and rough mountain work, the deer for the forest and green grassy meadows.

Looking at them I often cried out, 'That was good!'... I exulted in the power and sufficiency of Nature, and felt like saying aloud to God as to a man, 'Well done!' Our horses roll end over end, and our much-vaunted mules, so surefooted and mountain-capable, roll like barrels. I must so often judge for my horse, and, considering him as a kind of machine, I know the slope he can bear and how he will be likely to come off in making various rock steps.... These noble fellows — I would like their company! Where are the alp-loving ibex or chamois that can outleap them? Not long ago I saw their tracks on the Matterhorn. Clark said one must kill mountain sheep. But I'm glad these were not killed. They were as fearless of water and the roar of foaming cascades as the ouzel that sang for me today....

September 28.

I camped last night in a wide valley above the uppermost (San Joaquin) yosemite in full view of Mount Humphreys.[1]...

[1] Not the present Mount Humphreys; it was probably Mount Darwin. There was much confusion in Muir's time as to individual peaks and their names.

I saw three dead wild sheep that had been snowbound. . . . Lake Millar, fourteen hundred yards long, fifty to one hundred and fifty yards wide, has waters of a bright green, and lies along the Gothic front of Mount Millar on the south side. . . .

In ascending the main Mount Millar Fork of Joaquin, the first tributary of any size is on the right south side about one mile above the first meadow. It is a bright active stream coming down in a foamy cascade of one thousand feet. . . .

Mount Millar.

. . . Had a glorious view from the top of Mount Millar of the Owens River and Valley, and of the Sierra, one broad field of peaks upon no one of which can the eye rest. They are gothic near the axis — a mass of ice-sculpture. Mount Emerson is imposing with its evenly balanced crest and far-reaching snowy wings. Emerson and Millar are each about thirteen thousand five hundred feet high.

Millar is planted with ivesia to the top, also with Polemonium and with yellow Compositæ. At the base grow larkspurs, columbine, spiræa, dodecatheon. Birds on top and bees. Bush Epilobium far up and ferns Cystopteris and Pellæa.

Camp at Big Meadows.
September 29.

The direction of the slate cleavage is nearly vertical where it meets the granite at the fork of the river. . . .

A bright white silver strip cascade comes into the south South Fork just above its junction with the east South Fork on the west side. . . .

Deer, grouse, squirrels. . . .

Camp on South Fork, San Joaquin.
September 30.

In camp again with the main party. Up early and went with Clark to a point on the divide to view the landscape and plan the route. The view is awful — a vast wilderness of rocks and canyons. Clark groaned and went home.

North Fork, Kings River.
October 1.

Our tenth camp was on a small grass meadow, on the edge of the North Fork of Kings River (altitude eighty-seven hundred feet). From the divide of Kings River and the San Joaquin we entered a long straight valley remarkable for its many domes on the east side, from whose curved bases had come many avalanches of snow, cutting gaps in the timber and making hillocks on the west side. Some of these hillocks of former avalanches had crops of trees two feet in diameter.... The middle portion of this valley is blocked up and roughened in a remarkable way by rock-heaps swept down by avalanches. When a valley stream is dammed by débris, it sorts out the finer material if it has not too much velocity, and thus sand-heaps are formed, as at Big Tuolumne Meadows. If the valley be narrow, then the force of the stream carries all things away and sorts them out at long distances....

October 2.

Our twelfth camp is in a very interesting yosemite valley of North Fork Kings River, nearly circular at the bottom and timbered with sugar and yellow pines of rare beauty, some very large. Most of the bottom surface is oversown with boulders, and washed and spread by flood water. The upper end has a few acres of sandy flats on which are grouped pines and oaks....

A small mirror-lake occupies the extreme head of the valley where the river issues from a narrow and tortuous gorge in cascades. So narrow and tortuous is it that the upper end seems to be completely closed, and it is impossible to know the course of the river above for a single mile.

The walls of the valley begin at the head in a distinct yosemite style, but they extend without any marked change far below the bottom. At the head on the west side is an El Capitan rock, about one thousand feet high, and all the rocks which circle around the head are separate from the general wall. They give evidence of great glacial force, are still bright and polished,

and present many precipitous fronts with level tablets on which
are grouped small pines and live-oaks. . . .

The lake of this valley is a perfect mirror. Its banks, or rather
those of the calm deep lagoon-like outlet, are willowed and flowered superbly and banked with old logs, leaning oaks, and noble
pines. . . .

As night came on, the gray starry sky, meeting the moonlit
wall of El Capitan, seemed to blend with it. The granite was
luminous, as ethereal as the sky. A noble company of pines
reared their brown columns and spread their curving boughs
above us in impressive majesty. Only the trees near by were
in the circle of our campfire sun. Beyond was another circle
entirely distinct and distant — a black sharp-angled line of
tree-writing along the base of the sky. Who shall read it for
us? . . . Roar of the river in the night. . .

> Camp on divide between North
> and Middle Fork, Kings River.
> *October* 4.

Our camp last night was in dense fir, on the edge of a small
sloping meadow. This morning we ascended the ridge, and,
following it, found a trail descending near the junction of the
North Fork and main stream. Camped fifteen hundred feet
above the river. . . . The view is lovely from this dark, rocky
ridge nobly designed. . . . Before us lies the river canyon with
innumerable folds and groves sweeping down to it in front, swell
over swell, boss over boss, oak-dotted below, pine-feathered
above, all hazed with the atmosphere of Indian summer.

> Camp at Thomas Mills.
> Altitude 5700 feet.
> *October* 6.

Coming here we had a long weary uphill climb. Up from the
valley over a hill five hundred feet, then down two thousand
feet, to the wide-open, level-bottomed, oak-dotted valley of
Mill Creek, past an Indian town, then up the valley six miles

to a ranch and wagon road and six miles more to the mills. Beautiful valleys, brown as the plains.

The ice-action is not clear in the foothills. Did the ice sculpture these sharp ridges just before the breaking-up of the ice winter, as it sculptured the summits which they resemble? ...

October 7.

Sunrise, beams pouring through gaps in the rocky crest, warm level beams brushing across the shoulders in so fitting a way they seem to have worn channels for themselves as glaciers do. The oaks' stems are set aglow. The ground is colored purple with Gilia, and gray with *Plantago patagonica*, and redbrown with buckeye, dark live-oak, and sycamore.

Among the Big Trees.
Altitude 6500 feet.

The girth of 'General Grant' is one hundred and six feet near the ground. It is a tree much like the 'Grizzly Giant' of Mariposa. I saw a great many fine trees, fifteen to twenty-five feet in diameter, bulging moderately at the base, and holding their diameter in noble simplicity and symmetry. The 'General Grant' is burned near the ground on the east side, and bulges in huge gnarly waves and crags on the north and west. Its bole above the base thirty or forty feet is smooth and round.

The grand old tree has been barbarously destroyed by visitors hacking off chips and engraving their names in all styles. Men residing in the grove, shingle-making, say that in the last six weeks as many as fifty visitors have been in the grove. It is easy of access by a wagon-road between Thomas Mills and the Big Trees.

The great Kings River Canyon is just a few days beyond, with all that is most sublime in the mountain scenery of America, and as the S.P.R.R. is completed beyond Visalia, this whole region is now comparatively accessible to tourists.

The general appearance of the grove is so like that of Mariposa that one familiar with the latter grove can hardly know that this Kings River grove is not the Mariposa. I could not choose between the two. The trees are the same in size and general form, so also the firs and yellow and sugar pine are mingled with them in the same proportions. Underbrush of Cornus and willow grow along the streams, Rubus and a few ferns on the hillside. This grove on both sides is quite small, and grows on granite débris and moraine material well rotted, the soil not having been much removed from the roots since growth. In many places quite near the trees I observed the granite bedrock fast crumbling.

The Big Sequoias are always near the big yosemite canyons. A man told me he had seen an old barkless stump forty feet in diameter north of this grove.... Also another man told of another 'old snag,' 'bigger'n Grant,' he said. Many fine groves grow about Kaweah's streams....

> Camp on top of divide between Kaweah and Kings River Basins.

We camp on top of the divide between Kaweah and Kings River water, by the first large stream above Thomas Mills, in a lovely grove of *Pinus amabilis* and *P. grandis*.[1] We are fairly free in the mountains once more. The weather all gold.

> Camp where the trail crosses the south fork of the South Fork of Kings River. Altitude 6550 feet.
> *October 8.*

The scenery from the top of the divide between Kaweah and Kings River is sublime on both sides, the canyons of Kaweah showing beyond and yet beyond, till blue and faint in the distance.... On the left, the yosemite scenery about the many forks of Kings River presents sublime combinations of cliff and canyon and bossy dome, with high, sharp boulder peaks in the distance. A high conic peak, black with a small snow-patch,

[1] *Abies amabilis* and *grandis*, two species of white fir. Muir here uses an old generic name. — F. H. A.

towards the head of Joaquin was well seen from the trail today.
Was it Goddard?...

Camp in Yosemite,
South Fork, Kings River.
October 10.

We camped opposite the first Tissiack on a large mead....
Delightful river reaches, picturesquely treed and bushed and
flowered and ferned....

The Frémont pine [1] is common in the valley. Glacial polish
on the walls in many places.... The south wall from the fall to
the upper end of the valley is very picturesque, more so than
any portion of the (Merced) yosemite wall of equal extent.
Gothic peaks are well developed. Tissiack is very impressive
and like that of the other yosemite in general form....

Still another Tissiack, a magnificent rock, rises higher where
the large north tributary equivalent to Tenaya Fork comes in.
The view up this North Fork from the trail is exactly like the
view up Tenaya Canyon. A Washington Column is much
grander than that of Yosemite. Thus this Kings River Yosemite
has two Washington Columns and nearly two North Domes,
but no lakes and few meadows. The main portion is filled with
gravel, washed boulders, and sandy flats, well groved. It has
not many acorn-bearing oaks, therefore is not a favorite place
with Indians. They find more and better acorns below, also
fish. No Indian camps here....

At mouth of first tributary
of South Fork Kings River.
October 11.

Set out for Mount Tyndall [2] alone, leaving Billy and Dr.
Kellogg at camp with the animals....

It is hard traveling along this portion of the stream, the avalanche material having been planted with poplars and all kind

---

[1] *Pinus Fremontiana*, an old name (1847) for the nut pine, or pinyon. See
Sargent's *Silva of North America*, vol. xi. — F. H. A.

[2] Not the present Mount Tyndall.   Perhaps it was a peak on the Kings-
Kern divide, or perhaps Mount Brewer.

of chaparral. Even the bears seem at times to be at fault in making their trails. The colors of the poplars and Epilobium and mountain ash are glorious in combination. The light rain and snow of the previous days brightened all colors. The dogwood does most of the red, the aspen the yellow, and the sedges and a multitude of smaller plants the other tints. The bracken is splendidly colored now, and is in great size and abundance.

The falls on the North Fork are only a series of moderately inclined cascades, far inferior in power to those of Tuolumne. The South Fork rises fast but steadily, not doing or saying anything specially emphatic. . . .

At an elevation of ninety-seven hundred feet, the South Fork divides into many branches that run up to the glaciers and névés of a noble amphitheater of lofty mountains, forming fountains for the brave young river. . . .

I ascended two peaks in the afternoon. Clouds gathered about the brows, now dissolving, now thickening and shooting down into and filling up the canyons with wonderful rapidity. A great display of cloud motion about and above and beneath me.

Hurrying down from amid a thicket of stone spires to the tree-line and water, I reached both at dark. A grand mountain towers above my camp. A rushing stream brawls past its base. Willows are on one side, dwarf *flexilis* on the other. The moon is doing marvels in whitening the peaks with a pearly luster, as if each mountain contained a moon. I have leveled a little spot on the mountain-side where I may nap by my fireside. The altitude of my camp is eleven thousand five hundred feet and I am blanketless.

October 12.

Set out early for Mount Tyndall and reached the summit about 9 A.M. Had grand views of the valley of the Kern and the Greenhorn Mountains and north and south along the axis of the range, and out over the Inyo Range and the Great Basin. Descended and pushed back to the main camp. Arrived about noon to find Billy and Dr. Kellogg gone, though they promised to wait three days for me. They left me neither horse nor provisions. I pushed on after them, following their tracks on the trail towards Kearsarge Pass, by long stretches of dry meadows and many lakes, surrounded by savage and desolate scenery. The pass is over twelve thousand feet high. *Primula suffrutescens* is abundant on the granite sand up to the head of the pass. The scenery of the summit is grand. *Pinus flexilis* abundant. I overtook the runaway train at sunset, a mile over the divide, just as they were looking for a camping-ground. When asked why they had left me, they said they feared I would not return. Strange that in the mountains people from cities should so surely lose their heads.

October 13.

We descended the long pass, which is one steep declivity scarce broken from top to bottom. In a few hours we passed from ice and snow to the torrid plain. I took some provisions and my horse and left the party at the foot of the pass to make an excursion to Mount Whitney, while the rest of the party went to Independence to wait for me there. I skirted the base of the range past Lone Pine, and over into the valley of the Kern by the Hockett Trail. My horse got mired in a floating bog. Thought I should lose him, but got him out, though he sank over his back. He whinnied with joy at his escape. I camped on the edge of a sedge meadow. A cold, windy night; the wind blew the coals out of my fire, overturned my dishes, and took my hat.

October 15.

I left my horse on the meadow, and set out for the summit afoot. Soon I gained the top of old Mount Whitney,[1] about fourteen thousand feet high. Found a mule trail to the summit. I leveled to another summit, five or six miles north and five hundred feet higher, and set out to climb it also. The way was very rough, up and down canyons. I reached the base of the highest peak near sunset at the edge of a small lake. No wood

[1] Now Mount Langley.

was within four or five miles. Therefore, though tired, I made up my mind to spend the night climbing, as I could not sleep. I took bearings by the stars. By midnight I was among the summit needles.[1] There I had to dance all night to keep from freezing, and was feeble and starving next morning.

October 16.

I had to turn back without gaining the top. Was exhausted ere I reached horse and camp and food.

October 17 and 18.

Set out for Independence and reached it at night. Ate and slept all next day.

October 19.

Set out afoot for the summit by direct course up the east side.[2] Camped in the sage at a small spring the first night.

October 20

I pushed up the canyon which leads past the north shoulder of the mountain. Camped at the timber-line. . . .

October 21.

I climb to the summit [3] by 8 A.M., sketch and gain glorious views, and descend to the foot of the range.

[1] Muir was on the crest near what is now known as Mount Muir, about five hundred feet below the summit of Mount Whitney.

[2] John Muir was the first man to ascend Mount Whitney by a direct approach from the east side. See Farquhar's 'Story of Mount Whitney,' *Sierra Club Bulletin,* 1935, vol. 20, no. 1, pp. 83–85.

[3] On this summit John Muir found in a yeast-powder can two notes and a half-dollar. The notes copied into his journal are as follows:

'Sept. 19, 1873. This peak, Mt. Whitney, was this day climbed by Clarence King, U.S. Geologist, and Frank F. Knowles, of Tule River. On Sept. 1st, in New York, I first learned that the high peak south of here, which I climbed in 1871, was not Mt. Whitney, and I immediately came here. Clouds and storms prevented me from recognizing this in 1871, or I should have come here then.

'All honor to those who came here before me.

'C. KING.'

'Notice. Gentlemen, the looky finder of this half a dollar is wellkome to it.

'CARL RABE
'*Sep.* 6th, 1873.'

See Farquhar's 'Story of Mount Whitney,' *Sierra Club Bulletin,* 1929, vol. 14, no. 1, p. 45.

Independence, *October* 22.

I found *Pinus flexilis* at an elevation of ten thousand feet, forty feet high, ten inches in diameter at base, with cones three to four inches long, tapered and opening.

It is almost one unbroken slope from the Owens plain to the top of Fisherman's Peak.[1] First, the canyon is narrow and precipitous, walled at the bottom, then opens gradually. Half-way to the top the canyon forms and, of course, gives birth to a yosemite with high, bare walls. The canyon is filled with birch and willow and yellow pine and boulders from the sleepiness of streams. Higher still is a lake, and walls yet higher and exceedingly bare are planted with lichens and Primula, bush Epilobium, sedges and ferns (*Pellæa Brewerii*). Hush of stream.

October 23. Sunrise.

Orange on the sky. Mountains beyond of different shades of blue, then crimson, the near granite being chocolate-colored from the thick air of the desert....

[1] The people of Owens Valley unsuccessfully attempted to fix this name upon Mount Whitney. — Editor.

Chapter IV : 1874-1877

'Pines and waters and deep-singing winds.'

SHOUTING, 'I'm wild once more,' John Muir left Oakland in September, 1874, and climbed the Sierra to his glad 'reunion with the winds and pines.' A month later, feeling his intensive Yosemite studies were finished, he was trudging northward along the old California–Oregon stage-road from Redding on his way to Mount Shasta. At sight of its snowy grandeur he wrote in a letter, 'All my blood turned to wine, and I have not been weary since!'

On November 2, he scaled the summit of Shasta, although seasoned mountaineers said it could not be done so late in the year. 'But I did it,' he said, 'because I loved snow.'

Returning to Sisson's Station, he again set forth with a group of hunters to visit the haunts of the wild sheep among the Lava Beds. The month of December he spent encircling the mountain and exploring the adjacent region. January and February (1875) found him exulting in storms and trees on the Feather and Yuba Rivers. During the next two months he lived in San Francisco at the home of John and Mary Swett, putting his recent experiences in writing.

June, 1875, found him once more in the High Sierra guiding his friends William Keith, the artist, and Schoolmasters Swett and McChesney on a trip from the Yosemite to Mono Lake, and thence down the range to the Mount Whitney region, he himself preoccupied with trees and storms by the way. Upon their return he again set his face to the south, this time going

alone with Brownie, 'a small wild mule,' to make his first intensive survey of the Sequoia Belt. His tree studies continued
throughout the following year, and as his knowledge grew of the
widespread ravaging of the mountains by sheep, fire, and sawmills, he was aroused to launch his crusade for the Federal control
of forests.

In the early summer of 1877, John Muir visited Utah, begin-
ning his study of the vegetation and the glacial records of the Great
Basin. Since this excursion is well covered in his book 'Steep
Trails,' Chapters VI to IX, the journal is omitted here.

Autumn found him botanizing on the slopes of Mount Shasta
with his friends Asa Gray and Sir Joseph Hooker. Returning
from the mountain, Mr. Muir was the honored guest of General
and Mrs. John Bidwell at the Rancho Chico. Standing one day
with his host at Bidwell Landing on the Sacramento, he expressed
a desire to float down the river and explore its banks. The ranch
carpenter was thereupon instructed to build him a boat, and in
this craft he set out upon his voyage.

Accompanying the United States Coast and Geodetic Survey in
the summer of 1878, John Muir visited Nevada. The narrative
of this journey, told at the time in newspaper letters, is to be
found in 'Steep Trails,' Chapters XII to XVI.

- - - -

1. *Going Home to the Mountains* [1]

On way to Yosemite Valley.
September, 1874.

To rest and roll about for a week in the wintry freshness.
Smell the rosiny forests and feel the keen, chaste, caressing wind
on my cheek. Into the woods where the world has no entrance.
... Winter blows the fog out of our heads. Nature is not a mirror
for the moods of the mind.

[1] See Badè's *Life and Letters of John Muir*, vol. II, pp. 10–29.

Black's Ranch,[1] between
Coulterville and the Yo-
semite. Night.

Luxurious relaxation of toe-toasting and jokes while the
spectacled gudewife plies her needles and the rosiny fragrant
pine fire sheds a perfect sun-glow on the quaint furniture and
nippy faces. There is the well-known host, a shining mark for
the light, big-hearted and bodied, and what a shadow he casts
on the logs and against guns and buck-horn brackets!...

Every weary one seeking with damaged instinct the high
founts of nature, when he chances into ... the mountains, if
accustomed to philosophize at all, if not too far gone in civilization,
will ask, Whence comes? What is the secret of the mysterious
enjoyment felt here — the strange calm, the divine frenzy?
Whence comes the annihilation of bonds that seemed everlasting? ...

Tell me what you will of the benefactions of city civilization,
of the sweet security of streets — all as part of the natural
upgrowth of man towards the high destiny we hear so much of.
I know that our bodies were made to thrive only in pure air,
and the scenes in which pure air is found. If the death exhalations that brood the broad towns in which we so fondly compact
ourselves were made visible, we should flee as from a plague.
All are more or less sick; there is not a perfectly sane man in
San Francisco.

Go now and then for fresh life — if most of humanity must
go through this town stage of development — just as divers
hold their breath and come ever and anon to the surface to
breathe.... Go whether or not you have faith.... Form parties,
if you must be social, to go to the snow-flowers in winter, to
the sun-flowers in summer.... Anyway, go up and away for life;
be fleet!

I know some will heed the warning. Most will not, so full of

---

[1] Black's Ranch is located at the end of the wagon-road on the old Coulterville
Trail, and is approximately five miles east of Bower Cave. Muir's host and hostess
on this occasion were Mr. and Mrs. A. G. Black, owners of the ranch and of Black's
Hotel in the Yosemite Valley.

pagan slavery is the boasted freedom of the town, and those who need rest and clean snow and sky the most will be the last to move.

Once I was let down into a deep well into which choke-damp had settled, and nearly lost my life. The deeper I was immersed in the invisible poison, the less capable I became of willing measures of escape from it.[1] And in just this condition are those who toil or dawdle or dissipate in crowded towns, in the sinks of commerce or pleasure.

When I first came down to the city from my mountain home, I began to wither, and wish instinctively for the vital woods and high sky. Yet I lingered month after month, plodding at 'duty.' At length I chanced to see a lovely goldenrod in bloom in a weedy spot alongside one of the less frequented sidewalks there. Suddenly I was aware of the ending of summer and fled. Then, once away, I saw how shrunken and lean I was, and how glad I was I had gone. . . .

> Black's Hotel, Yosemite Valley.
> *September 25.*

In the mountain light the Oakland winter seems all a dream. Only the kindness of friends stands out clear. Pine trees, granite, and water-ouzels now. How true and pure and immortal they seem to me! yet somehow I feel satisfied to leave all and labor in other fields.

The very first evening I came in, the brave owl that was not afraid of the earthquake too-hooed unnervously as ever. Having heard him so many years, his voice seemed charmingly familiar.

The morning light is streaming in between the domes, and the sculpture of the arches is splendidly brought out. How eloquently they speak of the icy past! How marvelous the richness and delicacy of the sculpture wrought by so simple and blunt and unwieldy a tool!

I've been out sauntering on the meadows, and along the sleepy river. Two ouzels came glinting on the crisp, bright water, and after nodding recognizingly, and doing all their dainty manners,

[1] See *The Story of My Boyhood and Youth*, chapter VI.

began wading and ducking in the shallows. They are little dun nuggets of water music, as if the brown pebbles over which the river has sung for ages had at length been overgrown with feathers instead of mosses, and flown away.

2. *Hunting Shasta Wild Sheep* [1]

Sisson's Station. *November* 29, 1874.

On my return from the summit of Shasta I found four British hunters, three from bonnie Scotland, one from England — all sterling good fellows. Brown and Hepburn, the latter of whom had been here fishing and deer-hunting for several weeks, were eagerly bent on hunting the wild sheep, not only for the sport of the thing, but to learn their habits and to see their wild homes, and get specimens with which to adorn their halls.

Sisson, who is himself a keen hunter and an excellent guide to all kinds of game, soon made the necessary arrangements. Blankets, provisions, and rifles were heaped into a wagon, and on the morning of November 8 we set out for 'Sheep Rock.'

The party consisted of Sisson as guide and hunter-in-chief; Jerome Fay, an enthusiastic hunter and capital shot, who is in the employ of Sisson; Brown, the Englishman; Hepburn, the Scotchman, and myself. Sisson and Jerome carried Remingtons, Brown and Hepburn double-barreled breech-loaders. I went unarmed.

Winter had come early. The first week of November the frosts and rain and snows had stripped off most of the fine autumn leaves and bleached all that were left. The robins fled to warmer climes before the wild cherries were half done. The deer came down out of the snow and stood about in groves and thickets as if unable to decide what to do. The bears, notwithstanding their warm clothing, also began to shuffle down from the high founts of Shasta firs into the brushy valleys, and while on Shasta I saw that the brave wild sheep were also disturbed as if contemplating removal to their winter quarters. A few had weathered the storm,

[1] See Muir's *Steep Trails*, pp. 90–92.

in lee of flexilis pines a few hundred feet above my storm-nest, and it was hoped that the week of subsequent storm would certainly drive them down to lower pastures where they would come within the reach of rifles.

Sheep Rock, which we reached before night, lies to the north of Mount Shasta, about twenty miles from Sisson's. It is the principal resort of Shasta wild sheep during the winter months, and presents to the gray sage slopes of Shasta Valley a bold, craggy, precipitous front of two thousand feet. Here we hunted only one day. Brown got a good view of an old ram, grand and massive as a buffalo. It came within twenty yards, then bounded majestically away without Brown's getting a shot.

As signs of them were nowhere plentiful, it was clear they had not yet come down from Mount Shasta. We therefore resolved to set out next morning for the Van Bremers' mountain, one of the most noted strongholds of sheep in the whole Shasta region. Large flocks, we were assured, abide there both winter and summer. This journey of about thirty miles took a day and a half through wide stretches of sage plains, interrupted by rough lava mostly timbered with yellow pine and juniper.

On our way we reached Butte Creek about two o'clock of the first day. This is a favorite haunt of the antelope. After lunch, Brown, Hepburn, and Jerome went out from camp to hunt them, and Hepburn shot a fine buck antelope, which was brought in after dark. We brought him forward to the campfire and held up his head and steadied him upon his feet. The light fell on the beautiful stranger, and all his features stood out in startling impressiveness. His large eyes, even in death, were still beautiful. His limbs, slender and graceful, expressed abundance of strength.

The antelope is quite abundant in the plains and open timber to the north of Shasta. One of the fleetest and most graceful of all wild animals, he ranges not only the open valleys but the pine woods, and feeds upon grasses. In flocks of a hundred or more they are still seen almost any day by the vaqueros of the region.

The Van Bremers, whose cabin and cattle ranch we reached, are three — all hunters who, at length tired of hunting and trapping, have settled to raising stock in this wilderness. Their camp is situated at the base of the wild sheep mountain about five miles from the south shore of Lower Klamath Lake.

In answer to our inquiries they informed us that beyond rough walking we should have no difficulty in finding sheep; that perhaps two hundred or more made their home on the mountain and raised their lambs there, frequently issuing to the plains; but that when pursued on their favorite mountain they fled to the Modoc Lava Beds, which lie at the southeast base of the mountain. The Modoc Indians having been removed, the sheep have not been disturbed for two years. The Van Bremers said that when they first settled here, six years ago, they ascended the mountain with rifles and hounds to make a grand sheep hunt, intending to kill twenty or thirty. But after pursuing their noble game for a week, they wore out their boots and clothes, and wearied and lamed their hounds. On open level or ascending-ground spots the hounds would gain on them, but on jagged lava crags and rough loose slopes they fell far behind, and though they pursued day and night, could not capture a single one, so excellent is the sheep's endurance. So they were compelled to abandon the hunt, and have never attempted to hunt them since.

The morning after our arrival was delightfully crisp and exhilarating; frost crystals covered the sagebrush, and the snows of Shasta glowed rosily in the sunrise. The hunters strode up the bulging slopes of Mount Bremer, full of eager hope.

I spent the day examining a bluff of fossiliferous sandstone on the shore of Klamath Lake. On returning to camp at sundown, I found Brown and Hepburn sheepless and weary, declaring that for roughness and inaccessibility these habitations of the Modoc sheep surpassed all the highland crags they had ever beheld. Brown, who had hunted elephants in tropic jungles, and who had descended Bloody Canyon, declared this was infinitely rougher. ... However, I was glad to hear the sheep had so good a home.

It seems that some sixty or eighty head in different flocks were seen during the day, and a few bullets from three-hundred-dollar rifles were scattered among them without effect.

Jerome came in a few minutes later, and reported he had killed a magnificent specimen of the mule deer. This was packed into camp the next day on Bob, a limp old hunt horse. It weighed two hundred and twenty-five pounds without the viscera. The mule deer is quite abundant here, and is much larger and rougher and stronger than the black-tailed species. . . . Hepburn declared our specimen was about as grand and shaggy and noble as the red deer of Scotland.

Next day all the rifles were again carried up the wild mountain and many sheep were seen, but only one was killed, and that by Jerome. It was a bonnie yearling lamb whose horns were only small spikes. It was a shame to kill it. After being wounded, it still ran nimbly through brush and over lava blocks, followed by our one dog, Guy, who, according to Jerome, 'treed it on a rock,' where it was killed by a second shot. Brown and Hepburn were more than ever impressed with the excellence of Mount Bremer as a sheep castle, and with the nobleness of the sheep, several of which they saw leaping over jagged lava blocks, with horns thrown back.

So much hard hunting for so little mutton was rather trying, and when it came to facing the mountain a third time, both Brown and Hepburn shied. It was determined to try the Lava Beds, where, according to Van Bremer, sheep were often found, and as one of the Vans joined in the hunt, we were sanguine of success. But this day only wolves and sage hens were seen.

To get to Rhett Lake, where we were to camp that night, we rode past many a cliff and grassy knoll to Lower Klamath Lake, and along the shore, scaring ducks and geese, and through miles of plains where the sage hen delights to feed. Here and there stands a juniper dry and white in the alkali. Finally we came to our camp on a bluff about five hundred feet high.

In the sunset I walked down to the edge of the bluff. It was one of those delicious purple evenings of November when Indian summer is fairly ripe, and when the colors are richer and more pervading than at other times, and evening stillness is more profound. The air was filled with gossamer films of spider. Rhett

Lake lay to the left; to the right the Lava Beds, gray and black and pitted, with sage in sparse singed patches, the whole expanse broken only by the smooth sides of craters, gradually rising on the south to dark forests. Beyond all towered icy Shasta, a glorious white pyramid, while snowy mountains with immense purple belts of timber stretched southwestward. Lengthening reflections began to reach over the calm and glassy lake, so responsive to the sky in color and texture it seemed a sky itself. It was surrounded by a broad edging of meadow jutting out into the pale yellow and blue water in long rich brown promontories. The sky above was brushed and filmed with gray and purple.

Night came on, and in the still brooding light I crept back to camp. A raven and an owl croaked and hooted on the plain below. A robin sang in a cedar near-by....

Next day we returned to Van's ranch, but Sisson, who thus far had been occupied mostly about the camp, set out alone for what he called a square day's hunting all by himself, declaring he would 'kill a ram before night.' From Rhett Lake, where we camped, he struck directly across the lava plains for the main sheep mountain, while Jerome drove around with the wagon. Van Bremer, Brown, Hepburn, and myself rode over the sage plains, leaving the mountain on the left, hoping to find our game on the way; nor were we disappointed.

While we were all riding single-file in silence through the rough lava and sage plain, Van's keen eyes discovered a flock of fifty or more rams, ewes, and a few lambs. I was gazing at Mount Bremer, and the first intimation I had was Van's dismounting and handing me his halter. The noble game were about three hundred yards distant and stood gazing at us. To the right was a jagged battlement of lava, to the left their grand mountain stronghold, and they evidently were undecided as to which they should attempt to reach. The latter was the safer, but it was farther off. They turned this way and that, evidently frightened, feeling caught on account of the levelness of the plain in which they stood. Meanwhile, Van and Hepburn ran towards them crouching in the sage, and taking advantage of a slight swell in the ground. Hunter Brown, who was always doing unheard-of things, had taken his rifle apart, locked it into its pine box, and sent it back to Van's in the wagon.

As soon as the hunters' heads began to appear above the swell, the watchful game saw the absolute need of moving somewhere, and an old ram led off towards the mountain, all the others following slowly in single file, about fifty in the block, and as they bounded on at right angles it formed a very exciting scene. The hunters were now about two hundred and fifty yards from them, and just as the sheep got under full headway they drew up and took deliberate aim. Van, as he sighted his heavy Colt rifle, looked exactly like the figures one sees on powder-flasks, while the tall, manly form of Hepburn, slanting back and taking aim, resembled the mast of a clipper ship. They fired, and Hepburn's ram fell, a noble old fellow, broad and ponderous as a buffalo. Then bang went his other barrel, and his second sheep, a ewe, fell so suddenly that in the excitement she was not observed. Judge of the feelings of poor gunless Brown, witnessing the fray, outwardly cool as icy Shasta, yet doubtless, like that old volcano, hot within.

The brave sheep were now bounding wildly over the gray plain in a direct line for their castle mountain, and a bright thought flashed into the brain of Brown: he would head off the flying game and drive it back to be shot. So he gathered up his reins, and, as if he were riding steeplechase, dashed his spurless heels into his horse's flank. But the lazy mustang had no enthusiasm and made but a feeble response to Brown's ardor, so after galloping madly through the sage at the rate of about ten miles an hour, Brown drew rein in despair. Meanwhile, Hepburn's ram arose, and after staggering a few rods, while the hunters were reloading, ran firm and erect again with his huge horns thrown back over his shoulders. A second shot missed him and he fled like the wind to the shelter of the lava cliffs, the bullet probably having grazed his skull without inflicting any permanent injury. This was a fine specimen of a full-grown ram, broad and massive and probably weighing three hundred and fifty pounds. Just before he went down back of the cliff, he halted. There his form and noble horns were clearly outlined against the sky.

We little know how much wildness there is in us. Only a few generations separate us from our grandfathers that were savage as wolves. This is the secret of our love for the hunt. Savageness is natural, civilization is strained and unnatural. It required centuries to tame men as we find them, but if turned loose they would return to killing and bloody barbarism in as many years.

In the excitement and savage exhilaration of the pursuit of the wounded, I, who have never killed any mountain life, felt like a wolf chasing the flying flock. But all this ferocity soon passed away, and we were Christians again. We went up to the ewe. She was all that was left of them — left of the fifty. She was breathing still, but helpless, and I pitied her. A moment before, unarmed as I was, I could have worried her like a wolf, but helpless, and with so gentle an eye, she inspired pity as if she were human. Poor woman sheep! When bounding along with her neck curved high, she was shot through the head and never knew what hurt her. Bremer drew a big knife and coolly shed her blood, which formed a crimson pool in a hollow of the gray lava.

It was near sundown and we were five miles from camp. The stars came out, and every trace of excitement faded from our minds. Van Bremer tied the ewe skillfully on old Bob's back, and we hastened on.

Sisson reached camp just as we did and reported more blood. He had killed a ram on the mountain and a couple of mule deer. The aggregate of today's hunt was two sheep and two deer. Both the ram and ewe were said by Van Bremer to be considerably below the average size....

On the fifth and last day of the hunt, the fastnesses of Mount Bremer were invaded once more, but no blood was shed. Brown's 'luck' was as unique as ever. He had shot elephants in Ceylon, yet no one of these Modoc sheep seems to have suspected him of being a hunter, and whether crashing through the brushwood or hammering over the lava blocks with his ironshod shoes, they still seemed to welcome his approach. Today, after laying his gun beyond his reach, he sat down on the lava in a lonely place and deliberately took off his shoes. Presently he heard a footstep, and, looking round, there stood a ram as if for sacrifice. The grounds of this animal trust, so conspicuous throughout the hunt, are not easily guessed. Perhaps the secret lay in color and general brightness. For everything about Brown was bright. His coat, of glossy moleskin, was nearly the color of their own. His gun, also unnaturally bright, lay shining on the frosty ground like an icicle. And the nails in his English shoes glittered like crystals of feldspar.

There stood the ram; there sat the hunter. He dared not move towards his rifle for fear of breaking up the meeting. Big Horns therefore gazed on the brightness undisturbed, then quietly disappeared in a thicket. Brown, however, sat still four hours longer waiting for another, until the evening shadows grew out over the plain. Then he returned to camp declaring that the shooting of a wild sheep was only a matter of time, and that 'still hunting' was easier and better. After describing the gestures and immense horns of his visitor, he added with great animation, 'I would give twenty English sovereigns to shoot one of these noble animals.' Someone hinted that an ounce of lead was price enough if he only knew how to pay it.

This had been a fine sunful day, crystals on the sage and a rosy glow on Shasta. And the sun went down in that delicious purple so common in 'deserts.' But the next morning the wind blew stormily, and the air was dark with snow-flowers. We had intended hunting two days longer to allow Brown's 'luck' to arrive, that he might get horns wherewith to adorn his English halls, but, having a pass six thousand feet to cross, with the danger of being winterbound, we hastened our departure. Therefore all our game — sheep, deer, antelope, fox, geese, and sage hens — were packed and crammed into the wagon, and our hunt was done.

3. *A June Storm enjoyed on a Trip to the High Sierra with John Swett, J. B. McChesney, and William Keith*

June 17, 1875.

On the morning of the fifteenth, we set out on our camping excursion to the High Sierra. Cool, snow-tempered air was blowing on our faces, and I noticed that a massive storm-cloud muffled Half Dome like a huge ruff, and that the pines rocked in the wind and waved their tasseled boughs just as they do in winter. Also that the Upper Yosemite Fall was torn into long shreds and streamers as in winter by gusty storm-winds. I therefore interpreted the signals to Swett and Keith, but they were eager to go on. Accordingly we began to climb out of the valley in a long cavalcade, with the inevitable mules, mustangs, canned meat, and other camp débris. Before we had fairly passed the giant brow of El Capitan, rain fell, and the storm began shrouding many a beetling cliff and brow in dark rapid-moving clouds, and sending down rain into the valley and snow upon the mountains. We rode on through the dripping pines and snow-flakes and hailstones, Swett with a subdued clerical composure, and Mack in his abundant clothing snug as a beetle.

On reaching Gentry's Station, eighteen hundred feet above the wall, we deliberated whether to camp or advance to Eagle Meadows on the north side of the valley, where we could picket the horses. The clouds were at this moment somewhat broken and Mack and Swett were inclined to advance, but Keith broke up the weather council, declaring with a scowl and flash of savage wit blacker than any cloud in the sky, and with a voice like thunder, that it was 'perfect madness for poets, painters, and mountaineers to seek the darksome, dripping, snow-dusted woods in such wild, woeful weather.' Keith contains a poem whose appearance is momentarily expected, which fact explains the waving rhythm of his prose, and the sudden gushes of 'Oh, thou vast,' etc., in view of El Capitan, and when watching the snowy comets of Yosemite Falls. He has painted many a poem gathered from the dells of the Coast Range; now he proposes to take to the pen.

Keith's storm advice was followed, and we speedily found ourselves beneath a sugar pine roof and around a blazing fire. After supper Keith showed us to our rooms in the abandoned hotel, with obsequious bows and good nights. It was a wild night; rain alternated with wind-driven hail and snow, and vivid lightning flashes seemed completely to fill the river canyon beneath, and the thunder rolled in heavy reverberations from cliff to cliff.

Next morning all the ground was white with snow. Here and there a green fern frond or tall wandy grasses were seen above the snow. Not a violet or a gilia to be seen. The firs and pines showed a silver gray. In the forenoon we left camp to climb a dome down whose smooth unchanneled brow poured a fine cascade about three hundred feet high. This nameless cascade, enlivened by the fresh-fallen snow now melting (the sun appeared in fitful gleams among low-brooding clouds), swayed and glided, its waters delicately woven into a graceful lacelike pattern sprinkled with glassy crystal beads. Truly one of the finest nameless cascades of the region, its only want a want of size. The view of the sunny landscape from the top of the dome was exceedingly fine. The storm, merely halting, was massing its restless, flitting clouds for further rain and snow and hail. Suddenly the full sun broke forth, lighting up the south wall of Yosemite with the grand sweep of the Merced Canyon and the rich forest of sugar pine in front.

More snow and hail fell during the sixteenth, but about 5 P.M. the storm broke, the wind died away, the clouds disappeared, and the robins ventured from their cover in search of food, while after-storm clouds, so characteristic of these altitudes, floated above the dripping pines, purpled and silvered into bright glowing masses in the sun. The boughs were loaded with crystal beads that gave them a silver-gray appearance, and the irised sunshine straining through the green with dazzling brilliancy made a scene no words will describe. The storm was done. The birds came out and shook the raindrops from their feathers, chirped and sang, and searched about for food.

These storms, falling among the tender spring flowers, have a destructive harsh look as if Nature was blind and heedless. I noticed where large hailstones had broken off some of the corollas of the blue pentstemon and Mariposa tulip, and the ferns wore a pitiful expression as they lay with their fronds more than half buried. Yet Nature loves her gardens, and all that we call destruction is creation when viewed in its true relations. . . .

These Sierra gardens were all planted subsequent to the recession of the glaciers, which was one of the most recent of geological events. Tens of thousands of storms have fallen upon them, yet their beauty is perfect. Could we have come here during the main glacial period, we should have found only wastes of fathomless ice, with here and there trains of red and black slate borne down from the shattered summits. Yet the glaciers were the implements of all this gardening and forestry, furnishing soil, ploughing the solid granite, and spreading it out in long curving moraines and field-like beds. And since the planting of the first hardy pines and frost-enduring sedges, there has been a consistent development towards higher and yet higher beauty.

We are camped on the right lateral moraine of the trunk glacier of Yosemite Creek, which swept through it and flowed far below. And the excellence of the sugar pines, two hundred feet high, eight feet in diameter, without a decayed fiber shows the excellence of the processes of forestry. . . .

Paterfamilias Swett stands around waving his arms like a pine tree. . . . Mack catches Coleoptera, washes his fingers, and cooks. Mack is a rigid disciplinarian, as many an Oakland scholar and teacher will testify. It is shown here in the rigid scouring he gives the cups of the camp. Whether contact with the summits, glaciers, and granite will make him more tense or more lax will soon be determined. . . .

June is too early to make camping excursions into the high mountains — feed is scarce. July is better. October is made up of almost uninterrupted sunshine. Yet June snows are not to be

greatly feared; they last but a day or two, and disappear in the
sun as by magic.

Yesterday the firs were leaved, the violets and sphagnums out
of sight, the pine boughs laden and drooped. Today, except on
high summits, not a flake or hailstone is left. More than fifty
butterflies have passed me where I sit on the hillside among the
pines. The flies seem almost uninjured, and bees drone and
zigzag from flower to flower. A squirrel is within twenty feet
of me as I write, standing at this moment on his hind legs, pulling
down the seed-filled pods of the rock cress with his paws like a
bear eating manzanita berries. Not a cloud in the sky, just
breeze stir and shimmer; the bent needles and flower stalks are
erect again.... And gentians enamel the soft glacial meadows.

Cascade Creek goes brawling by, pouring from pool to pool
its waters, more than doubled by the melting snow. Doubtless
the storm will considerably increase the volume of Yosemite Falls.

The lowest point reached by this snow was about fifty-five
hundred feet above the level of the sea. Since morning this
summer snow-line has receded more than a thousand feet. The
breaking-up of the storm last evening was a brilliant affair.
The empty clouds changed to purple and pure snowy white shot
through and through with sun, and the dripping trees were laden
with flashing, irised crystals that burned on every leaf. The
clouds moved hither and thither, now down among the canyon
rocks, now up among the rejoicing forests, as if reviewing their
accomplished work.... Tomorrow we ride through many a mile
of silver fir, cross Yosemite Creek two miles from where it leaps
into the valley, and before sunset we shall camp at Lake Tenaya,
one of the brightest glacier lakes in the whole Merced region....

Additional Notes upon the Trip to the High Sierra

In Grand Tuolumne Canyon.
June 19 (?).

Tuolumne is a sleepy hollow. Here the river rests itself in a
curving, mazy channel emerging from the mountain, after

gurgling from the glacier on the north side of Lyell and flowing through many a granite canyon down falls, open meadows, and cascades. .... Here all sounds and sights conduce to repose — domes in refrain, the linnets' song, the rounded flowing falls, the curved, drooping oaks, sunny and hazy, and the tropic heat. ....

Mono Lake
June 20.

A Windstorm.[1] The party of three that I led through the pass to Mono Lake, were eager to sail its heavy waters and visit the islands. We borrowed an old waterlogged boat from a nomad who was stopping there for a while. He cautioned us against delaying our return, as stormy winds often raised a heavy sea. ... The lake was calm, lying like a sheet of molten metal — a dead lake in every sense. We paddled gently, rambling along the white shore curves, careful not to overbalance the clumsy craft that seemed trying to turn turtle.

A sail on the lake develops a group of pictures of rare beauty and grandeur. Long ranks of snowy swans on the dark water, clouds of ducks enveloped in silvery spangles. The mighty barren Sierra rising abruptly from the waters to a height of seven thousand feet, and stretching north and south for twenty miles with rows of snowy peaks. Ranges of cumulus clouds swelling in massive bosses of pearl — cloud-mountains and rock-mountains equally grand and substantial in appearance. Snowfields and ice in the higher hollows, white torrents dashing down shadowy groves, and smooth moraine slopes drawn out upon the gray sage plains along the base of the range, with silvery streams descending in bright cheery song to vanish in the dry desert.

The larger island in the lake is about two and a half miles long, and is composed of hard lava and loose ashes. The smaller, half a mile long with a cone two hundred and forty feet high, is of hard black lava, quite recent. Boiling springs and hot jets of gas boil up from the lake-bottom near its shore. ....

[1] Although found elsewhere in undated fragments, this Mono Lake narrative probably has to do with the camping trip of June–August, 1875, which Muir made in company with Swett, McChesney, and Keith.

In two hours we gained the larger island and wandered about looking at the cone basalt, lavas, hot springs, and vegetation.... A few white and blue gulls slowly winnowed the air on the way to their homes, while here and there the swift wing of a swallow was seen....

Then clouds began to settle low on the dark cluster of peaks about Mount Ritter and at the head of Rush Creek, allowing the hacked summits to appear free above them. Mount Dana had a round gray cloud-cap which at first lightly touched, then gently clasped his snowy head. Heavy cumuli gathered and grew to the northeastward toward Aurora, and shadows crept across the gray levels about the lake.

Suddenly, at noon, a breeze fell from the mountains which soon roused the sleeping waters into white-crested waves....

Then we made haste to get back to the mainland, a distance of about seven miles from the island. After rowing hard and going about a third of the way, the wind began to blow in heavy surges that lashed the lake into a fierce roaring tempest. Water poured over the boat, faster than we could bail it out. Fearing we should sink, we turned back to the island, glad to get ashore anywhere. The waves broke unweariedly on the sloping beach, waking memories of my boyhood on the seashore of Scotland.... I advised waiting patiently until the next day, but my comrades, hungry and without blankets, began to murmur. However, they submitted and went to bed, but the fire of slight brush could not be kept up and the cold made them shiver, causing them to roam in the darkness along the desolate shore, back and forth, like restless ghouls. They seemed to fear their lives were in danger, and they must escape at any risk, and gradually worked themselves up to a determined struggle for life, though the only danger lay in seeking to leave the island. At length, after midnight, they could no longer be restrained, and as the wind had slightly abated, I consented to try. We launched in safety and on we sped, keeping the boat squarely in the face of the wind to avoid upsetting. I, as pilot, sat in the bows to give warning of larger waves, one of my companions steered, and the other two rowed. . . . Seven miles to go. . . . The range to the south loomed dim and vast, and along the shore the Indians had built large encircling brush fire fences, for it was sage rabbit hunting time. These made a livid illumination, and dense black clouds of smoke were seen rising, making one think that after centuries of repose the volcanic cones were again bursting into action.

The wind howled, the waves broke repeatedly, and we had to bail to keep afloat. I sat with shoes unlaced, ready to swim, and feared not for life, as the water was not cold, though the dashing of bitter spray would be trying to the eyes. . . . Towards morning we got ashore and back to our camp in an old abandoned hut in the possession of wood rats. Yet it was a house, and all city visitors must have a roof over their heads. . . .

Summit of Mount Joaquin.
June 23.

A butterfly flew eight or ten times around the summit of the mountain on vigorous wing, as if rejoicing in strength. When it alighted on the warm granite near the glacier, it opened and shut its wings as if in a lowland flower garden.

Wallflowers are abundant here. I saw a woodchuck on the rocks, a thousand feet above the meadows. Also I heard and saw a small marmot that lives among granite rocks, chee-cheeping plaintively.[1] He was in company with the bright, frisky, confiding, inquisitive Tamias, or squirrel-sparrow.[2] . . . A Clark crow feeding her young.

Owens River Canyon.
June 26.

I saw young sage hens seven inches long from bill to tail, with twelve-inch wing expanse. They can just barely fly. Gray, blunt-billed fellows; when caught, they utter a clear piping like the sound of a boy's willow whistle. . . .

[1] Doubtless the pika, or cony, or little chief hare, not a true marmot. — F. H. A.
[2] A chipmunk. — F. H. A.

The scenery of the second great valley south of Red Slate Mountain on the east side of the range is the most sublime and most varied I have anywhere seen. It contains rocks for a dozen Yosemites.

No synonym for God is so perfect as Beauty. Whether as seen carving the lines of the mountains with glaciers, or gathering matter into stars, or planning the movements of water, or gardening — still all is Beauty!

Mountain streams vary greatly in speed, not only from declivity, but because of roughness of channel. Streams are bridled by having boulders in their paths. Also by being planted with birches, willows, and alders. Many of the steep-descending streams of the east flank are the most effectually held in check. Roots and branches lean and grow out, and many of the main stems are bent low over them with snow-pressure, and thus, in case of flood, they are submerged, and by their stiff opposition, coupled with flexibility, offer an admirable check to violent flood speed, thus protecting moraine deposits. Heavy streams, three or four feet deep and twenty feet wide, have thus been prevented from eroding their channels in moraines more than a few feet, subsequent to the glacial period. Also lakes act as checks against sudden and violent floods, and as heaters and coolers of waters.

Beside spray, many beadlike drops or angular water-masses are thrown into the air by the explosion of air bubbles contained in the white, churned stream water. These drops rise in curves, often meeting at the top just before they begin to descend, and forming a series of interlacing arches. And when the sun is right, they are seen to be beautifully irised, shining like colored glass beads.... A stream thus regarded is mightily exalted in beauty.

No flower in the mountain seems so truly to speak the heart love of Nature as does the violet. The wallflower,[1] large, yellow, and fragrant, is very abundant at almost all altitudes.... The

---

[1] *Erysimum asperum.* — W. L. J.

robin is an early riser, and his homelike call is delightful to hear. The hummingbird is also up before sunrise among his gilias and castillejas.

4. *Exploring the Sequoia Belt from the Yosemite to the White River, Tulare County*, 1875 [1]

Camp at Wawona Falls.[2]
August, 1875.

With a small brown mule I set out from the Yosemite to explore the sequoia woods. . . .

After purling around the roots and dogwood fringes of the Big Trees, the creek pours down a picturesque canyon filled with boulders and rosetted with giant saxifrage,[3] scarlet Mimulus,[4] azalea,[5] lily, lupine, Hosackia, chinquapin, Libocedrus, yellow pine, and groves of silver fir. . . .

In a nook filled with gray glacial boulders and shaded with a dense wall of highest spruce, this icy, crystal stream comes welling, its current made to pulse and waver with a scarce audible tinkle over small mossy pebbles. Boulders hardly a foot across, lined with mosses, lie along the water's edge. Above these is a zone of verdure made of short, slender grasses enameled with musk Mimulus, blue daisies, and three or four tall spikes of lupine.

Beyond the green zone, on the upper bank, two leafy bushes of mountain currant lean protectingly over, and above them a wild cherry bush. As we approach, we surprise a jay at his morning toilet. We are within a yard ere he sees us. Then he flutters off in such headlong haste he flies against a bush. Recovering, he darts confusedly into the recesses of the nearest fir. I seat myself,

---

[1] See Muir's *Our National Parks*, chap. IX: 'The Sequoia and General Grant National Parks.'

[2] The name 'Wawona Falls' was apparently given by John Muir to one of the many waterfalls among the forks of Big Creek. The exact location of the falls is not known.

[3] *Pelliphyllum peltatum.* — W. L. J.

[4] *Mimulus cardinalis.* — W. L. J.

[5] *Rhododendron occidentale.* — W. L. J.

and presently we are discovered by a bird who regards us as
intruders — a little dun nugget of a fellow with round gray bosom.
He comes nearer and nearer, uttering a scolding chur-r-r-r. His
mate soon answers with a nervous whit-whit-whit, perhaps intended for 'What?' but with feminine caution keeps out of sight.

A few moments later along come a pair of cheery, confiding
canaries, flying within a yard of our heads, from one currant bush
to the other, as if inquiring, 'What's a' the steer, kimmer?' ¹ Such
eyes! And such orange bosoms! How exquisitely modeled are
these pieces of bird beauty in form and color!.... Their fears sub-
side in the belief that we don't know their young's hiding-places,
and are not likely to harm them; soon they go about their affairs
among the fir boughs....

Evening.

For a while not a sound. Then the creak of myriad voices fills
the night with soothing, slumberous stir — all one subdued tone.
Yet above the general level of sound, like ripples on a woodland
lake, a few notes are heard — tiny cricket-like musical creaks
and chirps, infinitely sweet.

The woodpecker is latest at work, not like the hasty laborer
stopping before his task is done....

Beetles drone and boom, then drop into silence. Bats, winging
on easy whirls, circle in bays and deep pools of air among the trees.

A meteor flashes athwart the sky, startling us into a sense of
the majestic movements of other worlds. Stars, though bright,
are far less brilliant than on the heights.

Ant-lions flutter in campfire light. And, circling round, the grand
tree shafts are seen, the eye being confined to a few, not roaming
loosely over all the woods.... A soft, plaintive note like that of a
bird is heard frequently, but I have not yet traced it to its
source.... Then the owl unmistakable, and how cheery!

Stars glow brighter, for the moon is still below the horizon.
Here and there one is seen among the branches, like a white
incense lily; or past the black boles, alone over the hills.

¹ *Anglice,* 'What's all the stir (excitement), comrade?'

Mayhap one's mind will wander to other woods where the sun still shines. But that is not our affair, and if quite healthy, we shall be full of our own night.

A rustling is heard — the sound of a timid wood mouse....

The moon-day rises, and the deer, who have been sleeping and hiding in chaparral dens, come out to wander over the well-known pastures.

Noon, next day.

We go forth in the noon to walk. Our steps crackle over pine needles and empty brown burrs, through mazes of chaparral where only skilled mountaineers find their way.

The great California buzzard is sailing, whirling over game — a dead sheep or bear shot by a hunter and lost. Go where we will, we hear the sound of dropping cones — the harvest of the squirrel. Around the pines and firs are brown heaps of nibbled cone-scales and purple seed-wings. Here and there they are gnawed into shreds by bears.

Camp in Upper Fresno Basin. 5 P.M.

Now an old brown log is glorified with the evening sun-glow. Two bars of mellow light shoot up the meadow; both margins are in shadow, with scarce a flower panicle stirring in the hush. Here and there a willow-tuft glows against the gray shade. One grand promontory of firs stands full in the light, the long branches clad in yellow lichen. Farther back the brown trunks are flecked with sunshine, and on the north side one young pine towers transfigured, while all its companions are in shadow.

Tamias is frisking, now near, now retired and sitting on the top of a stake, calmly watching and listening. A big cone falls near-by with a heavy thud, but he has heard that before and knows it well enough. Now the logcock clucks, but he has heard that also, and the woodpecker's rap. But to the tiniest uncommon sound he listens attentively....

A few rays slant into the shadowy amber deep. Now the highest tops of trees are in night. Far up the mountain the slopes are

still steeped in thick unshining purple, and we think of the alpen-
glow still farther beyond, inspiring the snowy peaks.

Night.

A few moments later the day is done, and all is changed to
dull gray. Fainter, fainter grows the twilight. . . . The brownest
trunks lose their color. . . . How solemn the hush, the rest! Not
a squirrel note now. Every one is at home motionless, sleeping in
a rolled ball, all his lightning-filled limbs wrapt about with his
tail for warmth. The dew is falling. The violets and daisies are
drinking. Not a breath stirs the innumerable plumes of fir. Go
out into the latest twilight on the mead and see even the airy
panicles of Agrostis scarce moving. Yet the heart of Nature is
still beating.

Now the architecture of the forest is seen in all its grandeur.
In the daylight we see too much — more than we can attend to or
appreciate. But at night all is massed, and the spires and towers
of black shoot up to the gray sky along the meadow, forming a
wall, a street of trees. . . . How palpably this studied arrangement
strikes the dullest eye!

Morning.

A whiz and a swoop as of a bolt from the heavens. We look
and see a hawk pouncing on a buzzard. . . . The forest opens and
a gray dome rises into the sky.

There before us is the grand wide-open basin of the Upper
Fresno, with innumerable trees — spiry firs, yellow pines, and sugar
pines with outstretched arms, and on the distant hills the kingly
sequoias. . . . Now the sound of rushing water-cascades mazing
ceaselessly to the sea. . . .

We come upon the highways and byways of deer. Young
squirrels watch from tree limbs. If you cannot see the squirrels,
don't go to seek them, but bide a wee, and they will come per-
chance on their own errands, or from sheer curiosity. You hear
a strange note of questioning, of wonderment, and lo! there he is,
gazing with fearless eyes. A moment more and he darts upon

you, running across your legs, if you have the nerve to remain
motionless — electrifying you like a stream of lightning. ...
What a world of expression in his eye, as of the woods condensed.
No eye so bold, so unflinching. Perhaps he tries to drive you
away with a sudden onset, screaming 'Pyow, pyow, pyow,' like
a bubble bursting in a laugh. Or perhaps he leaves without a
word, and returns to nutting. His whole flesh exhales the odor
of balsam, and tastes of pine needles and rosin — every fiber
leavened. ... No man incapable of calm waiting will see wood-
dwellers, winged or footed.

In seeds what plans for future centuries! ... In meadow pools,
with beetles and skaters, what a world of faerie!

As we walk, nature in the noon glow lies beautified. Passive,
yet active, immortal.

Cool coverts for deer in the chaparral, pressed and outspread by
winter snow, then rising elastic a few feet, forming a bower and
dim retreat.

> Wah Mello (Fresno Dome).
> Headwaters of North Fork
> of San Joaquin River.

Coming in sight of the massive dome, rounded and bare, it
seems so ethereal after the still terrestrial woods that our thoughts
undergo a change. New landscapes span the far horizon. A mile
away is a ridge of pre-glacial lava, the residual mass of fiery
floods. ... And over the meadows an avalanche of water, rocks,
and logs swept a few years ago — a terrible manifestation of
Nature's power. But the law of these things how few can see! ...
The cooled lava is forested now. The sun shines lovingly upon it,
and all is joyous life. New flowers are already planted on the
flood belts, showing Nature's modes of working towards beauty
and joy. ... Over all came floods of glacial ice, bringing all landscapes, forests, and gardens with their tender loveliness. We
read our Bibles and remain fearful and uncomfortable amid
Nature's loving destructions, her beautiful deaths. Talk of immortality! After a whole day in the woods, we are already immortal. When is the end of such a day? ...

Sunset.

There is no rankness now in the flower stalks, or lushness in the grass. No fiery tropic splendor, but warm mellow lights on rock and tree and mead — subdued tones and gray transparent shades, and from the west an amber flood of glory!...

Walking these woodland paths we find ourselves following deer and bear. Fragrance from beautiful mountain carpets greets us in open places where the pines stand well apart. Here is the snow-flower gone to seed, yet wonderful in crimson color. And Ceanothus still in bloom, and sweet hawthorn fragrance.

We are heated on the open hills, and soon descend to the valley, where water runs cool amid saxifrages, brawling calmly and leisurely now... from pool to pool, over boulders once flood-rolled. And we enjoy the flood-music once again, for it is all well written on channels of rock and boulder walls, and stranded, battered logs.

Every flower, every needle is exhaling odor. Amid such innumerable fragrance fountains, how wonderful that Nature keeps so admirable a balance! The air is never gross, but subtle essences combine to give health and pleasure. So also the streams of our meadows are mixed with the juices of a thousand flowers — aye, and minerals too, for water is a universal solvent.... Yet how rich and pure and exhilarating — a drink for gods!

Dawn.

Morning comes again, hallowed with all the deeds of night. Here it is six or seven thousand feet above the sea, yet in all this tranquil scene we feel no remoteness, no rest from care and chafing duties because here they have no existence. Every sense is satisfied. For us there is no past, no future. We live in the present and are full. No room for hungry hopes, none for regrets, none for exultation, none for fear.

Down in the willow wilderness are found the red-stemmed cornel, and giant larkspur eight feet high, interspersed with Castilleiæ and lupine, and boat-leaved Veratrum. And Leersia, finest of mountain grasses. And among drier woods the mottle-leaved Goodyera, and thickets of tall arching grasses.

A cow comes through the woods down into the meadow, and I know by her tracks she has been here before. Will all this garden be made into beef and mutton pastures, and be delved by the hog-herd and ditcher's spade? I often wonder what man will do with the mountains — that is, with their utilizable, destructible garments. Will he cut down all the trees to make ships and houses? If so, what will be the final and far upshot? Will human destructions like those of Nature — fire and flood and avalanche — work out a higher good, a finer beauty? Will a better civilization come in accord with obvious nature, and all this wild beauty be set to human poetry and song? Another universal outpouring of lava, or the coming of a glacial period, could scarce wipe out the flowers and shrubs more effectually than do the sheep. And what then is coming? What is the human part of the mountains' destiny?

The Sierra crop of conifers is ripe, and will no doubt be speedily harvested. New lumber companies are being created almost every year, and a flume is being rapidly pushed to completion to extend to the railroad, when the magnificent firs and pines of the Fresno Valley — not excepting the sequoias — will be lumbered and floated to market.

Night.

Now a pale spirit light broods over the meadow and willow-fringed bosses. The trees are bordered in white, their trunks clearly outlined against the intense jet of the darkness beyond....

Morning.

An owl, prince of lunatics. Health in his soft, angleless 'too-whoo'-hoo-hoo'.' Sometimes he is heard an hour after sunrise. A yellow flicker, a noble pileated woodpecker, and a robin. The jay is not so vociferous here as east of the plain. He chaffers pleasantly enough, but his scream seems out of tune, as if a pine needle or a butterfly wing were in his throat.

Linnets and nuthatches are below, pecking the moss of tree trunks, and flycatchers with silvery bosoms and wings black as wood-shadows.

Here and there sounds the tap-a-rap of a woodpecker, and presently a hawk sails majestically over all.

Forest shadows fall across the mead in front of us. The brown woodland slope beyond the trees is half in shadow. The fine brown trunks — some are wholly in light on one side, others are flecked and mottled. Brown tones of meadow sedges, grayish green of willows, still grayer huckleberry patches and dark green of alder.

Compare walking on dead planks with walking on living rock where a distinct electric flash seems to attend each step. Then there's the soothing softness of mossy bogs, and brushing past lily stalks and columbines in ravines. . . . There is no danger in night walking.

A garden with Senecio and yarrow, dense mosses, Camassia, and Viola with purple-striped lip, oval opposite petals turning back, delicate spurs seen between short stems, every hair tipped with dew. The young buds look like the bills of gorblings,[1] and the heart-shaped round leaves mingle with the primrose and Mimulus.

Dainty rosettes of liverwort pasted down on the ground. Alpine clove white-flowered. Farther out are long leaves of Dodecatheon, and taller brown mosses. Long-stemmed Calamagrostis waves in the faintest breeze. Towers of Spiranthes, and daisy-like dandelions, and sprawling rushes. A crooked stream with black mud bottom, bays of shadow, and promontories of moss, bossy and rich and lustrous.

Around crinkled willows grow the tall crimson paintbrush, and *Hosackia grandiflora*, Helenium, and towering spikes of mint. Brooklime with running stems and blue flowers, lupine and Epilobium, Polygonum and long-leaved runnel and tall rue, also pretty beds of Galium.

On the dry meadow-margin ferny leaves and flowers make a fine carpet. St. John's-wort, yellow-starred, makes the softest mats of all. Monardella, Gayophytum, musk Mimulus, pink Gilia, and blue-curls in moist shadows, with a margin of ferns

[1] A Scotch name for unfledged birds.

and life-everlasting. *Rubus nutkanus* under the trees. Potentilla, snowberry, and purple Eunanus, purple-flowered Malva, and a violet like a hairy wood rush.

Camp between two forks of
Big Creek. *August.*

A Forest Dawn. Bird time of day is the morning when the sunbeams begin to sift through the treetops. Lie down in a silver fir thicket at night and wait for their coming in the wakening day. Fifty or sixty visited my grove this morning on the edge of a green forest meadow where white violets grow all the year.

The night wind was a mere soft breathing, and the meadow brook was heard plainly speaking and singing its pebbly words and songs. The stars made themselves felt like flowers with exciting fragrance. The great moon looked down into the recesses of the shadowy wood as if giving all her attention to its concerns. Some bird — I regret a stranger to me — uttered a sweet low note, simple and unrelated, at intervals of three or four seconds. Then a broad-voiced owl hooted across the meadow. Soon these became silent, and all the night was given to the moon and stars. Only the brook spoke more and more earnestly and eloquently.

At dawn a multitude of bird voices were heard aloft in the tall firs and sugar pines far and near. Soon they came to my grove, perching above my head, looking down with merry morning twittering, pecking at the fir buds and burrs for breakfast.... One little full-breasted nubbin with white belly and dark back and wings.... Also a brown wren following the curves of furrows in bark of fir, and a dainty canary with orange bosom uttering sweet spicules of music that filled the air like frost crystals on a frosty morning. Steller jay was here too, and the woodpecker. But by nine o'clock every wing was still.

Then came butterflies on the meadow, and dragonflies and buzzing bluebottles and a few small gray mosquitoes.

And the wind waked the sleeping firs, which threw tremulous and warm shadows on the green meadow ground, and tall stumps barred it with shadows black and straight as if ruled. Squirrel notes were mingled with the birds' earliest. Their first note is a fine musical sparrowy half chatter, half chirp. They began their work, and soon were heard the thumping sounds of falling cones, for they are all ripe now, and the squirrels are cutting them off to store away for winter beneath logs and leaves.

One fine confiding and bold fellow eyed me for a time, then came towards me in nervous birdlike dartings along the small fir boughs above my head, then, descending, ran across my coat skirts and hastened away, looking over his shoulders as if filled with unsatisfied curiosity and astonishment.

The forest edgings are here intensely and excitingly beautiful. Tall spires of fir are mingled with sugar pines with outstretched arms, fringed along the base with chestnut and dogwood and with shadows shimmering and waving on the smooth mead....

Night.

A Campfire. The glories of a mountain campfire are far greater than may be guessed.... One can make a day of any size, and regulate the rising and setting of his own sun and the brightness of its shining. You gaze around at the illumined trees as if you never saw trees before. How marvelously the plumy fronds of the fir show out their beauty, as if the tree had ferns for branches. And each grass and daisy, now the attention is directed, may be seen for what it is, the shining corolla and panicles waving and nodding in sympathy with the flashing flames.... The bossy boles and branches ascend in fire to heaven, the light slowly gathered from the suns of centuries going again to the sun, in clear eddying sparks and flames of ever-changing motion, the very type of unweariable, elemental power.... Sparks stream off like comets or in round starlike worlds from a sun. They fly into space in milky ways of lavishness, then fall in white flakes feathery and pure as snow.

Camp at Lower North Fork of
San Joaquin.   Altitude about
8000 feet.   *August,* about last day [*sic*].

This day wore the bluest vesture of the sky I have beheld. It was the ordinary mountain blue intensified ten times or more....

It was as if the air were steeped in indigo, fairly dyed, yet of mountain-sky transparency and tension. At evening, the sun blazed in glorious splendor of purple and thick fiery gold, fairly igniting the forests — a most impressive sunset of the still, hushed species.

No meadow plant is more glorified than the little alpine Calamagrostis ... with its fine glossy stem and floating featheriness. ... Yarrow and Ivesia also bring you to your knees to gaze at them. ...

There is scarce a human being in existence that would not shout with excitement on seeing the silver fir in campfire sunshine. Such towers of sunful whorls — no tree, fern, or palm in the world may rival it in sharp lights and shades, every leaf seeming filled with still rapt enthusiasm. ...

The sparks of my fire are tonight echoed clear and sharp quite near-by, producing a remarkable effect, like the popping of muskets let off irregularly by practicing recruits.

The fragrance with which one is feasted in the woods is, like the music, derived from a thousand untraceable sources. In music there are not only birds, main wind-tones, the frogs, a flutter of leaves like the clapping of small hands, squirrels, waterfalls, and the rush and trill of rivers and small brooks, but the whole air vibrates with myriad voices blended that we cannot analyze. So also we breathe fragrant violets, the rosiny pine and spicy fir, the rich, invigorating aroma of plushy bogs in which a thousand herbs are soaked, ... and the air is laden with a multitude of scents gathered from ocean wave-tops, from pine forests and gardens, making a combination so marvelously poised we scarce notice it on account of its excellence. Yet it varies every moment, this vast scent-flood, and is not the same in two portions of the current, as when the central plain is in bloom or withered, or the lower woods are putting out young leaves or making balsam. ...

Water also ever varies, and is remarkably compounded. Miles

of drip is distilled from fern moss or minerals. No two streams
are alike. I fancy I could discriminate between Merced water
and all others. Merced water is one thing, Tuolumne another,
Kings River another, while town water, deadened and lost, is
nothing — not water at all. . . .

There is a sound of ah-ing in the woods. You hear it, and know
a storm is nigh, and every tree knows it, and every waving
branch. Look down and consider the grasses. They, too, with
every panicle swaying this way and that, reveal the thousand
minor ripples and eddies into which all wind is broken.

The impetuous rush of flame from a dead pine would seem to
show that not only the gas[?], but also all the storm-winds that
had ever beat and surged through its boughs, had been imprisoned and stored up in its cells.

Where the crowns of five or six trees come together is the spot
for a camp bed.

Though Nature in her green, tranquil woods heals and soothes
all afflictions, yet their prime uses are not for healing and conso-
lation, but for food for the healthy; and the healthiest robust
minds and bodies will enjoy them most. . . . The woods are made
for the wise and strong. In their very essence they are the
counterpart of man. Their beauty — all their forms and voices
and scents — seem, as they really are, reminiscences of something
already experienced. . . . Let an imprisoned man see the grand
woods for the first time . . . he will enjoy their beauty and feel
their fitness as if he had learned of them from childhood.

How little note is taken of the deeds of Nature! What paper
publishes her reports? If one pine were placed in a town square,
what admiration it would excite! Yet who is conscious of the
pine-tree multitudes in the free woods, though open to everybody?
Who publishes the sheet-music of winds, or the written music of
water written in river-lines? Who reports the works and ways

of the clouds, those wondrous creations coming into being every day like freshly upheaved mountains? And what record is kept of Nature's colors — the clothes she wears — of her birds, her beasts — her live-stock?

Hawks live on beautiful food, as bats on finely painted moths and beetles, squirrels on nuts with fairy wings, and the deer on blooming shrubs and dainty flowers. But the hawk devours beautiful birds, the very darlings of Nature. Today I saw a hawk watching a blue jay in the chaparral where he was driven for shelter.

The foliage of *Pinus contorta* is much yellower in mass than others. No tree needles are more silvery than these long brushes all shining on one side where the sun strikes.... No other conifer comes together in groups of three or four to make so perfect a union. Nor does any other fork into two and three so airily and elegantly — all the separate heads forming one slender, wandy spire.

Libocedrus approaches the wet meadow most closely. It can grow in wet or dry soil. It is the most angular and uncircular in bole. The foliage is transparently warm yellow like that of Sequoia. It is no match for the pine in the struggle for soil and light, when these conditions as well as temperature are suitable for its rivals. It has the knottiest trunk, sharply and deeply furrowed and ridged. Infinitely less noble in gestures and presence, it has less to say to humanity, and it is a very old inhabitant of the planet....

How infinite the nooks where only the loving eye can approach, the holy recesses in grove, mead, rock-mossy dells and cups made of ... stones wedged together by some torrent! Here Nature does the very dearest things. Here her sweet unutterable serenity is most clearly manifested and felt. How lovely the shimmering of the sun on pine needles! Pines are more interesting than firs. Fir trees never move, only their branches move, but it is worth all we can pay to witness their grace.

On south side of Joaquin
River, near mouth of
Chiquito River. Undated.

The common purity of Nature is something wonderful — how she does so vast a number of different things cleanly without waste or dirt. I have often wondered by what means bears, wild sheep, and other large animals were so hidden at death as seldom to be visible. One may walk these woods from year to year without even snuffing a single tainted smell. Pollution, defilement, squalor are words that never would have been created had man lived conformably to Nature. Birds, insects, bears die as cleanly and are disposed of as beautifully as flies. The woods are full of dead and dying trees, yet needed for their beauty to complete the beauty of the living. . . . How beautiful is all Death!

One would never think of removing a single dead limb or log from these woods were the thing not suggested by man foresters, such is the sense of fitness and completeness. In contemplating some lovely grove, I have wondered how if this dead stump or white mast were removed, would it be bettered. But I never could see room for even such paltry improvement. See the fineness of finish, how each object catches the light. Look at this dead forest, burned — its branches down-curved around the trunks like a white fog or cloud, or overgrown with lichens as if living. There is a dead stump with a woodpecker on it, and alive with mosses and lichens — homes, too, for beetles and ants. And so, when we walk the aisle-like defiles of the woods over ridges, through meadows, and still, cool glens, we find each in perfect beauty, as if God had everywhere done His best in putting it in order that very day.

Some days ago I came drifting through the gorges and woods from Mariposa Trees, arriving in the Fresno Basin. Then the grove was full of noon sunshine, and on sauntering from tree to tree, marking their form and condition, making my way through hazels and dogwood, and over huge fallen trunks, I came suddenly upon a handsome cottage with quaint old-fashioned chimney and gables, so new and fresh it still smelled of balsam and gum like a new-felled tree. At the door I found a gray hermit [1] wholly unlike the ordinary California mountaineer, sad-looking and unspeculative, living a true hermit existence in the woods. Bit by bit he gave me his history, romantic but in the main only a typical example of eventful pioneer life full of intense experiences during the Gold period — now up in exciting success, now down in profound reverses. Finally, the day of life wearing on into the afternoon and long shadows stretching before, health and gold gone, the game played and lost, he had crept into these solitudes to await the coming on of night. . . . How sad the undercurrent of many a life, and now that the clang and excitement of the gold battles are over, what wrecks of hope and health remain, and how interesting are some of the wrecks! No country is so full of unique and rare men. . . . This old man of fine breeding and intuitions gazes back at the home he hoped to make for his children, as a dream.

Being a true tree-lover, his eye brightens as he gazes on the grand sequoia kings that stand guard around his cabin, which seems as much a part of the woods as a squirrel nest in the bark. He is finely alive to the silent influences of the forest pets, the mountain quail and the squirrels, talking to them as to friends, and stroking the tender sequoias a foot high, hoping they will yet become giants and rule the woods.

September 1.

The mountain meadows are now being painted in delightful blendings of brown, yellow, and green upon which the mellow sunbeams love to fall. . . .

The Sierra can hardly be said to have any remnants of an ancient flora, for the whole range has been lately ploughed up by the glaciers. Some of the meadow-edgings have been so lately made that we can positively say that these forests are the first ever grown upon them.

[1] See Muir's *Our National Parks*, pp. 312–14.

I camped the other day upon a meadow sloping to the San Joaquin at an angle of thirty degrees. Everywhere beneath the flowery sod I could hear the rush and swirl of running water. An ancient landslip had choked the gorge with boulders which gradually acquired this fine sod of vegetation, and beneath it the stream still ran ... forming a covered cascade. I slept on it for the sake of the music, which was sweetly and rarely varied.

Springs of the Sierra occur mostly in the middle region, where moraine matter is abundant, and where rocks have so disintegrated as to form veins for rain and snow-water to percolate through and be absorbed. These springs often give rise to sloping meadows which are surrounded with beautiful firs, and adown whose verdant, flowery bosoms the sunbeams pour, making a creation of sun and shade that causes one's whole being to glow with sympathy.

The cool crystal springs that well forth know better than to sing loud in such places of hushed and sacred repose. The springs themselves are always edged with moss, no matter whether issuing from rock-heaps or sandy banks. These mosses swell forward over the water in rounded capes and headlands, like the glaciated bosses of alpine lakes — green and yellow and brown, marvelously blended, and with here and there a stalk of overarching grasses and a few violet tufts. Here the robins and larks come to wash and drink, as well as the great brown bears.

Birds. Now a wide-winged hawk heaves in sight — sailor of the air, fish of the upper sea, with pectoral fins ten times as big as his body — so high you scarce hear his fearless scream.

Now comes a cloud of cranes with loud uproar — 'coor-r-r, coor-r-r' — breaking the crisp air into greater waves with their voices than with their broad brown wings, their necks outstretched as if eager to see farther and go faster, their legs folded and projecting back like the handle of an umbrella. Looking down as they go, they see the woods below dappled with meadows and glistening with streams, and know the location of all the frog bogs for hundreds of miles.

A little dusky crested bird dwells among the willows, keeping the twigs in tremor, though seldom seen. Now a linnet flits across the open and lights on willow sprays, making shimmer of shining leaves like the beautiful disturbance made by ducks plashing down from the sky into a sunny mirror-lake.

Jays with guttural notes hop from limb to limb, leaving stiff dead twigs in fine vibration like the fibers of a violin.

Woodpecker is drumming on hollow logs, tapping dead spars. Then comes the way-cup with golden wings colored like October leaves, clad in perpetual autumn, the dearest of the woodpeckers, elegant in form notwithstanding his short barbed tail. He moves gracefully on the ground and sits well on slender sprays, and climbs as easily and fast as any of his tribe.

Now we hear the loud cackle and chuckle of the logcock, prince of Sierra woodpeckers, larger than a pigeon, with ivory bill,[1] crimson head and jet wings, making the woods ring, loving the deepest dells where the sugar pine and sequoia grow tallest and cast dim shadows. Astonishing how far they are heard in calm weather drumming on dead sequoia tops.

Now a hummingbird as big as a bee alights wing-weary on a twig, and begins to smooth his feathers. He has flown many a mile since early morning, and touched more flowers than the botanist could gather in a week.

The squirrels send down showers of burr scales and purple seed-wings and bark that flicker and alight like snowflakes.... The Douglas squirrel gives forth more appreciable life than all the birds, bears, and humming insects taken together. His movements are perfect jets and flashes of energy, as if surcharged with the refined fire and spice of the woods in which he feeds. He cuts off his food cones with one or two snips of his keen chisel teeth, and without waiting to see what becomes of them, cuts off another and another, keeping up a dripping, bumping shower for hours together. Then, after three or four bushels are

---

[1] The bill of the pileated woodpecker, or logcock, is better described as lead-gray in color. The ivory-billed woodpecker, which Muir may have seen in Florida, now almost extinct, was never found in California. — F. H. A.

thus harvested, he comes down to gather them, carrying them away patiently one by one in his mouth, with jaws grotesquely stretched, storing them in hollows beneath logs or under the roots of standing trees, in many different places, so that when his many granaries are full, his bread is indeed sure. Some demand has sprung up for sequoia seeds in foreign and American markets, and several thousand dollars' worth is annually collected, most of which is stolen from the squirrels.[1]

How infinitely superior to our physical senses are those of the mind! The spiritual eye sees not only rivers of water but of air. It sees the crystals of the rock in rapid sympathetic motion, giving enthusiastic obedience to the sun's rays, then sinking back to rest in the night. The whole world is in motion to the center. So also sounds. We hear only woodpeckers and squirrels and the rush of turbulent streams. But imagination gives us the sweet music of tiniest insect wings, enables us to hear, all round the world, the vibration of every needle, the waving of every bole and branch, the sound of stars in circulation like particles in the blood. The Sierra canyons are full of avalanche débris — we hear them boom again, for we read past sounds from present conditions. Again we hear the earthquake rock-falls. Imagination is usually regarded as a synonym for the unreal. Yet is true imagination healthful and real, no more likely to mislead than the coarser senses. Indeed, the power of imagination makes us infinite.

The *Pinus contorta* stands rigid on rigid rocks, but wandy along meadow-edges and in rich alluvial basins, where many grow close together and wave as one.

Middle Fork Kings River.
September 10.

Huckleberries are ripe here at nine thousand feet.
Nature makes beautiful use of smoke. During September and

[1] This paragraph is taken from Muir's 'The New Sequoia Forests of California,' *Harper's Magazine*, vol. 57, November, 1878, pp. 816–17.

October the Indians fire dead logs in hunting the deer, and shepherds do the same in making ways for their sheep. Great smoke springs are thus started, which, oozing and curling forth into the still Indian-summer air, make whole skies of smoke. The sun, especially in the morning, fires this new sky and burns it white, producing a truly glorious effect.

The canyon of the Middle Fork of Kings River is this morning full of smoke from bank to bank as when ice-filled during the glacial epoch, the sun burning the fading edges, the deeps slightly purple. The bold sheer headlands, facing the canyon on either side, stand out into the fairy smoke flood, as into a boundless sea. The pines in the foreground are finely relieved upon it. And far beyond on the southern slope, the rich woods — some half submerged, other standing clear — seem enchanted and hushed in the crimson light of sunset.

From the Middle and North Forks' divide glorious views are obtained of all the Kings River Kingdom — the wideness of the valleys grassed with pines, the grandeur of their architecture on canyon-edges and all along their fountains, and the sweet, gentle beauty of their meads and gardens.

I have yet to see the man who has caught the rhythm of the big, slow pulse-beats of Nature.

Last eve I heard a night bird I would gladly lie awake a week to know. Its note was very musical, flutelike, very soft and sweet, yet brave, cheerful, and clear — 'Ka-wu′kuk, ka-wu′kuk.'

Camp near South Fork,
Kings River. Undated.

Sequoias. While camped recently in a fir grove near the head of a tributary of the Merced, I caught sight of a commanding granite dome looming above the trees, called Wah-Mello by the Indians, and, though now studying trees, I could not resist running to its summit. Here I obtained glorious views of the forests filling the Fresno Basin, vast expanses of yellow pine stretching many a mile, forests of sugar pine with outstretched feathery arms, and, towards the southwest, the kingly sequoias rising high

in massive, imposing congregations. There is something wonderfully impressive in sequoias at a distance. Producing foliage in
dense masses, they can easily be recognized miles away. One is
seen crowning a ridge, rising head and shoulders above companion pines, with inexpressible majesty on his massive crown,
or they are beheld in dense, close-together companies, their fine
outline curves exceedingly distinctive....

A supremely noble kind of tree. Redwood was once more
widely distributed, but not the *Sequoia gigantea*. The sequoias
are the most venerable-looking of all the Sierra giants, standing
erect and true, in poise so perfect they seem to make no effort —
their strength so perfect it is invisible. Trees weighing one
thousand tons are yet to all appearance imponderable as clouds,
as the light which clothes them, so fine is their beauty. Huge
limbs six feet in diameter, of heaviest wood, give no look of heavy
sagging, but take the slant which gives the most perfect form.
Brown and gray and yellow-lichened, with indestructible vitality,
they stand sound and serene after the hardships of wind and
weather of five thousand years. They are antediluvian monuments, through which we gaze in contemplation as through windows into the deeps of primeval time.

Sequoia is a serious-looking tree, but not so serious as the
juniper. Instead of standing silent and immovable with only its
light outer sprays like the tentacles of barnacles, sensitive and
full of motion, the sequoia waves and sings gloriously in the great
winds and leads all the forest choirs.

'Towering to the dimness of a cathedral spire,' no other tree
has seen so much. No other is so full of other days — scores of
centuries of sunshine are in it. Some are still standing older than
the Pyramids.

The Kings River Sequoia Belt extends from Old Mill Flat north
northeastward, almost unbrokenly, a distance of ten or twelve
miles, to near the south South Fork of Kings River. In some
places it is two miles wide, and forms the bulk of the woods.
Beautiful meadow-edgings are in many places. Here are no evi-
dences of decadence. For every old and dying tree is one or more

in prime, and for every one in prime, many young trees, saplings and seedlings. Here as elsewhere they seem to follow streams, small cool oozing brooks in which they dip their roots. However, they mostly make those streams. . . .

From the southwest end of the Kings River Sequoias there is a break of about half a mile to those of the Kaweah at Hyde's Mill. Here they attain full possession of the forest for several miles, covering the hill south of the mill in magnificent order. The sky outline . . . I shall never forget — such swelling domes of verdure so effortlessly poised in the cool blue sky.

Hyde's Mill, booming and moaning like a bad ghost, has destroyed many a fine tree from this wood — two million feet of lumber this year. And it has been running three years. When felled, the sequoia breaks like glass — from twenty-five to fifty per cent unfit for the mill. This is not true of the *Sequoia sempervirens.*

From Hyde's Mill across all the North Forks, the Sequoia Belt is broken only by deep canyons and dry ridges. On all the ridges from five to seven thousand feet high, where the soil is sufficiently moist, they grow thriftily with no visible thought of dying, there being an abundance of young and middle-aged to follow the fathers to the highest deeps of the air and sun. Near Bald Mountain Dome is a lone sequoia about three feet in diameter, young and vigorous, at an elevation of seventy-seven hundred feet. This is the highest I have yet met.

After crossing three large streams I found a fine grove in primitive beauty about two miles long from northwest to southeast and a mile wide, standing at an elevation of sixty-five hundred to seventy-three hundred feet high. Here to the southward another break occurs.

The sequoia is never found in any valley exposed to the rush of floods, nor on any hillside so steep and unporous as to shed its soil and rain. It grows always where the deep sandy or loamy soil is capable of holding the winter moisture all the year, or where the rock is full of innumerable fissures and shaded and cool and moist. It thrives better than elsewhere upon low passes between partial tributary divides, where the sides of the pass possess sufficient drainage to supply moisture. Also the largest trees are always the oldest, and therefore are found upon ridge-tops isolated from fire by rocky bareness or by streams. Yet not so high but that water may be reached by sending roots down perhaps hundreds of feet.

> Camp in a hollow sequoia in
> the midst of a burning forest
> on divide between Middle and
> East Forks of Kaweah River.

Sequoia Forest Fire. Varied beauty of fire effects: fire grazing, nibbling on the floor among old close-packed leaves; spinning into thousands of little jets — lamps of pure flame on twigs hung loosely, and taller spurts of flame; big bonfires blazing where heavy branches are smashed in heaps; old prostrate trunks glow-ing like red-hot bars. . . . Smoke and showers of white fluffy ashes from the fire boring out trunks, rills of violet fire running up the furrows swiftly, lighting huge torches flaming overhead two hun-dred feet, on tops of pillars dried and fractured by lightning strokes. Down below working among arches of roots and burn-ing whole trunks hollow into huge tubes as they stand up, which you may look through as telescopes and see the stars at noon-day. . . . Smoke fragrant like incense ascending, browsing on fallen twigs and tiny rosebushes and Chamæbatia, flames advanc-ing in long bent lines like a flock of sheep grazing, rushing in a roaring storm of energy like devouring lions, burning with fierce fateful roar and stormy booming; black and lurid smoke surges streaming through the trees, the columns of which look like masts of ships obscured in scud and flying clouds. Height and hollow filled with red surges, billows roaring uphill in ragged-edged flapping cataracts. Every living thing flaming.

The destruction, in great conflagrations, of fine buildings on which loving art has been lavished, sad as it is, seems less deplor-able than the burning of these majestic living temples, the grand-est of Gothic cathedrals.

In two-leaved pine groves thousands burn at once in one continuous flame, flying like storm-clouds with terrific grandeur —
an ocean of billowy flame reddening the sky at night.

> Camp in a mountain garden,
> South Fork of Kaweah River.
> Undated.[1]

The river goes foaming past two thousand feet below, while the
sequoia forest rises shadowy along the ridge on the north. This
little garden is only about half an acre in size, full of goldenrods
and Eriogona and tall vaselike tufts of waving grasses with silky
panicles, not crowded like a field of grain, but planted apart
among the flowers, each tuft with plenty of space to manifest its
own beauty in form, color, and wind-waving, while the plant-
less spots between are covered with dry leaves and burrs, making
a fine brown ground for both grasses and flowers. The whole is
fenced in by a close hedgelike growth of wild cherry, mingled
with Ceanothus and glossy evergreen manzanita, not drawn
around in strict lines, but waving in and out in a succession of
bays and swelling bosses exquisitely painted with the best
Indian-summer light, and making a perfect paradise of color.
I found a small silver fir near-by, from which I cut plushy boughs
for a bed, and spent a delightful night sleeping away all canyon-
climbing weariness.

Next morning shortly after sunrise, just as the light was beginning to come streaming through the trees, while I lay leaning
on my elbow taking my bread and tea, and looking down across
the canyon, tracing the dip of the granite headlands, and trying
to plan a way to the river at a point likely to be fordable, suddenly I caught the big bright eyes of a deer gazing at me through
the garden hedge. The expressive eyes, the slim black-tipped
muzzle, and the large ears were as perfectly visible as if placed
there at just the right distance to be seen, like a picture on a wall.
She continued to gaze, while I gazed back with equal steadiness,

[1] This entire entry has been taken from Muir's 'The New Sequoia Forests of California,' *Harper's Magazine*, vol. 57, November, 1878, pp. 823–24. The passage has
been changed from the published form to include Muir's marginal revisions.

motionless as a rock. In a few minutes she ventured forward a step, exposing her fine arching neck and forelegs, then snorted and withdrew.

This alone was a fine picture — the beautiful eyes framed in colored cherry leaves, the topmost sprays lightly atremble, and just touched by the level sun-rays, all the rest in shadow.

But more anon. Gaining confidence, and evidently piqued by curiosity, the trembling sprays of the hedge indicated her return, and her head again came into view; then another and another step, and she stood wholly exposed inside the garden hedge, gazed eagerly around, and again withdrew, but returned a moment afterward, this time advancing into the middle of the garden; and behind her I noticed a second pair of eyes, not fixed on me, but on her companion in front, as if eagerly questioning, 'What in the world do you see?' Then more rustling in the hedge, and another head came slipping past the second, the two heads touching; while the first animal that had ventured inside came within a few steps of me, walking with inimitable grace expressed in every limb. My picture was being enriched and enlivened every minute; but even this was not all. After another timid little snort, as if testing my good intentions, all three disappeared; but I kept perfectly still, and my wild beauties emerged once more, one, two, three, four, slipping through the dense hedge without snapping a twig, and all four came forward into the garden, grouping themselves picturesquely, moving, changing, lifting their smooth polished limbs with charming grace — the perfect embodiment of poetic form and motion. I have often-times remarked in meeting with deer under various circumstances that curiosity was sufficiently strong to carry them dangerously near hunters; but in this instance they seemed to have satisfied their curiosity, and began to feel so much at ease in my company that they all commenced feeding in the garden — eating break-fast with me, like gentle sheep around a shepherd — while I observed keenly, to learn their gestures and what plants they fed on. They are dainty feeders, and no wonder the Indians esteem the contents of the stomachs a great delicacy. They did not eat any of the grasses, as far as I noticed, only aromatic shrubs and mints. The Ceanothus and cherry seemed their favorites. They would cull a single cherry leaf with the utmost delicacy, then one of Ceanothus, now and then stalking across the garden to snip off a leaf or two of mint, their sharp muzzles enabling them to cull out the daintiest leaves one at a time. It was delightful to feel how perfectly the most timid wild animals may confide in man. They no longer required that I should remain motionless, taking no alarm when I shifted from one elbow to the other, and even allowed me to rise and stand erect.

It then occurred to me that I might possibly steal up to one of them and catch it, not with any intention of killing it, for that was far indeed from my thoughts. I only wanted to run my hand along its beautiful curving limbs. But no sooner had I made a little advance on this line than, giving a searching look, they seemed to penetrate my conceit, bounded off with loud shrill snorts, and vanished in the forest.

> Sequoia forest,
> Middle Fork Tule River.
> *October* 20.

Here is temple music, the very heart-gladness of the earth going on forever. On the Middle Fork of Tule I found a sequoia forest eight miles long, six wide, and wedge-shaped. . . . I saw flocks of ladybirds going into winter quarters.

Sequoias fall mostly uphill because leaves and branches fall and pile against the upper side, and burn off the roots and trunks on that side, throwing the ponderous trees out of balance. Also the squareness or angularity of trunks is controlled by the quality of the soil. Sometimes the soil is good and deep and encouraging on one side, with bare rock on the other. Sequoias often seat themselves on bed rock if moist soil be near.

As soon as any mishap befalls the main axis, as being burned off at the top or broken down with snow or storm-winds, every branch beneath, no matter how situated, at once seems to become excited and anxious that the onward growth be continued, and branches which before grew contentedly outward now turn upward — rush to the front to take the fallen leader's place....

I found a sequoia struck by lightning this summer. Forty or fifty feet were shattered and stricken off, the bright red and brown-black fragments mingled with strips of bark and green branches, making a magnificent heap of ruins. When a pine or fir is struck by a powerful charge it is split into long angular rails and slivers and smithereens, but sequoia, being very brash, breaks up like blasted granite. When a giant falls, it makes a regular trough in the ground, four or five feet deep and thirteen or twenty feet wide, like a rounded ditch. When the fallen trunk is burned, young trees grow in the ditch thus formed and reach a height of a hundred and fifty feet, with a diameter of four or five feet....

November 12.

Pausing in my studies this peaceful afternoon, I chance to think of the thousands needing rest — the weary in soul and limb, toilers in town and plain, dying for want of what these grand old woods can give. And though I suppose it may be of no avail, I yet shout: 'Ho, come to the Sierra forests. The King is waiting for you — King Sequoia!' There is health and life in his very looks, in the air he breathes, in the birds he keeps, in the squirrels that gambol in his arms, and the flowers that blow and the streams that flow at his feet....

Our crude civilization engenders a multitude of wants, and lawgivers are ever at their wits' end devising. The hall and the theater and the church have been invented, and compulsory education. Why not add compulsory recreation?... Our forefathers forged chains of duty and habit, which bind us notwithstanding our boasted freedom, and we ourselves in desperation add link to link, groaning and making medicinal laws for relief. Yet few think of pure rest or of the healing power of Nature. How hard to pull or shake people out of town! Earthquakes cannot do it, nor even plagues. These only cause the civilized to pray and ring bells and cower in corners of bedrooms and churches.

The form of a perfect sequoia is an ellipse — the aged, blunt and domelike at top; the younger, sharp and slender, but not aspiring or arrowy like the fir or yellow pine. The colossal brown trunks are tapered with infinite care and beauty — often branchless to a height of two hundred feet, yet not altogether leafless, for slender sprays issue at intervals, flecking the brown pillars with green as if pinned on for the sake of ornament only. The cones measure two inches long by a quarter in diameter, bright green in color, and are made up of about forty diamond-shaped scales lined round and round inside with a thin tissue of rich purple color, each containing from five to eight seeds, making from two to three hundred in a cone. . . .

No one of the conifers seeds so profusely. In a single day one could count every seed on the most fruitful of sugar pines. Not so with the sequoia. It has the smallest seed, and there are enough in this grove to plant the globe. Winged . . . wafted on the breeze, like a boy's kite . . . alighting silently, lightly as flakes of snow, they grow in silence, making only one grand sound, rock-shaking, when they fall. Sequoias are dying, but so are all forest trees. They live their appointed time, like mastodons of the vegetable kingdom, and like other mastodons will be known only as fossils. . . .

Undated.

Come to the woods, for here is rest. There is no repose like that of the green deep woods. . . . Here grow the wallflower and the violet. . . . The squirrel will come and sit upon your knee, the logcock will wake you in the morning. . . . Sleep in forgetfulness of all ill. . . . Of all the upness accessible to mortals, there is no upness comparable to the mountains.

5. *A Night Scene at Lake Tenaya*

Camping at Lake Tenaya
August 1 (?), 1876.

On three sides the mighty mountains, all the sculpture massed in blackness, yet the white glaciers visible, and here and there bosses glowing in the cool light. The lake with its rocky bays and promontories well defined, its depth pictured with the reflected mountains, its surface just sufficiently tremulous to make the mirrored stars warm like water-lilies in a woodland pond. The starry sky twinkling over from cliff to cliff; Ursa Major above the trail. . . . A crescent moon. . . . All hushed save the roar and rush of the glacier-stream descending a thousand feet into the lake.

Huge glaciated bosses with adamantine surfaces sweetly adorned by the blooming Bryanthus. The flap and flick of my fire illumining a clump of mountain pine. . . . Mysterious impressiveness of every sight and sound. The sinking of the so-called busy world, the vital sympathy of the very God world. . . .

This is my old haunt where I first began my studies. I camped on this very spot. No foot seems to have neared it. There is the mark of my fire, and the brown pine tassels of my bed, close to the beveled bank of the lake.

Constancy of the feelings which Nature excites is the most enduring of all. Love is only intensified with absence. 'Familiarity breeds contempt' is the meanest of all aphorisms.

The Milky Way is like a moraine of stars.

Later: The night wind begins to flow and sigh over rocks and through the clumpy trees. The rush of the waterfall blends with the wind and the fire.

6. *Canoeing Down Three Rivers* [1]

On Sacramento River.
October 3, 1877. Night.

Sailed yesterday from Bidwell's Landing, seven miles west of Chico, at 3 P.M. Camped the first night on an island beneath a large sycamore. Enjoyed the strange sounds, the busy owls, the rush of the river, the planets reflected in the water, the seven stars of Ursa Major, the effect of my camp light on the willows

[1] See Badè's *Life and Letters of John Muir*, vol. II, pp. 73–94.

and sycamores.... A lovely dawn.... Great numbers of birds — cranes, geese, and ducks.... Rapids are numerous. I fear lest my precious boat, 'The Spoonbill' [1] — a lively memento of the Bidwells' kindness — be stolen....

The river is full of snags. The banks are high — twenty to thirty feet; sand and gravel bars are common. Steamers are hard tried — a steamer going uphill is one of the most distressful sights imaginable.

Birds, especially brants, are abundant; I disturbed a flock of a thousand or more. They rose in a dome and gradually parted, then wheeled in a loud uproar, settled again into one flock, and alighted farther down.

Curlews are numerous along the gravelly, sandy open beaches; and small birds, such as sandpipers and killdeers. Also flocks of mallards, blue-wings, cranes, and wood ducks.

Cattle coming to drink are afraid of my craft where I am camped on a low sloping bank between two clumps of willow. My boat sprang a leak from the swelling of ill-nailed bottom boards.

October 4.

I turned the boat upside down and removed one of the boards, and with my jack-knife whittled it smaller to fit, replaced and nailed it with a stone for a hammer, then lightened it by throwing unnecessary lumber overboard, and sailed on more buoyantly and swiftly in the glorious purple sunrise.... Struck snag while rowing and nearly upset. Began to think of climbing a tree to look for Marysville Buttes, when, at the end of a lovely reach, they appeared in full imposing view, exceedingly jagged in outline.

An hour before sunset I reached Colusa, and then sailed four or five miles farther over a charming bit of river of gradual sweep, with banks magnificently treed and vine-hung. The water was glassy with just a perceptible flow, and now and then a murmur of rippling current around the roots or dipping sprays. Camped under a huge old arching sycamore on a narrow ledge widened

---

[1] Because of 'Spoonbill's' efficiency in surmounting snags, she was soon rechristened 'Snagjumper.'

with my paddle. A strange animal grunted like a pig — I saw
the track in the mud, probably a coon, or badger or skunk. . . .
Plashing and plumping of fish.

October 5.

The sunrise on the Buttes this morning was a glorious rose
purple. I sailed a few miles farther, passed the mouth of Butte
Creek, and when the Buttes bore a little south of east, landed,
hid my goods in the viny trees secure from pigs and civilized men,
and set off cross-lots for the Buttes. Soon came into a road which
led off eastward past the base of the southmost and highest of
the group — found it nineteen hundred and fifty feet above the
river, or in round numbers two thousand feet above the sea.

It is chiefly composed of old trappean lava with tilted stratified
beds around the bases, chiefly conglomerate, with polished quartz
peebles [1] abundant. The whole group evidently consists of mere
remnants of a much higher and more united mass.

Butte Creek, a fine stream luxuriantly treed and lined, flows
past on the south; the distance from river to base of westmost
(Butte), three miles; to southmost and highest, six miles; from
Marysville, fourteen miles. I left my boat at 8 A.M., and reached
it again at 3 P.M. The south sides of all the five main summits are
dotted with small oaks and laurels, the latter more abundant.
The north sides are clothed with dwarf oak, Salvia, Carduaceæ
common, also Spiræa, Mahonia, and Baeria. Gilias, pentstemons, and eriogonums. . . .

The Buttes for fifty miles or so form a striking feature in the
river scenery. In most places only the banks are seen with
their luxuriant vine-festooned trees, but from many a reach
opposite them, they loom grandly and effectively as a mountain-range many times higher.

I camped on a sandy bar on the edge of a fine growth of young
cottonwoods.

[1] The Scotch word for pebbles.

The river-bank is becoming gradually lower, and the colors are becoming riper — glorious reds and purples on the vines, the sycamore a rusty brown, the oaks gray and blue green, the Cephalanthus a fine yellow and falling with beautiful curves in long sprays into the water notwithstanding the lowness of the present levels. The Cephalanthus is more patient under submersion than any other bank plant, not excepting even the willow. The fine-leaved willow of the bank sends down long tresses of rootlets into the water, giving the trunks and branches up to the high-water mark a peculiar appearance.

Beneath the successive strata of flood sediment is a blue-gray deposit of fine material evidently belonging to an earlier chapter of the valley's history — probably deposited prior to the existence of the Sacramento as a distinct one-channeled stream.

Below Colusa, there are no rapids. The river becomes stately and calm, flowing on in reaches of surpassing loveliness. The bends become longer and more abundant, now north, now south, east, and west.

Of all the domestic animals, sheep are the only ones not greatly excited and terrified by my strange boat. The birds are curious without much fear, and often wait until very near. One came out and fluttered above the bow as if intending to alight. It is interesting to witness their morning bathings. The buzzards are abundant along the river, being well supplied with drowned animals. They walk as if tenderfooted and waddle like geese. Often they sit along the stream on the ground, always in open places. Frequently they perch on dead oaks or sycamores, especially in the morning after heavy dew-falls, with outspread wings held open to the sun to warm and dry. The blue loon,[1] a long-necked heavy-billed fellow, also alights on trees, and rises from the water by beating his wing-tips and paddling with his feet for a hundred yards or so. Even when perched on high snags, in taking to flight he swoops to the water and beats his wings as if rising from it in the first place.

[1] Probably a local name for some unidentified bird. Loons are not known ever to perch on trees, and no loon is blue. — F. H. A.

Although not in large flocks, wood ducks are quite abundant, especially on the lower river, where there are few open bars. Geese and mallards, however, are far more numerous on the upper river. The curlew is not found at all on the lower reaches. Blue cranes [1] — the large species with a tremendous coarse, hoarse squawk — are common, also the smaller kind, all leg and wing, and a big kink in the neck. Sand-hill cranes are here and there — noble birds, flying in harrow-shaped flocks.

Camped in a little sandy cove.

October 7.

Passed Knight's Landing. The river is intensely lovely, slow and stately in flow. Camped on top of the bank. . . .

October 8.

Just before reaching the confluence of the Feather, the river is more than usually crooked. Below the Feather it is doubled in width, but shallower. The difference in temperature of the river between Chico and Sacramento is only about two and a half degrees — sixty-four to sixty-six and a half — though the distance is nearly two hundred miles.

- - - - -

John Muir beached his *Snagjumper* at Sacramento October 8 or 9, and took a steamer for San Francisco. From there he went by train to Visalia, and thence to the Middle Fork of the Kings River, where he explored for the first time 'the new Yosemite' of that tributary. His extant notes upon the canyon are brief, but they include the following half-effaced description of a night storm.

- - - - -

[1] Herons. — F. H. A.

Middle Fork Kings River.

At midnight the storm songs of the wind began to sound on the mountain-tops, and, looking up twenty-five hundred feet, I saw a cloud ragged with streamers on the edges. In a few moments snow-flowers began to fly with the awful deliberation... of a mountain storm. The gray light of the coming dawn began to mingle with the dull starlight. Then a thin metallic white clearness appeared on the canyon cliffs, below which the storm wrapped the peaks in gray.... Then a yet stranger light came on. The cliffs here and there caught a glow of yellow, changing the white gradually to rose. One beheld a thousand views of clouds — a new climate, a new world. The mighty pines glowed like flames, and the falling waters rejoiced as if conscious of the treasures being heaped in their fountains.... A grand reproduction of Egypt's Sphinx on the south wall... its head turned north.

Shafts of glad, warm sunlight came streaming through the pines....

- - - - -

Returning to Hopeton, Mr. Muir 'built a little unpretentious successor to *Snag* out of some gnarled, sun-twisted fencing, launched it in the Merced opposite the village, and rowed down into the San Joaquin, thence down the San Joaquin past Stockton, and through the tule region into the bay near Martinez.'

- - - - -

On the Merced River.
November 10.

Setting out from Hopeton at noon, I made the first day of my voyage to a camp in oak timber eight miles below. Had considerable difficulty in shoving the boat over cobbly bars. Fish abundant in deep pools — salmon, trout, and suckers.

November 11.

Lunched at Neil's Ranch and camped beneath overhanging willows on the bank. Rain fell early the next morning.

November 12.

I rowed to a house a couple of miles and remained all day. Rain nearly constant. From this point (fifteen miles below Hopeton) are no cobblestones and scarce any pebbles — sand only is carried into the San Joaquin by this stream.

November 13.

I rowed to Collier's and stopped with his family overnight. He has twelve happy, hearty children. Met a man here who had been shot in the head. He is an inventor and showed me a working model of a seed-sower and harrow. He complained that the shot had scattered a portion of his brain, and affected his mind. Just above this ranch, at Turner's, is an artesian well, seven-inch bore, three hundred and five feet deep. The bottom is probably one hundred feet below sea-level, but the water, after passing through sand and blue clay, a fine estuary sediment, is still brackish.

From here to the San Joaquin it is four miles — by the river ten or more. The lower river courses are always more sinuous. The Merced is a charming river from its icy fountains all the way to its end. This lower portion is beautifully vine-draped — the leaves now ripe and glowing in red and yellow. Many charming reaches with oak and willow and Cephalanthus on the banks. ...

The channel of the Merced from its debouchment into the plain is from one to two miles wide, though seldom covering it all, not even in flood time. This channel, eroded in sedimentary rock, sandstone, and conglomerate, forms the fertile bottom lands on which most of the best ranches are located. ...

Beaver are still common. They make no dams here but dwell in holes along the bank, and eat willow and Cephalanthus.

San Joaquin River.
November 15.

The amount of water now in the river — a dry year — is a current about ten feet wide and two feet deep, flowing three miles per hour. A month ago, while evaporation was much greater, the amount was less than half as much — nearly dry. The San Joaquin was, in most places, a series of currentless pools separated wholly by sandbars; now it is quite a current even above the Merced.

November 16.

Camped about five miles below Hills Ferry in a hog pasture. I fired a big oak trunk. Heavy dew and fog.

November 17.

I rowed about seventeen miles in a singularly dull, snaggy, worn-out looking portion of the river. Camped in a fertile grassy field in a willow thicket on the Patterson Grant. Made a fire of drift.

November 18.

I rowed eighteen miles, and camped just above Graysonville. Enjoyed some fine reaches with glassy, slippery current. Had grand views of the Coast Mountains. Cold and foggy until noon, then hot.

I passed over two pebbly bars. Found a salmon trout new killed and dressed and laid out on the bank for me by fish hawks. Saw pelicans in large flocks, three hundred or more, white with black wing-tips. Magnificent birds with strong deliberate strokes, imposing as they fly, their wings making a fine silky hiss and rustle. . . . They fly in harrows and perch on trees. I noticed gulls, white and gray, here for the first time. Also cranes blue and white [1] — the white less common.

I saw an eagle with a white head, his body black and white, and a great variety of ducks and geese, but no curlews, so common on the upper Sacramento.

[1] Herons and egrets. — F. H. A.

Salmon in great numbers are making their way up the river
for the first time this season, low water having prevented their
earlier appearance.

November 19.

Passed the mouth of the Tuolumne about noon. It is not wide
but has a rapid current. The waters are brown with mining mud.
Above the confluence the San Joaquin is clear, just as the Sacramento is clear above the confluence of the Feather.

Camped in a small opening among brier roses with strong cat-
claws, as well as prickles. A fine erect species stood ten or twelve
feet high with handsome flowers.

Great numbers of dead cattle float in the river. I saw a pair
with horns interlocked. They had probably met on a narrow shelf
and disputed the way. Many sheep also fall in and struggle to
climb the steep bank, until exhausted. Hogs are wiser and better
swimmers. They shoot across the river instead of wasting
strength.

November 20.

Passed the mouth of the Stanislaus at noon, much smaller than
Tuolumne. Its waters are clear.

A high wind all day and dead ahead raised whitecaps against
the current and made rowing nearly impossible. I camped among
wide-spreading willows on the east bank. . . .

- - - - -

The Journal account ends here. The following notation is on
the fly leaf: 'November 27, 1877. Arrive at Strentzel's.' This
records what was probable Muir's first visit to the Alhambra
Valley Ranch and the family of which he later became a member. (See Badè's *Life and Letters of John Muir*, vol. II, pp. 73–98.)

Chapter V : 1879-1881

'A new world is opened — a world of ice with new-made mountains standing vast and solemn in the blue distance roundabout it.'

IN 1879, John Muir made his first journey to Alaska, sailing up the Archipelago to Sitka, returning to Fort Wrangell to explore the near-by coast and the inland reaches of the Stickeen River, then, late in the fall, traveling north by canoe with Indians and that 'adventurous evangelist,' S. Hall Young, to the discovery of Glacier Bay. The lateness of the season, however, and the unwillingness of his companions to penetrate farther into the realms of storms and ice, prevented John Muir from gaining more than a glimpse of the vast region he had come so far to see.

He returned to California, and on April 14, 1880, was married to Miss Louise Strentzel, daughter of Dr. John Strentzel, the pioneer landowner and horticulturist of the Alhambra Valley, Contra Costa County. The gift of the bride's father to the young couple was the original ranch home with twenty acres surrounding it, planted to orchards and vineyards.

Late in the following July, John Muir sailed north once more to resume his explorations, this time visiting by canoe Sum Dum Bay and its tributaries, Taku Inlet, Glacier Bay, and Taylor Bay. While tramping on Taylor Glacier he had the memorable adventure with the dog Stickeen.

Most of the journals of Alaskan travel, including those of the cruise of 1881 to the coasts of Siberia and Wrangell Land, have been absorbed into the two books, 'Travels in Alaska' and 'The Cruise of the Corwin.' The first journal of the series, however, is presented here in its original form. Portions of it run parallel with the formerly published version, but the two are by no means identical. Muir's Alaskan journals abound in drawings made by himself of glaciers, mountains, trees, and islands. A few of these are reproduced in this text.

The 'Memories of the Dog Stickeen' here published are selected from a large notebook filled with random thoughts concerning the little Indian cur that gave him so much insight into the kinship existing between man and his 'horizontal brothers.'

· · · · ·

1. *First Journey to Alaska*, 1879

On the mail steamer *California*
en route to Victoria, B.C.
July 10, 1879.

Left Portland this morning at 3 o'clock. Fine trip down the river. The Columbia is a broad massive flood of brownish, opaque color that tells but little of its eventful journey down from its far fountains in the Rocky ranges. It seems rather like an arm of the sea. The shores are forested fresh and have scarcely been touched, excepting here and there on low tables where clearings have been made on a small scale for agriculture.... The trees are smaller and more difficult of access than about the low shores of the Sound on account of rocky, sheer banks, which, however, are not high. Many of the bank cliffs are bare of trees, covered only with moss; streams tumbling into the muddy river here and there, handsome, ferny-edged, invite one back to the woody valleys whence they draw their clear waters.

The harder outstanding rocks of the banks show plain traces of glaciation, so abundant was the ice flowing down the Columbia and Willamette supplied in the early portion of the glacial period from the great Utah and Nevada Basin, and from the region to the north of it, and throughout the autumn of the period from the Cascade Range, many of whose founts are still fruitful. The

ice once flowed in the Columbia Valley as far as the sea. Therefore the traces are far clearer and fresher here than in most other
valleys hereabouts at the same altitude. That the Columbia
River had cut its way through the Cascade Range, as is so freely
and thoughtlessly asserted even by geologists, is wholly untrue —
a mistake of the first water — and shows how comprehensive is
the ignorance of that portion of geology which from its nature,
above all others, admits of demonstration. The Columbia is
newborn and has as yet only commenced to flow. But however
long it may flow, it never can excoriate and model a valley in the
least like the one it now occupies. So far from eroding this lower
section of its valley, it is filling it up with material brought from
the higher and more rapid sections.... Coming geologists may
find bones in abundance in the formation being deposited here....

The Columbia bar the day we crossed it was about as gloomy
and forbidding as may well be conceived even by seamen. A
stiff wind was blowing in and a thick, muddy fog, thin and thick
in masses ragged and draggled. Our steamer is small and smooth-
propelled by a screw, which gives her fine rolling capacity, which
she had ample opportunities to prove in the rough sea all the way
up to Victoria.

Along the coast.

Here ... I saw the glacial rocks and traces which seemed yet
more fresh and telling in all that relates to the action of the ice-
sheet — its course, the way it deposits the so-called glacial drift,
excavates harbors and fiords, and brings landscape features in
general into relief. Yet, strange to say, man, with his reason,
builds his houses, grades streets, tills the glacial soil on the
ground prepared by this mighty agent, where the phenomena are
so strange and so striking as to attract and arrest the attention of
animals, without once attracting his. So truly blind is lord man;
so pathetically employed in his little jobs of town-building,
church-building, bread-getting, the study of the spirits and
heaven, etc., that he can see nothing of the heaven he is in. Place
people who sing heaven and explore it so zealously here, and they

would still be seeking it without guessing for a moment their present whereabouts.

The sail over to Port Townsend is interesting chiefly from the fine open display of the Olympic Range made to us from the deck nearly all the way across. At Departure Bay, a short distance to the north of Nanaimo, we went to coal. Here are located the Wellington coal mines, which have sent so much excellent fuel to San Francisco. Coal is said to have been discovered at Nanaimo by the Indians about twenty-four years ago, and through them became known to the Hudson's Bay Company, who began to work it, but in 1861 sold it to an English company now known as the Vancouver Colliery. The Vancouver coal measures (cretaceous) are said to rest in a narrow trough extending from a point about fifteen miles from Victoria nearly to Cape Mudge, a distance of one hundred and thirty miles. The rocks associated with the coal are sandstones, conglomerates, and shales, abounding in fossil plants and marine shells, and resembling in general appearance and degree of metamorphism the true carboniferous of some parts of eastern America. . . .

Weather constant rain all day — a tepid, drizzling, leaf-making day; the foggy clouds trailing slowly through the treetops and at times descending among the bushes and mosses, fondling and nursing every leaf. Leaving the bay, our trip to Alaska fairly began. Day after day we seemed to sail in true fairyland, every view of islands and mountains seeming ever more and more beautiful; the one we chanced to have before us seeming the loveliest, the most surpassingly beautiful of all. I never before had scenery before me so hopelessly, over-abundantly beautiful for description. One grand master view of mountains beheld from some definite point, some vantage ground gained after passing through what is common and generally known, or some lovely bonnie bit of picture definitely bounded, as for example a lake in the woods — one of those glacier meads walled in by trees or rocky moraines, or even one's first grand view of the High Sierra after climbing from height to height through the veiling woods — these may be attempted, and some kind of description, some picture more or less telling, made of them; for in them there is definite aim, something with a beginning and end to enable one to make an effort with hope of getting done. But here it is so nearly endless, and so varied and at the same time so similar, the lines of one seem graduating so delicately into the next, while the whole is so tender, so fine, so ethereal, any penwork seems coarse and utterly unavailing. This enchanted land of lake and fiord, forest and waterfall, mountain and island, begins to appear in full force just below Departure Bay, and ends far north if it ends at all. It seems as if surely, following this shining way, we should finally reach heaven. What can the heart of man conceive more divine?

I suppose, however, now I come to think of it, that this sheet of scenery rock, of embroidery coast lace, fine spun to grace the rugged, mountain-laden continent, ends beneath the ice at the north, uniting there, no doubt, with another extending down the east coast. Some would say that it is all one grand monotone of beauty, but it is not. The general type is the same, but no two landscapes are alike. Here you behold a long stretch like a river, lofty beveled walls on both sides, the view open back and ahead but rigidly closed on either hand, stretching on and on for a hundred miles or more. Then you may come to a lot of islands sown broadcast in some wide sound, big and little in endless combinations, sprinkled or clustered, some sheer-faced, plunging deep into the blue prairie; others rounding in fine convex brows or with a hollow curve terminating in a long promontory, all timbered. Only the lofty summits that rise to a height of four or five thousand feet are bare or patched with clouds of snow. Some are so small they show no curving of the woods; seem like handfuls of trees set in the water to be kept fresh, and spreading slightly as if leaning out against the rim of a vase. Rarely, comparatively so, we see small bare rocks like black dots, mere specks, punctuating the end of a grand, eloquent, on-swelling sentence of islands tree-laden; all reflected in mirror-blue water, forms and meaning doubled.

The variety is caused chiefly by differences in the structure and the composition of the rocks, and also by the pre-glacial features to some slight extent; differences also in the amount of glaciation other portions of the landscape have received, some sections having been profoundly influenced by the influx of vast, steeply inclined, onthrusting glaciers from the mountains of the mainland. Especially heavy was this influence towards the end of the period when the main sheet flowing parallel with the coast was beginning to fail. And again the higher mountain islands nourished local glaciers, some of them of considerable size, which sculptured the summits and sides too, often quite deeply, making wide round shell-shaped amphitheaters at top with canyons leading down from them to the water's edge. These causes produced endless variety, but in one particular these landscapes all agree. They all have a rounded, over-rubbed, sandpapered appearance, an exquisite finish caused by the one wide, all-embracing hand of the ice.

Saying what little we can about it, then, in a general way, it is an exquisitely wrought web of embroidery of islands and water, graduating in a fine fringe out into the ocean expanse on the water edge, and in a heavier, more massive, mountainous margin to the mainland with its lofty glacier-laden heights. Some of the islands are continents in appearance and in effect from any view to be had of them save only on the map; but by far the greater number are small and appearable as such; hundreds of them less than a mile long dotting the glassy levels in charming combinations and compositions, their direction of extension complying with the main waving lines of the coast, especially those farthest away from the coast; those near the coast having been in great part eroded and shaped by the ice descending from the mountains of the mainland, and therefore extending in a diagonal direction or at right angles to the coast line. In this generalization a most delicate harmony is everywhere apparent.

The channels, passages, canals, straits, sounds, etc., are subject, of course, to the same law. Many are like rivers, not only in separate reaches, but continuously so for hundreds of miles. The tide currents, the floating driftwood, fresh and leafy, the luxuriant, overleaning foliage of the banks make this resemblance all the more complete.

In the midst of the thickest island clusters, the impression is that derived from wide lakes, however much fretted by the island abundance; and in the thickest-sown archipelagoes the water seems always and everywhere deep, never fretted away into shallow, dabbled pools. And however bewildering any attempt to describe the whole sheet of this ravishingly lovely landscape, the eye easily takes in and dwells with ever fresh delight on the smaller of the individual islands. Though in their relations to each other the members of a group are evidently derived from the same source, one rock-mass hewn from one stone, yet they never seem broken or abridged as to their individual lines of contour, whether plunging at once abrupt and sheer with round, projecting brows, or in calm hollow lines sweeping out in low points tipped with sedge. Viewed one by one they seem detached beauties, like extracts from a fine poem; while from the way the lines sweep over from side to side and the way the trees are put on, each seems in itself a finished stanza.

The arrangement of the trees is such, the most marked specimens matching and harmonizing with each other, that a distinct impression of their having been sorted from the common forest and rearranged is felt. In some of these tufted islets a central group is planted in the middle, and two smaller groups that evidently balance each other are planted at about equal distances on the ends. Or the whole appears as one handful, a marked tree leaning out from each side. These relations to harmony are so constant that I believe they are as much the result of design as are the beauty arrangements in the painting of flowers and the counting of their petals and their occurrence in whorls, or the arrangement of feathers on a bird.

Their beauty is in every sense the beauty of youth, for though freshness of verdure and the remarkable universality of the woods may be attributed to copious moisture from the warm, steamy, all-bathing, all-embracing ocean river from the sunny fountains of Japan, the very existence of the islands, their forms, finish, and peculiar distribution are all immediately referable to the creative action of the ice during the period here just closed.

The sky above this young fairyland is in every particular becoming to the landscapes it mantles; usually pale, tender blue, with pearl clouds hovering in calm, fleecy, filmy masses, combed out on the luminous edges, but never garish-colored. Morning and evening are in orange, purple, and red, and at noon in summer the whole bland atmosphere palpitates with an intense subdued passion of light, in which the tranquil sheets of water shimmer and spangle, and the evergreens of the islands dip their spires with inimitable grace and repose.

I should like to sketch one of these Alaska summer days, however imperfect the sketch must be. It is a day without night, for it begins and ends at midnight, which is the low noon of the great round day. The sky is red and orange then, for clouds more or less distinct are almost always present. The day opens slowly, the center of greatest light insensibly increasing and circling round the horizon's rim; and when at length the sun appears, it is without much of that stirring, impressive pomp, that flashing awakening energy so suggestive of the Bible image of a strong man coming forward to run a race. The colored clouds with their dissolving edges seem to vanish as their color leaves them, sinking into a hazy dimness around the horizon. The islands, some of them with ruffs of mist about their bases, cast black ill-defined shadows over the glistening water, and the whole dome of the sky becomes pale, whitish gray. For three or four hours after sunrise there is no striking feature to be felt or seen. The sun may be looked in the face though seemingly unclouded, and the islands full in the light, and the mainland mountains are seen in the distance. Yet in all their beauty of form and wealth of woods they seem to be yet asleep, rather dull and uncommunicative. As the day advances towards high noon, the light of the sun shining down in full power lights the water levels to silver. Brightly play the ripples about the bushy edges of the warm shores, inzoning every island with a white-glowing girdle. The bland air beats now and makes itself felt as a life-giving ocean of energy through the all-pervading, sifting, drenching, luminous mist of pearl sunshine. Now we may think of the life all about us. It comes to mind of itself: the marvelous abundance of fishes feeding beneath, the myriads of trees and bushes drinking the moist light and heat, the glaciers — the ice-mills of God — on the mountains, making meal for ever. Through the long afternoon, the whole creative way down to sunset, the day grows in appreciable beauty and impressiveness. The light seems to thicken and become more generously fruitful. Everything is in conscious repose, winds only breathing gently or wholly asleep; the few clouds downy and luminous, scarce casting a shadow; a white gull here and there winnowing the warm air on easy wing; no singing birds even to stir and sweeten the air and keep one's senses separately active and awake; sky, water, islands, and mountains blending in one inseparable scene of brooding enchantment.

Then comes the sunset with its exciting colors, mingled purple and gold; not a narrow arch on the horizon, but filling half the sky, its focal point well round to the north. I have seen far more gorgeous sunsets than any I have yet witnessed here, but never any more impressive. The clouds that usually bar the horizon are fired on the edges, while the spaces between glow in yellow and green, and a soft, mellow purple flushes the sky to the zenith and fills the air, fairly drenching and dissolving the islands in its rich, glowing floods.

The crimson and gold soon vanish after the sun goes beneath the horizon; but instead of going straight down, it sinks on a curve nearly concentric with it, so that even this glowing portion of the display lasts much longer than in more southern latitudes. The sun, never sinking but a few degrees below the horizon, has even at midnight power to reach and color the low-lying clouds and mists that hang along the sky. The colors, then, of the sunset circle around to the northward and eastward and unite with those of the sunrise.

The most extravagantly beautiful of all the richly colored sunsets I have seen in this cool moist northland was one painted on a late July day when we were about halfway between Nanaimo

and Fort Wrangell in the midst of one of the thickly sown archipelagoes. The most of the day had been rainy, but during the
latter part of the afternoon the clouds cleared away — all save
a few that lay along the horizon like islands in overlapping bars
and belts, and in misty rolls on the mountains and down in some
of the larger canyons of the mainland. It was a calm evening, a
calmness that felt its way pervadingly back into one's soul, and
the color came on gradually increasing in extent of area and
richness of tone (like a bank of roses coming into bloom in slow
devices as if taking long to ripen); then it faded in the same way,
though there was a marked dying-out of the more glowing fires at
the moment of sunset. At a height of about thirty degrees, there
was a heavy bank of cloud with dark gray sky above it, and its
lower edge deeply tinged with red; below this were three horizontal bars of purple edged with gold, and pale yellow-green sky
between them; while a spreading fan of flame radiated upward
across these bars of color, fading on the edge of the separating
radii in dull, lurid red. But beautiful and impressive as was this
painting on the sky, the most novel and exciting effect was found
in the atmosphere itself, which was so loaded with moisture that
it was painted, or rather became, one mass of color, a thin translucent haze of wine purple, in which the islands seemed to float in
a half-dissolved condition. A narrow strip of water, also red,
seemed drawn like a border to the isles of the blessed. Luminous
rolls of mist lay in the troughs between the mountains. Snow-
fields, glaciers, peaks were not simply steeped in the rosy atmosphere, but dissolved in purple flame. The main effect was seen
in looking directly into the heart of the sunset as a focus. The
three gold and purple bars, with yellow-green sky between; a
dark cloud-bank overhead with heavy, lurid caves; then the whole
mass of the air a soft, purple mist, with innumerable purple islands
seen more and more faintly beyond and beyond, formed a vision
of the isles of the blessed, spiritualized in faint, or rather, tender
lines, dissolving yet distinct, more glorious in reality than ever
the heart of poet conceived, a realization of a more extravagant
dreamland than the poets of any nation have ever dared to write.

Some people on deck were transfigured — doctors of divinity, tarry sailors, and all — and worked in as an inseparable portion of the one grand effect.

It is not generally known that during the long summer days in Alaska it is never dark, for though the sun sets here about nine o'clock and is therefore some five or six hours below the horizon, it is so short a distance below even at midnight that there still is light enough to read, and the moist clouds that usually lie around the base of the sky are colored orange and red, so that the sunset in subdued tones may be said to last all night. In northern Alaska the sun scarce dips beneath the horizon.

Perhaps one-third of the summer days are rainy; but the rain is so warm, and the air is so calm as a general thing, and the rain falls so gently, that even this rainy weather is not stormy. Most of it would be called showery, as it seldom pours evenly through all the twenty-four hours. The clouds are usually gray, often pearly white, and at sunset always well colored. There are few raw, black, draggled days without some dashes of late or early color, or white illumination about the noon hours. I never before have seen so much rain fall with so little noise — no loud, rushing winds, no thunder — at least I have heard none as yet, and from what I can learn from residents here it is quite as rare a phenomenon as in California, a flash and clap of a faint far-away kind once in two or three years. The largest cluster of dry and sunny days was about two weeks long near the end of July.

There is a fresh wholesomeness about the wettest of this weather that seems conducive to health. There is no mildew in the houses, as far as I have seen, nor any tendency to eternal mouldiness in any nook, however hidden from the sun. Neither the people nor the plants have a flabby, dropsical appearance — the very land of health for man and beast. Ordinary noon temperature from 65° to 70°; in the warmest months, never beyond 80° as maximum. In September clear days are rare; most of them are all cloudy and rainy throughout the twenty-four hours. The rain is now often driven slant by winds moderately strong, and the clouds during the intervals between showers crawl and droop in a ragged, unsettled way without manifesting any tendency to violence such as one so often sees in the storm gestures of the California mountain clouds. While the rain is falling the general color of the clouds is dull leaden gray with darker masses ill-defined, the whole settling down on the islands, giving them a weird, ghostly look, their outlines melting in the gray gloom, although on account of their dark forests they are yet impressively visible in the foreground, while fading to fainter and fainter cloudlike masses in the distance where sky, land, and water are all beheld as one.

The lightest portions of these storm days, or rather rain days, are found between the showers, when there is usually an upper cloud system that admits the sunlight in a finely sifted reduced state as if coming through ground glass, while darker clouds float beneath, and gray downy islands of mist drift about over the woods, sometimes slipping back of an islet, throwing it forward in bold relief with its treetops charmingly distinct.

In winter, from what I can learn, the storms are mostly rain at a temperature of thirty-five or forty degrees, but driven by strong winds which, when sweeping the channels, lash them into waves torn into scud that is carried far into the woods. The long nights are then gloomy, as may well be conceived, and raise to the highest terms the value of a snug home with blazing, crackling yellow-cedar fire, and the lights of shaded lamps on book-covered tables.

Snow falls quite frequently too, but seldom to any great depth or to lie long; usually about three or four inches; once since the settlement of Wrangell, four feet. Between storms are single days and clusters of considerable size that are quite clear and brilliant. Lying alongside of the steaming ocean, the thermometer falls below the freezing-point only when the wind blows from the mainland. These remarks apply to the coast and islands only. Back in the mountains, and beyond the mountains on comparatively low ground, one thousand to three thousand feet above sea, the mercury falls to zero and far below, forty to sixty-five degrees at times — what they call a cold snap.

July 14.

Fort Wrangell at 6 A.M. Our whistle-screams and cannon-shot awakened the boggy village, and down came a score or two of Indians and a half-dozen whites to the end of the wharf ere we were alongside. The Captain assured us we should find it a miserable place built in a swamp, no good thing about or in it; only looked well to him over the stern of his ship when leaving it. Going ashore, we rambled through the squalid streets, hoping to escape to the woods and rising ground back of it, but found the ground everywhere boggy, though covered with beautiful mosses and alpine flowers. Even the stumps and logs, where most of the trees had been cleared away for fuel and building timber by the military, were overgrown with tufted grasses and mosses and bushes in an exceedingly picturesque manner.

In company with Mrs. Patou and Mr. Maynard, a photographer from Victoria, we visited the Indian southeast end of the town, where a number of handsome carved posts are set up in front of the larger Indian houses. The finest of these were photographed while the steamer was discharging freight for the Cassiar mines up the Stickeen River. Gathering a few plants, I escaped back into the cabin, feeling that this was the most repulsive of all the wildcat frontier towns I had yet seen. No house seemed to have a corner fresh enough to make it possible to spend a single night in the town, while I could see no spot in the surrounding wilderness dry enough or free enough of suspicious Indians and Indian dogs to camp on in any comfort.

In the afternoon we sailed for Sitka, lying at a distance of eighteen hours — one hundred and eighty miles — calling at Checan [1] on the way, for lumber, entering a beautiful bay glassy as a lake with a bold mountainous shore covered with good timber and bushes above of bright grass-green color, with now and then a mountain so steep that the snow descending into the bay in avalanches had kept it free of trees and allowed only bushes to grow. These last are often beautifully fretted with zigzags of silvery cascades. Some of these mountains on the larger islands here

[1] Doubtless the place now known as Shakan. — F. H. A.

and all along our way have well-formed glacial wombs, and of course they also show the effects of shadows on their sculpture. Several good specimens of sheer-north- and rounded-south-side mountains may be seen here; also a good specimen of half-dome. Glorious are the views we had of the mainland mountains with their glaciers and spiky black peaks, mere far-away glimpses but most telling to me.

The mill at which we called for lumber was built by the Government while the troops were located at Wrangell; a small steam circular, capacity four or five thousand feet per day. The timber along the shores is not yet marked by what has been cut. The trees are felled into the water and floated round to the mill: Menzies spruce[1] and yellow cedar;[2] the latter more valuable and less abundant — the former, a white wood easily worked — some of the logs five feet in diameter; apt to be knotty, making good deals and scantling. The cedar here is the best in the Territory for inside work, very durable, of fine mellow color, takes good polish and is fragrant as sandalwood. Will undoubtedly come to be sought for in other markets. Prostrate trees moss-grown, lying one hundred years on the damp ground, are still sound in the heart, retaining both their delicacy of color and fragrance and toughness. Though rather soft, it should make good ship timber. No cattle are about the mill. Wildness crowds close about it. Above eight or ten feet of mossy logs, roots, and humus the present forest is standing; some of the largest trees — the tallest about a hundred and fifty feet — with not a root reaching the rock beneath. Hemlock is the common tree in this charming wilderness of wood, water, and mountains.

July 15.

Reached Sitka next morning. Rough and stormy outside, but fine and calm in the bay, which is filled with islands and has a background of noble mountains boldly sculptured in capitals of

---

[1] *Picea Menziesii,* an old name for the tideland, or Sitka, spruce (*Picea sitchensis*). See Sargent's *Silva of North America,* vol. xii. — F. H. A.

[2] Yellow cypress (*Chamæcyparis nootkatensis*). — F. H. A.

ice. I walked back a mile or two into a charming bog; the vegetation so perfectly alpine, or rather, arctic. I seemed to be in one of the bogs of Canada about Georgian Bay, so many old friends did I meet, such as Linnæa, Ledum, Coptis, Pyrola, Vaccinium, cranberry, partridge-berry, Lycopodium, mosses, ferns, and a' that and a' that, and ... muckle of true bog beauties dear to the heart of every dweller of the cool North. There is a hard compact quartzite here which rises above the general level in smooth rocky mountains which retain the grooves and scratches and some of the polish of the ice which flowed west. Glacial traces are better preserved here than elsewhere on the coast as far as I have seen. So much moss and moisture seem to erode the surface in a short time, as well as hide it.

Sitka has a rusty, decaying look — a few stores, a few houses inhabited, many empty and rotten and falling down, a church of imposing size and architecture as if imported entire from Constantinople; ... cannon lying in the streets sinking like boulders in mud; dirty Indians loafing about; everybody of any character away at the mines or out a-fishing. It was the capital while the country was in the hands of the Russians and a place of considerable trade, but since the purchase it has been practically abandoned. The discovery of gold in the adjacent mountains has created quite a stir; should the mines prove productive, then it will rise from the dead and grow. Nothing of consequence has yet been discovered, but the possibilities are great.

From Sitka we went to a salmon cannery six miles from here to land freight. This and another one at Klanack are the only canneries yet established in the Territory; both have been established withir the last three years and are doing well, the fish being found in great abundance and costing nearly nothing. The Indians furnish them for half a cent a pound, and receive most of their pay in goods at a high rate. Many Indians are also employed in the cannery, together with a few Chinamen. They receive a dollar a day. When the Chinamen came up on the steamer, the Indians at first raised a great outcry and refused to let them land, claiming that the work rightly belonged to them; and it was only after being assured that no Chinamen would be employed as fishermen, and that after the Indians learned to make the cases they would all be sent back, that order prevailed. A more perfect settlement of the Chinese labor question on a small scale could not easily be found.

The Alaska salmon are said to be inferior to those of the Columbia, but in some other places, as at Chilcat, they are said to be equally good in every respect — size, flavor, and fatness; and since the Columbia fisheries and those of the Fraser and other rivers of British Columbia are beginning to fail, while the demand is increasing as the market expands, this immense fountain of fish must surely be drawn upon to a grand extent in the future, and create an extensive trade.

Cod and halibut are also found in abundance and some little attention has been given to them, and to herring also; but with the exception of a few barrels of salted salmon and packages of dried and smoked codfish, the salmon produce of the canneries is all that has yet gone to market. The difficulty in the way of those who regard every uneaten and unsold fish and every unsawn tree in the woods, and every dollar's worth of mineral in the mountains, as lost and worthless, is that all of these may yet be found in more accessible portions of our big country. But the time is coming when Alaska will seem nearer home, when her resources will be laid under contribution searching enough and intense enough for the most exacting of utilitarians. . . .

I saw some good green gardens both at Sitka and Wrangell — potatoes, cabbages, peas, beets, and similar products, requiring no great amount of sun-heat, but under temperate moisture doing well. So also the grasses. But the spots of alluvium and moraine beds along streams and lakes and low shores form but a very small portion of the area of the whole country. The fact is, most of the southeast coast of Alaska is of solid rock that has but just come to the light from beneath the ice sheet. The moraine soil is in the sea, and whatever the rock may become after the action of the rain and snow and air have disintegrated it to soil, it is in the meantime not yet ready for the husbandman. Not yet is there either the climate or the soil required for the growth of the cereals. The unborn lands about the North Pole will probably become a fine agricultural region, and so may Alaska, but mankind must bide a wee. Even if the soil and climate were ready for the farmer, the country would remain for many years unsettled while land nearer markets and centers of civilization may be had infinitely easier of improvement. A young man would be old, however industrious and strong, before he could clear and bring into a fair state of cultivation even a very small farm of, say, fifty acres of the forest lands near the coast anywhere on the sea side of the coast mountains....

Returning to Sitka the lovely scenery was enjoyed over again, but it all, while calling me, seemed beyond reach. There are no settlements, no base of supplies or communication in the wilderness, while I have been everywhere assured that the life of an adventurer away from the towns of Sitka and Wrangell would surely be short; but I must see something of the interior. Will therefore remain over a month or two and watch for opportunity; may at least go up the Stickeen.

Were the attractions of this north coast but half known, thousands of lovers of nature's beauties would come hither every year. I know of no excursion in any part of our vast country where so much is unfolded in so short a time and at so little cost. Without leaving the steamer from Victoria, one is moving silently and almost without wave motion through the finest and freshest landscape poetry on the face of the globe. The discomforts of a sea voyage are not felt; nearly all the long way is on still inland water. It is as if a hundred Lake Tahoes were united end to end, with banks and backgrounds multiplied in the same ratio as to sculpture and extent of range and refinement of waterlines. While we sail on and on through the infinite beauty enchanted, hard, money-gaining, material thoughts loosen and sink off and out of sight, and one is free from oneself and made captive to fresh wildness and beauty, obeying it as necessarily as unconscious sun-bathed plants. When, as often happens in still warm days, the islands are swung in the air with edges dissolved and outlines blurred in fairy features, then the enchantment is complete.

On the river steamer *Cassiar*,
exploring coast between Fort
Wrangell and Cape Fanshawe.
July 21.

Setting out from Wrangell [1] at 5.30 A.M., and steaming until 6 P.M., we had sailed only about sixty miles, the engineer running very slowly to save the sixty tons of fresh water he had aboard. It was all consumed an hour or two before stopping, and salt water had to be used, an effort made to obtain fresh from a stream near an Indian salmon fishery having failed. The boat was chartered for twelve persons at the rate of sixty dollars per day to make an excursion to Chilcat, at the head of Chatham Straits, to visit the Chilcat Indians — for missionary purposes by Drs. Jackson, Lindley, and Kendel and Mr. Young, and by Mr. Vanderbilt for business and pleasure. Four wives, mostly of the missionaries, [were included in the party], and I went to see glaciers, etc. The distance was estimated at two hundred and fifty miles and the time four days, but at the rate our steamer was going we saw that we should be twice as long as the estimated time in making the trip. I would fain have gone on at any cost, but the divines, as they are called, esteemed the cost of reaching and saving the souls of the Chilcats as too great. Like the Macedonians, these savages so warlike and inflexible in their opposition to the entrance of miners into their mountains, called for the missionaries and invited them to establish a church and school in their midst; but the *Cassiar* sailed too slowly and therefore they turned back, leaving the Indians to die unsaved yet a while. Therefore we directed the Captain to return to Wrangell.

Anchored first night at ——— for fresh water. Eagerly embraced the chance to go ashore with Dr. Lindley. Was rowed to the mouth of small rill by an Indian; the tide was low; walked

---

[1] See Muir's *Travels in Alaska*, chapter v, 'A Cruise in the Cassiar.'

up the shingly, shelly, dulsy beach. The stones were blue slate, quartz, and granite (in the order of abundance); the first plant terrestrial was a noble grass nine feet high, forming a margin before the dense intertangled forest; then an alder thirty feet high; beyond that a noble shady wood composed of *Tsuga mertensiana*, and *Picea sitchensis*, with a few cypresses. The ferns were developed in noble beauty and size — two aspidiums, one six feet high, one Polypodium, one Gymnogramme. Also *Rubus nutkanus*, eight feet high, just out of flower; three vacciniums eight feet, Ledum, a woody prickly tropical-looking Aralia ten feet high, very warlike, a moderate number of young conifers, mosses indescribable, dwarf cornel, Pyrola, Coptis, and Solomon's seal. The trees were mossy at the foot and up the furrows a little way, with cushiony mosses a hundred feet up; the tallest trees a hundred and fifty feet, the limbs mostly horizontal or drooping in all the spaces.

Sounds. Not a leaf stirring; deep, hushed repose; one bird, a thrush, singing sweetly, lancing the silence with its cheery humming notes, as the sunshine sifts in thin sunbeams between the boughs, marvelously effective. The whole blessed scene coming into one's heart as to a home prepared for it. We seem to have known it always. Strange, how strange, is this untamed, untouched solitude of the wild free bosom of Alaska, yet how eternally and necessarily familiar! Then through all, penetrating, saturating all, is the awful hushing sweet-voiced monotone of the stream, like the very voice of God humanized and terrestrialized.

The stream is about five feet wide, sunken in the woods ten or twelve feet, crossed by innumerable log bridges. The trees on the banks lean over across from side to side, forming high gothic arches, very beautiful to see as one looks along the stream. The bridge I first crossed upon was the most richly ornate I ever saw, wild or artificial — a massive log densely plushed to a depth of six inches or more with golden mosses and a swath of bushes, mostly young Menzies or Sitka spruce, not rank or roughlooking but as if put on for beauty only. There was hemlock also with delicate flat plumes, and a few currant bushes and dogwoods.

Beneath bridges so beautiful does this cool wood stream flow and sing. Some are ferny and flowery also with saxifrage, smilax, and Pyrola. Sat on the bulging mossy roots of an old hemlock until nine o'clock; daylight still lingered and the stillness of the noble wood became yet more impressive....

Not a track is visible of bird or beast or man. Snow retains for a time at least the lightest track, even mine; the next best surface is fine dust in a thin stratum on a smooth surface; next, fine sand; but here is the most untrackable portion of the earth's surface I ever saw on mountain or plain, a covering of elastic moss over extensive areas. No bird's foot marks it, not even those of the deer which inhabit these woods, or the bear. So much obstructed are these forests with fallen trunks and bulging roots, the animals naturally seek the waterways or meadows or lakes, leaving the woods virgin; but where for any cause they are traversed, no track is left.

The commonest traces found along the shores of these bays and inlets are those of man and woodpeckers on the trees. The Indians have camped from time immemorial, and so have many white men in the last century, for these water lanes are the only open ways at the mouths of the many streams. Stumps of trees that have been felled with axes, or hacked drift-logs are common; only, however, on the immediate margin. Knife-marks, too, on the decayed punky portions of trees are still more common, light-wood being scarce and much sought for in so damp a climate.

The Indian deck-hands were carrying water aboard in a canoe, eighteen or twenty pailfuls at a time, carrying it some hundred yards or so from the mouth of a small creek that comes brawling from the hills beneath innumerable moss-embossed bridge logs. They are patient, garrulous, plodding fellows, never in a nervous civilized hurry and never at a loss to find out the means of doing whatever is required of them, their slow, cautious, complete methods always being found better than they promise, contrasting strikingly in this respect with those of the narrowly educated townsmen, educated only at the top....

The scenery along the sixty miles we sailed is not unlike that of the straits and channels passed through in coming from Victoria to Wrangell, lovely shores wooded close down to tide-level, seemingly untouched by man save, at long intervals, for some lonely cabin where a grassy margin allows an open space; islands, too, in endless variety and composition. The mountains, however, on the mainland are higher and glaciers larger and more numerous. The charm of all was the bland, mellow sunshine lying on the lovely shores and calm, glassy water, and in the distance the lofty mountains rising higher, hacked and worn into an imposing array of pinnacles and towers and black outstanding battlements, the jagged walls, mostly black, circling around the most fruitful in-wombs I ever beheld; then below these the wide névé fields sending up long white fingers into the dark recesses of the peaks and the bluish-gray ice-currents pouring from the base of these and coming down in beautiful curves into the forests, the largest of them to sea-level. These are the most striking and attractive of the features. Small glaciers, mostly buried in snow and seeming far smaller than they really are, attract but little attention from the ordinary observer; they are simply regarded as snowbanks, remnants of the general mantle of winter. But a great glacier, river-like, curving right and left around projecting bosses, sweeping past lofty rock-walls and down through the forests, streaked with trains of sand and angular boulders that are relieved against the gray and blue of the ice and mark its flow — such a glacier always fixes the eye and calls out the wondering attention of every beholder. Counting big and little, about one hundred came into view, the smallest perhaps a mile long, the largest fifteen or twenty. The snout of this one comes down to tide-water in a beautiful yosemite valley. It reaches across from wall to wall with a front of a mile or so and a height of five or six hundred feet. It is as if, taking the scenery into consideration, gables, battlements, huge outswelling bosses tree-crowned and guarded, the great glacier which formed Yosemite Valley were still in existence, stretching across from the Three Brothers to below the Sentinel, the ice crushing heavily against the forested rocks of the walls.

Yet strange as this may appear, to see glaciers in contact with flowery and tree vegetation, I have so long looked at the ancient glaciers of the Sierra that they have been seen as present in the flesh, so that I seem to be contemplating what I have long been familiar with. At low tide there stretches from the water to the snout of the glacier a nearly level sheet of moraine matter — granite, sand mud, and pebbles and cobbles, with comparatively few large or even moderate-sized boulders — gathered by the many small streams flowing from beneath the glacier and from countless rills issuing from the small tunnels in the snout from top to bottom, and also from those running over the beveled front wall in tiny cascades.

The large quantities of water-rounded granite deposited along the flank of the Sierra, and spoken of in connection with the glaciers of Switzerland by Agassiz, Guyot, and others, and also observed by myself on Vancouver Island, where the glacial phenomena are so fresh, had made a difficult problem for me in the Sierra, as there is little or almost no washed moraine material found in connection with the small and medium-sized glaciers. But here the problem is fully solved. Nearly all of the moraine matter of this glacier was not only rounded but roughly stratified before it traveled a single yard from the snout. This is evidently effected beneath or on the surface of the glacier by the streams, some of them torrents in size and force; on account of the rough crevassed condition of the glacier, the surface water finding its way to the bottom far back, cutting and keeping open channels of considerable size. Into these wells, mills, crevasses, and underchannels the surface moraine matter falls and is well ground over and over again before it is finally discharged, a thoroughly milled grist. This finest flour, superfine, is carried out to sea many miles, twenty or more from the mouth of the Stickeen, where it slowly settles and forms rocks of various kinds, the origin of which, with their varying texture and their imbedded fossils, is seldom guessed. The middlings of various grades are laid to rest alongshore in the abandoned channel of the retreating glacier, the coarse gravel and boulders in the moraine to be wasted out of shape in after time by the various post-glacial agents.

The main portion of all the material removed by the ice in mountain sculpture is here, as perhaps in most ranges, carried far off and spread abroad in the sea. This explains the inadequate quantity of visible traceable moraine matter found in connection with profound canyons and heavily degraded ranges. In the case of the glacier under consideration there appears to be little or no terminal moraine, only a bank eight or ten feet high close up against the foot of the ice-wall, extending in an almost level draggled slope to the bay; but were this glacier to come to rest on a mountain-side, the moraine would appear more nearly proportioned to its source, for the water is deep just beyond the moraine. The bottom of the bay or valley once occupied by the glacier may be deeply moraine-covered, and the slope, should the material be mostly fine, would be so gradual that it would be taken most likely for the natural slope of the bedrock. This view, based on the actual facts, explains the so-called objections to the glacial theory for the formation of the wonderful system of fiords always found along the northern glaciated coasts all around the world. The relationship between fiords and glacial founts is, when fully appreciated, exactly cause and effect. Every cross-fiord — that is, those that trend in a direction at right angles or approximately so to the general trend of the coast — conducts back into a system of tributary glacial founts. Those trending parallel with the coast are more obscure as to their origin only because they are on a scale of greater magnitude and therefore the distance between the cause and effect is greater. That these so-called channels, straits, reaches, inlets, bays, sounds, etc. were eroded from the solid by glacial action I have no more doubt than that the small hollows and canyons in the mountains at the head of which the glaciers are still seen at work were so eroded. They were eroded by the ice-sheet which came from the north at the period of its great extension. Their general parallelism with the coast, and with the well-marked scratches, striæ, and moutonnéed rocks on comparatively level ground where the flow was nearly uninfluenced by the topographical features, would of itself suggest this origin as to cause; while as to trend-variation the compliance to groups of mountains bearing ice corroborates it. The determining cause of the location of these parallel coast fiords was the pre-glacial mountain spurs, the modified, heavily overswept remnants of which form the walls of the fiords, and the highest of these towards the last of the period nourished and sent down small local glaciers, sculpturing them in some places quite deeply, the tributary canyons coming down to tide-water. All above two thousand feet nourished these local glaciers.

The islands, so remarkable a feature, are simply roches moutonnées brought into relief from the general mass as in inland glaciation from superior strength and favorable location. That this is the simple cause of this stupendous group of surface features will be plain to all who are capable and willing to make the necessary observations. The theory that the channels and corresponding bounding walls are corrugations effected by primary foldings of the rock, subsidences, elevations, etc., has not one fact for its support. These islets are fragments so evidently of one mass with stratification cleavage. General rock characters show this at a glance, while in the case of such resisting materials as gneiss, granite, metamorphic slates, etc., the waves beating upon them since they were created have not yet wasted them at all to any appreciable extent. They are newborn, and as plainly moutonnéed as any glacier-polished porphyrited boss in the Upper Sierra. The larger islands are only less so on account of after glaciation of local kind. Finally, in the small fiords on Vancouver Island we in many instances find the traces distinct where the bottom of the ice-sheet has gone down into and up out of them just as they do on high ground valleys, the water alone obscuring the phenomenon, which elsewhere would appear quite natural. Remove the water from these fiords and from all the archipelagoes along the coast, and the whole would seem a perfectly common piece of mountain scenery — canyon, valley, ridge, and mountain. Many detailed proof-facts will be required to compel the assent to this in the minds of most geologists on account of the defectiveness of glacial education in general and in special. But the glacial millennium will come.

Besides the large glacier which we visited I noticed two others nearly as large, their tributaries reaching back far and wide into the coast mountains and uniting into grand trunks reaching nearly to the sea, into which they at a comparatively recent time flowed, making fiords. The captain kindly sailed his boat up to within a few feet of the terminal moraine of the first and sent us ashore in a large canoe paddled by four Indians. By *us* I mean Doctor of Divinity Sheldon Jackson, Mr. Young, missionary at Fort Wrangell, and myself. The greater portion of the glacial water is discharged from beneath two grand arches, one at either side of the snout, the larger on the north side nearly as large as the Merced in Yosemite; the other, which Mr. Young and I forded easily, is about fifteen feet wide, and of a fine milky color. The outlets of glaciers are usually on the side where the pressure against the walls is less, allowing the underchannel to be kept open — that is, melted — as fast as the ice advances against it.

The trees on the left bank were for a height of eight or ten feet being carried away by the glacier, showing that for some cause it stood at a higher level this year at this particular point than at any time since the trees were planted. They were probably from fifty to a hundred and fifty years old. This would by no means, however, show any *general* advance or increase in depth of the glacier. Some great avalanche, shot down on its edge a little above this wasted point, would of itself deepen the glacier and account for the phenomenon. Other portions of the forests fringing the glacier for miles inland showed no trace of a like destruction, but rather that a gradual shallowing of the glacial current had been in progress. The same is corroborated by all the other glaciers I have seen in the region.

One of the most remarkable facts presented to me here was the want of polish on the fresh surfaces of even compact granite. The sides had evidently been rasped by moraine boulders, different portions being successively brought under their action as the surface of the glacier fell to a lower and lower level. The brighter portions of these canyons undoubtedly lie along the bottom. In general form and in grooved and scratched surfaces these rocks were as telling as any I ever saw, but not so freshly so as I should have expected.

> On Wrangell Island, four-
> teen miles from Wrangell.
> *Late July.*

Ruins of Old Village of the Stickeen Indians.[1] An outcurving piece of ground slopes slightly to the bay, two hundred yards long, seventy-five wide. A swath of fine rank grasses grows in front, a forest of spruce and hemlock behind, the trees sixty or seventy feet high, probably grown since the village was deserted. The whole of the ground is strewn with the immense timbers of the houses and overgrown with rank vegetation — grasses, ferns, elder bushes, nettles, raspberry, cowbane, etc. — while the boulders on the beach, piled in rows and submerged at high tide, show the ways up which the Indians pushed their canoes in coming and going on their fishing, war, and gossip expeditions.

The site of the village is marked most interestingly by carved totem pillars, now moss-grown, and some of them picturesquely planted with tufts of grasses and bushes. One carved post is surmounted by a bear, life size; others are carved along the whole column into human forms, and are said to have been the receptacles of the ashes of the dead.

A loon flew past as we lingered, screaming and making the solitary place more impressively solitary by his intensely lonely wild laugh. Two species of birds were calling excitedly about their nests. These and the low swash of the waves along the dulsy beach and the wind in the woods were the only sounds

The rafter beams of the houses are often seventy-five to a hundred feet long, of cedar, two feet in diameter, hewn perfectly round. The posts supporting these are curiously carved into animal and human forms. In front of many of the old houses

[1] See Muir's *Travels in Alaska*, pp. 87–94.

there are tall carvings, some of them over three feet in diameter
and twenty-five or thirty feet high, monuments of an historical
character, meant to keep the dead with their history and character in mind. A small cavity is usually found at the base of each
containing the charred bones and the ashes of the dead. Frequently these contain several boxes like a family vault. Some of
these monuments are tall posts smoothly rounded and with a
plain figure about life size on top, always in a sitting posture, and
said to resemble the deceased whose ashes they contain; others
have figures much larger than life, of men or of women or of both,
carved from one large trunk. Still others have the figure of a
bird or animal, such as a bear, crow, eagle, dog, or whale — the
totem of the family. Those of the old Stickeen village are now
much weather-beaten, grasses and bushes growing on top in a
very picturesque way, grass tufts in the ears, a bird on the top of
the head, colored lichens and mosses enlivening the whole surface. They are all made of cypress, or yellow cedar, as it is called,
and as some of them are much decayed, they must be quite
ancient — a hundred years or more.

In moving to the new village at Fort Wrangell most of the
carved family monuments were left in place, and a few of the
Tyees had new ones erected. The Shakes family, however, with
a civilized eye to economy, have moved most of theirs. The bear
is their totem; one six feet long, tolerably well carved, is placed
on top of a pillar twenty feet high in front of a new house at the
south end of the island. Bear tracks are carved on the pillar
from the ground to the top with a view to make it appear that
the bear at top had climbed to its present position. The effect is
quite ornamental. The porpoise is also common, judging from
the frequency of its appearance.

These totems are not only erected in front of the dwellings, but
also over the hut-like dead-houses, small hovels five or six feet
square. One of these has a dog or wolf on the roof, with head up
and mouth open as if howling for his dead master. When a
Stickeen Indian dies, the body is burned with feasting and chanting and bestowal of gifts. If these rites are properly celebrated,

the soul on reaching the bank of a great river is ferried over, given a body, and guided through the woods far inland to a fruitful, gameful country and is happy. But in case the death rites be imperfectly administered he is unable to get over the river and is compelled to wander, an unhappy ghost, in a wet, dark wood, and work mischief to the living. . . .

Standing on the shore of this beautiful bay looking out over the water, the forest dark and shadowy behind them, these grotesque memorials are remarkably striking and tell the history and character of their makers in durable language, while the decay suggests too well the fate that awaits every tribe on the continent.

Besides the carvings on wood, the Indians about Wrangell have worked in stone to some extent. A ledge of slate jutting out into the bay a mile from the town is covered with rude figures. . . . The head of a hog shows that they are not very ancient.

There are some twenty-five Indian tribes in Alaska, whose language and customs are more or less distinct, but they were all doubtless derived from the one Mongol stock. Their physiognomy — down-slanting oval eyes, wide cheek bones, thick outstanding impending upper lips — reveals the secret of their origin, while their sculpture and close geographical position would at once suggest their connection with the Chinese and Japanese. Many of them have rather short legs and are square-built, broad-backed, and stumpy-looking; not a single specimen compares at all favorably with the best of the Sioux, or, indeed, with almost any of the tribes to the east of the Rocky Mountains. They also differ from other Indians in being willing to work. They are industrious when free from the contamination of bad whites or those who are too good. They live chiefly on salmon along the coast and inlets and rivers, the density of the forests no doubt in great part determining this. There is one characteristic, however, that they have fully developed in common with all wild tribes, namely, superstition. They attribute to witchcraft all diseases and events of which they know not the cause. Witches are not hanged or burned as a general thing, but punished by beating and ducking in the sea until nearly or quite drowned. After repeated thrashings, etc., they are asked for the means or source of power of their craft, and if they deny the crime and fail to disclose their bad charm, then they are killed by cruelty of some kind. One was hanged a few years ago in Wrangell; another was saved this year by the interference of the captain of the *Jamestown*. A small boy eight or ten years old was in the habit of strolling about among the graves of his tribe, and was at once suspected of being a witch and feared accordingly. He seemed pleased with the importance thus acquired and turned it to good advantage, until finally driven from home and left to starve. He was taken into the cabin of a kind family on condition that he would not exert his mysterious power against them. Every crime, as far as I have heard, excepting only that of witchcraft, may be atoned for by blankets.... If one kills another, the victim's relatives must kill one of his or receive two hundred blankets more or less. If a trader sells hootchenoo to an Indian, and he gets drunk and kills himself or is killed in a row, the one who sold the whisky must buy blanket redemption. Blood in particular must be redeemed in blankets. They are the unit of wealth and unit of atonement. A man's wealth is measured in blankets. 'He is rich, he has five hundred blankets.' The raising of a carved heraldic monument is a grand occasion. Some of the monuments cost as high as three hundred blankets, plus the cost of a grand entertainment and potlatch, when all are fed and all receive presents. Occasionally a rich Indian holds a grand potlatch, giving away all the hard-earned savings of a lifetime. Then he becomes a chief, or Tyee. A good way to get rid of riches in old age when from their kind they are hard to keep. It is the common price of fame and power.

But notwithstanding the superstition in which all wild, or rather ignorant, peoples are sunk, that by which these Indians are afflicted is attended by and checked by more sound sense and natural reason than are found among the so-called enlight-

ened and religious of our own race.... As an illustration: A
Wrangell Indian had a grudge against a certain family. He indus-
triously scraped the dirt which was of potent age and kind and
ample in quantity to raise the dead or kill the living. This powerful medicine he slyly placed in the family vault in contact with
the bones and ashes of his enemy's dead, assured that it would
cause death to the living. This was quite reasonable; and it was
quite reasonable too that he should be punished for his bad intentions and compelled to go to the vault, remove the mighty medicine, and cast it away.

The presence of the *Jamestown*, with her dignity of cannon-
power and official buttons, is said to exert a most salutary influence on the behavior of the Indians. That the Government
should in some way have an unmistakable representation here is
plain. Then let power appear to whatever extent is requisite to
the enforcement of simple justice, such as every Indian under-
stands quite as well as whites. This may easily be accomplished.
One gunboat, first-class in everything save size, would be sufficient for the whole territory, because the Indians dwell in permanent villages on the edges of navigable waters, wedged in
between the woods and deep water at the mercy of such a ship,
seldom going back into the tangled forests. They are human;
therefore do what is wrong now and then, but a good thrashing,
prompt, heavy, wholesale, and whole-souled, discriminating only
as to the offending tribes, not as to individuals — such a punish-
ment will last on the average about one generation, and is ever
found to be productive of the best results, peace and good will.

It is a common saying that savages respect only power; these
seem literally to kiss the rod and love it, and therefore need it
seldom. The agency fort-and-soldier system begets only disease, crime, debauch, general demoralization, and death. These
people, contrary to the instincts of other Indians, are eager to
know and learn. They beg for teachers and missionaries, not
probably because they are predisposed to piety, but simply because Christian teachers are the only ones they ever see, and

because these from their comparatively unselfish devotion to the welfare of the Indians gain their confidence. It is too often found that in attempting to Christianize savages they become very nearly nothing, lose their wild instincts, and gain a hymnbook, without the means of living, being capable of taking nothing more. Then they mope and doze and die on the outskirts of civilization like tamed eagles in barnyard corners, with blunt talons, blunt bills, and clipped wings. These, however, are capable of civilization in a substantial sense. They are industrious, willing to give good and fair work for fair wages, and to adopt all the benefits of civilization within reach. They would compare favorably with the very best of the uneducated of any people under the sun. Unprincipled whisky-laden traders are their bane; common-sense Christian teachers their greatest blessing. A few good missionaries, a few good cannon with men behind them, and fair play, protection from whisky, is all the Alaska Indians require. Uncle Sam has no better subjects, white, black, or brown, or any more deserving his considerate care.

2. *Memories of the Dog Stickeen* [1] *and Our Adventure on Taylor Glacier, Alaska, August* 30, 1880

An Indian Cur. What labyrinth of relationship Stickeen had, I never tried to trace; probably he came from the Stickeen tribes.

Indian curs are all solemn, even when puppies, I suppose from having to work hard. They are made beasts of burden as soon as they can carry a pound or two. Those of the interior of the country are all thus used, old and young. Each has its pair of saddle-bags, according to size, marching solemnly. Hiding, Stickeen probably learned in trying to avoid his load.

When we came to the Indian villages, he hardly noticed other dogs, being wholly unlike those kingly, magisterial dogs that ever draw a company of dependents at their heels. Rather he was a hermit — a wandering star beyond the influence of any of his kind.

[1] See Muir's *Stickeen.* See also Muir's *Travels in Alaska,* pp. 277–311.

A good-natured sailor and vagabond, he was fond of wandering but without visible enthusiasm, nosing among logs and trees with sober, unhasting industry, now and then shaking the rain from his hair or the dew from the huckleberry bushes, heeding no kind of weather, never in a hurry or fuss, eating as if it were a sad duty.

Never cat-like with the small game he killed, without yelp or growl he did the job with official decorum, like a dull boy doing his chores.

He was smooth and glossy as a berry. His little round feet made no sound on the moss carpet of the woods, turf and moss, nor on the glacier that never before felt the foot of man or dog.

From most of dog sins he was free — didn't fuss, steal, whine, or get in the way. He never offered his head to be patted as an affectionate terrier would, or a loving collie; neither did he refuse a caress — he just didn't care. He was cold and unemotional as a glacier curled up in a blue shadow on the mountains. Yet his eyes puzzled me — aye, those eyes! They were deep, calm, fateful, suggesting the unfathomable wells of the glaciers.

But at first he seemed to me a small dim dull sluggish unromantic nobody . . . as unfussy as a tree. Even the minister [1] knew him not. He was just a little dog that followed him.

No mark of years was on him — he seemed neither old nor young. He always looked grave, no matter how playful you might be with him; scratch his ears, pat his head, set him up on hind legs, or cuddle him on your lap, he was still irresistibly grave — not a tail-wag. A little black horizontal philosopher, calm, pensive, silently watchful. . . .

He was particular about his beds, and no dog had better ones, for the woods down to the tide-line are not simply carpeted with mosses, but richly plushed with them, so that all the ground was a bed, fresher and finer far than any king could buy of fluffy down silk. Only sticks and roots woven into the yellow fronds had to be avoided. A mossy hollow — not too damp — at the foot of a spruce tree between bulging roots was the favorite place where

---

[1] The Reverend S. Hall Young, the 'adventurous evangelist,' Stickeen's owner, who accompanied John Muir on his canoe journey among the glaciers.

he curled up comfortably, himself like a little boss of moss, swelled up over a boulder.

He was an adventurous swimmer, as much at home on the waves as a seal. And he could make his way through any wind like a salmon through swift water.

Our Horizontal Brothers. This poor little wild apostle of Alaska, child-dog of the wilderness, taught me much. Anyhow, we were nearly killed and we both learned a lesson never to be forgotten, and are the better man and dog for it — learned that human love and animal love, hope and fear, are essentially the same, derived from the same source and fall on all alike like sunshine.

He enlarged my life, extended its boundaries. I always be-friended animals and have said many a good word for them. Even to the least-loved mosquitoes I gave many a meal, and told them to go in peace.

In all my wild walks, seldom have I had a more definite and useful message to bring back. Stickeen was the herald of a new gospel. . . . And it was in that dreadful crevasse that I tried so hard to avoid that I saw through him down into the depths of our common nature.

Any glimpse into the life of an animal quickens our own and makes it so much the larger and better every way. But this was more than a glimpse, it was a deep look 'ben in the heart,' as the Scotch say, and made all animals friends and fellow-mortals indeed.

Those who dwell in the wilderness are sure to learn their kin-ship with animals and gain some sympathy with them, in spite of the blinding instructions suffered in civilization. See the nar-row selfishness of the attitude of man in dealing with animals — selfish even in religion, for after stretching to the utmost his mean charity, he admits every vertical mammal, white, black, or brown, to his heaven, but shuts it against all the rest of his fellow-mortals. Indian dogs go to the Happy Hunting Grounds with their masters — are not shut out. . . .

Stickeen's homely clay was instinct with celestial fire, had in it a little of everything that is in man; he was a horizontal man-child, his heart beating in accord with the universal heart of Nature. He had his share of hopes, fears, joys, griefs, imagination, memory, soul as well as body — and surely a share of that immortality which cheers the best saint that ever walked on end. . . .

How thick a mass of sound, serviceable ignorance is comfortably covered up in the stupid word instinct! Simply by pronouncing this one fetish word, all the wisdom, all the mental powers of animals, lose their natural significance, are at once emptied. . . .

On the Taylor Glacier. Morning.

Strange that Stickeen should have ventured forth this stormy day when all the other animals were glad to stay at home in the rocks and woods, trying to curl up and keep warm in their damp dwellings — the lithe ermines about the roots of trees; marmots on bare mountain-tops, their burrows thatched with kind overlacing Bryanthus; wild goats huddled back of cliffs out of the wind, too cold to go out to feed; bears and wolves crouching in the best chance shelters they could find.

Having bread and rain for breakfast, we set forth in the big gray day, letting the storm's hearty tides roll over us. Never is the Creator more visible than in storms — a sublime vision.

In the majesty of this new morning with its triumphant storm-song, none exposed to its enchantments could be calm. It bore all common caution and comfort before it like the pine needles and drops of rain. We were pulsed along in the very blood of Nature and could feel every heart-beat. Storm days are holidays; every leaf is tuned and chanting in fine accord like a trained choir. No silence anywhere, everything singing — the drenched rocks, the ice, the trees, every bush and blade of grass as well as winds and streams. Small drops beating on stones, glinting on angles of crystals, and flat leaves, whispering, cooing, plashing, throbbing in one song — eternal necessary beauty following everywhere. . . .

The gashed wilderness of ice and the gloomy slow-crawling clouds made the sounds of the thrashing wind fearfully solemn. The glacier stretching away in the gloom lay huge and forbidding in the storm, like a load on the weary eye, a place where no men have dwelt nor son of man passed through — a huge malevolent creature crawling over the landscapes and blasting, crushing them, overthrowing forests, and damming up salmon streams.

By the Grand Crevasse.
Late afternoon.

The gray, fateful gloom of the stormy night was settling down. All the birds were driven from the sun and sky to seek shelter where they might. Not a wing was seen, all were folded; nor any of the multitude of 'wee, cowrin', sleekit, timorous beasties' that even in these hills found home and food. The roaring storm had all the sky to itself, yet was not sullen or destructive, but in fullness of joy rather, with life and gladness, only infinitely earnest.

The mountains, as the clouds lifted, loomed like ghosts in the distance, frowned gloriously, and vanished again, leaving only the gray glacier and the blue abyss. No sunbeam ever sounded its depths. Death seemed to lie brooding forever in the gloom of the chasm — a grave ready made, neither land nor sea, blue as the sky and as pure, capacious enough for the largest army that ever fought and bled. . . .

The poor child, Stickeen, was left alone on the side of that awful abyss, unspeakably lonely, so pitifully helpless and small, overshadowed with darkness and the dread of death. . . .

Crossing the Ice-Sliver Bridge. But in his little hairy body there was a strong heart, for notwithstanding his piercing recognition of deadly danger, he was able to hush his screaming fears and make firm his trembling limbs.

At length, as with the hushed, breathless courage of despair, he slipped down into the shadow of death, the storm was not heard or seen by either of us. I saw only those pathetic, feeble little

feet as he slid them over the round bank into the first step. Soon
the hind pair followed, and all four were bunched in it. Then he
worked them down into the next notch, and the next, and, hushed
and silent, lifted his feet slowly in exact measure; I, breathlessly
watching him walk along the long narrow sliver, waited on my
knees ready to help him up the cliff at the end; but when he
reached it, he hooked his feet into the steps of the ice-ladder, and
bounded past me in a rush. Then such a revulsion from fear to
joy! Such a gush of canine hallelujahs burst forth on the safe side
of the gulf of ice!

How eloquent he became, though so generally taciturn — a
perfect poet of misery, and triumphant joy! He rushed round
and round in crazy whirls of joy, rolled over and over, bounded
against my face, shrieked and yelled as if trying to say, 'Saved,
saved, saved!'

... And so, the Lord loving us both, we got back to camp. He
was indeed a fellow-creature — a little boy in distress in guise of
a dog.

Chapter VI : 1882-1889

'It is not now so easy a matter to wing hither and thither like a bird, for here is a wife and a baby and a home.'

AFTER his return in 1881 from the cruise of the Corwin in Alaskan and arctic waters, John Muir devoted the following six years almost exclusively to his family and ranch. Making up for the loneliness of his long wanderings, he reveled in 'dormer windows, open wood-fires, and perfectly happy babies.' Buying and leasing more land from his father-in-law, Doctor Strentzel, he applied his inventive ingenuity and his scientific knowledge to developing high grades of fruit and large productivity. He became one of the most successful horticulturists in California, harvesting annually 'hundreds of tons' of fruit that commanded top prices.

But during this period he kept no journals, and wrote nothing for publication. Even his correspondence was only with immediate relatives. Some of these relatives felt much easier now that John no longer risked his life on perilous journeys. True, he went to the mountains every summer, but he never strayed far from post office and telegraph station, and the grapes ripening on the vines were sure to hasten his return to the ranch.

However, his wife, Louie, wisely recognized that this giving up of his old wilderness freedom and the satisfactions of his literary and scientific work was fatal to his ultimate happiness and well-being. She saw him becoming 'nerve-shaken and lean as a crow' and racked with a cough. So while in August, 1885, he was camping for a few days on Mount Shasta, before going East on a hur-

ried visit to his mother, she wrote urging him to remain among
the trees and snow until he was well.

> Even your mother and sister would understand [she said]. My
> father and mother at last realize your need of the mountains.
> Then as for the old ranch, why it is here, and a few grapes more
> or less will not make much difference.

But John Muir was not easily relieved of his Scotch sense of
duty. It was difficult to find a capable foreman to handle the
ranch work. And Helen, his younger child, was exceedingly frail.
To go far from home was to be tortured by anxiety.

At last, however, in 1888, his wife persuaded him to join the
botanist Charles C. Parry on a week's camping-trip to Lake
Tahoe, and to go on from there with his friend William Keith to
the woods and mountains of Oregon and Washington.

When he had gone on this journey, Mrs. Muir sent after him
a letter in which she said:

> A ranch that needs and takes the sacrifice of a noble life ought
> to be flung away beyond all reach and power for harm. . . . The
> Alaska book and the Yosemite book, dear John, must be written,
> and you need to be your own self, well and strong, to make them
> worthy of you. There is nothing that has a right to be considered
> beside this except the welfare of our children.

So John Muir, in the fragmentary journals of these trips,
resumed his writing. And he climbed Mount Rainier with something like the old ecstasy. 'I did not mean to climb it,' he
wrote to his wife, 'but got excited and soon was on top.'

The following year in June he camped with Robert Underwood
Johnson near the Soda Springs in the Tuolumne Meadows, and
together beside the campfire they planned the campaign that
should bring about the creation of the Yosemite National Park,
and, with that success, inaugurate the establishment of millions
of acres of forest reserves and parks throughout the nation.

-     -   -   -   -

1. '*Fixing' Nevada Falls* [1]

Undated.

Nevada Fall is about six hundred and fifty feet high, and in general interest usually ranks next to the Yosemite Fall among the five main falls of the valley. A short distance above the head of the fall on the north side, the river gives off a small part of its waters, which forms a cascade in the narrow boulder-filled channel and finally meets the main stream again a few yards below the fall.

Sometime last year, the Commissioners came to regard these cascades as a waste of raw material, a damaging leak that ought to be stopped by a dam compelling all the water to tumble and sing together. Accordingly, the enterprising landlord [2] of the upper hotel was allowed a few hundred dollars to 'fix the falls,' as he says, and by building a rock dam he has well-nigh succeeded in abolishing the Liberty Cap Cascades, though no corresponding advantage is visible in the main fall.

'Where is the landlord?' the guests would inquire. 'Oh, Albert is up there fixing the falls,' his goodwife would reply, evidently considering her husband's importance greatly augmented by so stupendous a task. Tinkering the Yosemite waterworks would seem about the last branch of industry that even Yankee ingenuity would be likely to undertake. But that men such as the Commissioners should go into the business of improving Yosemite nature, trimming and taming the waterfalls properly to fit them for the summer tourist show, is truly marvelous — American enterprise with a vengeance. Perhaps we may yet hear of an appropriation to whitewash the face of El Capitan or correct the curves of the Domes.

The view of the head of Nevada Fall,[3] fifteen or twenty yards

---

[1] It has been impossible to fix the date of the cascade-damming episode, since the Park Commission Reports make no mention of it. Authorities upon Yosemite Park history, however, incline to the belief that it occurred in the early eighties.

[2] F. A. Snow was the landlord of the 'upper hotel,' known as the Casa Nevada. It stood on a flat tablerock just below Nevada Falls.

[3] See Muir's *Mountains of California,* vol. II, pp. 103–54.

back from the verge, is one of the most exciting I have yet seen. The channel is roughened, both on the bottom and sides, with ribs and elbows projecting, the object of which seems to be to toss and fret the stream so as to churn it into white foam to fit it for its grand display; and with what deep enthusiasm it goes to its fate!

The lateral dashing on side angles, the surging and churning in potholes, and upglancing in shallow curved basins make altogether the wildest of water movements. No other stream is so elaborately prepared for its falling as Nevada. No wonder it is the whitest.

The most exciting view of all, however, is gained from the point of an overhanging rock which projects beyond the brow of the fall, while the eager surging flood rushes past. Here we can look straight down into the middle of its intensely white glowing bosom where it is displayed by glancing upon a smooth, shelving apron. At the foot it cannot be approached on account of the heavy spray. In autumn, when the water is low, we may go close up to it. Then the fall is charmingly beautiful, all its wild outbounding, thundering, hurrying waters hushed and soothed to a sweet, whispering, tinkling web of lacelike embroidery outspread, fold over fold, as if on show, ever changing. Indeed, few would hesitate to say it was more interesting than the white triumphant passion of spring.

Apart from the fall, the view down the canyon from its head is one of the most wonderful in the valley. The river rushes in wild exultation down to Emerald Pool over bare granite folds, the majestic walls on either side of the sublime mass of Glacier Point ridge blocking it abruptly, making a basin shaped like an immense triangular hopper three thousand feet deep, filled with the roar of the waters and the winds as if it were some grand grinding Mill of the Gods.

Today the basin is full of clouds, some of them like smoke and spray, now hiding, now revealing one point after another of the mighty rocks, going hither and thither as if on urgent business. Some rush down with winglike arms outspread and held steadily like those of a hawk when swooping upon its prey, or leisurely descending from the sky to its nest.

Wildly the white and darker clouds commingle, trees, faintly outlined through their silky fringes, seeming to welcome the touch of the soothing messengers of the sky on whom they depend for life-giving snow and the rain.

2. *Camping at Lake Tahoe with Charles C. Parry* [1]

Along the Truckee River.
Late June, 1888.

The road from Truckee is dusty, the ground strewn with fallen burned trunks or tops of trees felled for lumber, but the freshness of the young trees springing up, and the chaparral and the beautiful river ever in sight, redeem the view from all the destructive action of man.

A warbler was singing on a nest in a cottonwood close to the road. Going down to the river, I heard the singing bird and searched for him. Found him on the nest, where he seemed to court attention rather than to seek concealment.

Lake Tahoe.

Wild rose abounds, and strawberry, spruce, alder, Rhamnus. All pines, *Jeffreyii* and *contorta* and *grandis*,[2] are but little touched as yet by the axe. Libocedrus grows to a large size, reaching its greatest perfection along the west shores of the lake, on a gravel beach. I saw one a hundred and sixty feet long, five feet in diameter four feet above ground, five hundred and eighty years old.

First camp, north
of Rubicon Point.
June 30.

Snowy mountains in the background, heavily timbered. The best timber has been cut, but it is still very fine with young growth

---

[1] Charles C. Parry, 1823–90, botanist, an authority upon the flora of Mexico, the Rocky Mountains, and California.
[2] Probably *Abies concolor* — W. L. J.

and the picturesque old Libocedrus, the yellow and the two-leaved pines, and the pines[1] *grandis*[2] and *amabilis*[3]. ... A large cedar stands in front of our bedroom on the lake-shore.

The flash and roar of waves on the shore begins suddenly — a hundred voices stirred by wind in the forest. Jays, ducks, gulls, yellow linnets. ... A lovely, calm evening; colors on the lake.

Emerald Bay, second camp.
July 1.

Arrived at Kirby's, 10 A.M. Talked with Mr. and Mrs. Kirby, both kind and intelligent. A charming retreat from lowland heats and lowland cares. A fine rough tangled canyon above, watered by Cascade Creek brawling among the rocks. Ferns, Allosorus, Pellæa, Cheilanthes, dogwood, alder, willows, Mentzelia, *Pentstemon confertus*, Spiræa, Pyrola, lilies very abundant — the white and small tiger.

Emerald Bay is about two miles long. Its mouth is nearly closed by a terminal moraine; the sides are formed by lateral moraines. The left lateral is very striking, well formed, three or four hundred feet high where it joins the shoulder of the mountain, timbered with pine and spruce sparsely on the grayish slopes.

As we look from the head of Emerald Bay, the distant mountains are blue and the lake is blue, with exactly the same tone, and the yellow summer sky is like the glow of the desert beyond the mountain-rim.

The trees — slender pines and firs — are firmly relimned and inverted in the mirror blue of the bay. In the foreground a strip of meadow, reddened with castilleia and fringed with willows, is reflected with fine effect. Also the shore of washed granite sand, and the one island, a hundred feet high and half a mile in circumference, adorned with pines and bossed with chaparral. ...

[1] Really firs, not pines.
[2] *Abies concolor* — W. L. J.
[3] *Abies magnifica.* — W. L. J.

Cascade Lake. Altitude 6675 feet.

Snowy mountains and a fall and a grand glacier basin, and well-timbered lateral moraines, make a fine setting for Cascade Lake.

I saw a duck with her young sailing and running on the lake, a fine, wild, happy sight. Also young robins and young grouse....

Landed at Dr. Brigham's cottage, bark-covered, neatly shingled, with wide verandahs. Flakes of bark on the sides in handsome patterns... very tastefully rustic....

Third camp. *July 2.*

We camped at the southeast end of the lake on a sandy beach and made a grand fire. The sage was beautiful in the firelight. Grand tea and potatoes — the pomp of kings ridiculous!

July 3.

I landed in a rocky cove three miles from camp at 8 A.M. to write these notes and rest and botanize. Ducks are here, and a small tamias with a voice like a robin. Saw a blackbird feeding its young down on the shore this morning. The sunrise lovely, calm, hushed. Yellow amber light touching the mountains of the western shore, tinting the purple lake yellow.

We set out towards Glenbrook at 10 A.M., sailing along the rugged shore, past many small indentations and long rocky points weathered into islets and single stones. Saw a brown-headed duck with young making frantic calls to her brood to make haste for fear of us. She led them along the shore half running, half swimming, making fast time; then, holding their heads low, they glided among the boulders and hid until we passed. Saw another flock of ducks, nearly grown, that could barely fly. They flew a few rods, then made haste out into the lake, running and wing-beating a mile or more without taking rest.

A heavy swell arose, making rowing difficult, though no wind was on this side. Parry nearly fell overboard backwards. We landed at 3 P.M. and camped beneath a large yellow pine a few feet above the low-water mark on a sandy beach, two miles from Glenbrook.

Fourth camp. *July* 4.

We had a rough cobble bed, but slept well nevertheless. Rose at 4 A.M., got breakfast — a stew of rice, beef, and deviled tongue — and set sail at 5.45 for Glenbrook to get provisions — milk, meat, etc. . . .

Noon camp, five miles
north of Glenbrook.

Most of the stores are closed. No meat; got milk, which soured. Mill hands are dawdling in their best clothes, with the dull semi-holiday gestures of men accustomed to routine hard work, evidently not at ease, and not enjoying rest or idleness as they think they should.

Saw a camp of Washoe Indians. They came up to see the Fourth festivities. Women in gay calicoes . . . with white, perfect teeth. Fat and careless. . . . Some went fishing.

We set sail with chocolate and sardines and sour milk added to our stores. Marvelous reflection in calm mirror-waters, of trees and rocks flapping and wallowing yet distinct in the swells raised by the oars. Color pale emerald — from water and from sky.

The lake was calm until towards evening. In sailing across the mouth of Crystal Bay we met a heavy swell. Found a fine safe, quiet anchorage and harbor at an old log-slide on the east side of a rocky point on the west shore of Crystal Bay. Just enough of sand to sleep on between boulders. Had a good fire, and a good bed of dwarf oak and Vaccinifolium.[1]

[1] *Quercus vaccinifolia.* — W. L. J.

3. *A Journey to Mount Rainier* [1]

July 11, 1888.

Start for Oregon in a Pullman car.

At Sisson's Station on flank
of Mount Shasta, California.
July 12.

At the station meet Sisson, who is getting feeble.... He deplores the destruction of the forest about Shasta. The axe and saw are heard more often in the Shasta woods, and the glory is departing.

The view from the back of Sisson's Hotel: In the immediate foreground, a meadow yellow-green, fringed with pea-green willows, then more yellow meadow, and more willow fringe beyond, leading to a spiry dark-green forest that sweeps to the snow in one grand curve. The trees get darker, until beneath a white ruff of cloud they are deep black. Their spires, each distinct in the middle ground on the edge of the main meadow, gradually become indistinct until they form one solid wash of black.

Just in front of the forest, in a strip long drawn out, is the raw shingle town of Sisson, glaring in the Shasta meadows. It has a good schoolhouse; some buildings are tasteful, mostly they are boxy. The station eating-house is large and well finished.

A cloud-ruff through which sections of the mountain above the tree-line show intensely luminous.... Snow fell last night....

Flowers here are in full bloom — the Washington lily more purple than in the north, Pyrola, Spiræa, Brodiæa. Huckleberries are ripe, as are raspberries and blackberries....

July 13.

Shasta is glorious in clouds and sun-fire — a white filmy, satiny, fibrous texture crowning the summit with delicate touch and adjustment. All the rest of the mountain, with its moraine glacial

[1] See Muir's *Steep Trails*, pp. 204–346. John Muir climbed Mount Rainier accompanied by Keith, the artist, Professor Ingraham, and five ambitious young climbers from Seattle.

sculpture, is fully revealed and standing out in glowing floods of clear sunshine — all the colors washed in the frequent rains. ...

A heavy shower in the afternoon, then a clear and glorious sunset with magical changes of light and shade and waftings of fragrance from the piney woods. The summit looming immensely high and clear above cloud-fringes — castles in the air, but substantial as rock and ice and snow can make them.

On the train going north.
Undated.

From the summit of the Siskiyou Mountains, Shasta towers majestic and imposing even at this distance.

After passing through the tunnel into the Rogue River Valley, there is a marked increase in the freshness and lushness of vegetation. ... It is one bed of fertile soil from Ashland to Portland. Man and beast will be fed.

Portland, Oregon. Undated.

Views from the Portland Heights extend over a great width of green hills and forests to the Cascade Range, with the white cones of Hood, Adams, St. Helens, and the tip of Rainier. In the rosy divine glow on the mountains at sunset all the onlooking landscape seems in sympathy to yield an indefinable subtle response.

The city in the foreground with leafy avenues of elm and maple, its outgrowing edges fading in loose straggling points among the cultivated fields and orchards, and these again graduating and losing themselves in the dark woods which rise in billowy hills to higher, higher mountains; the crest of the Cascades with its grand beacon lights of volcanic cones, now cool in ice and snow eternal.

The woods on the heights opening to the river make glorious vistas from which all this vast panorama may be studied. Ash, maple, red and yellow tinged, and wild cherry a foot in diameter and forty feet high — a beautiful leafy tree, translucent in the down-sifting light. Also hazel, dogwood, rose, arborvitæ, and Douglas spruce.

Overlooking Puget Sound,
near Seattle, Washington.
July 20.

A calm morning. The Sound silvery smooth of spangles. Detail of mountain sculpture lost in a fine fading-edged haze of white. The summit of Mount Rainier clear but fine with glaciers glistening. The Cascade Range to right and left just visible, the dark ridges in the mid-ground feathered black and spiry. Not a cloud in the whole blue sky.

In the foreground a red-shirted Indian in a canoe, his oars flashing silver.

July 21.

In the dark woods primeval, a dull cloudy day. A hushed awe-ing sound among the tops of tall cedars and hemlocks — one with gray fibrous bark, the other red and flaky. The hemlock has fine yellow foliage, and the young trees are elegant, drooping at the tops. So also the cedars, level-topped in age with lower branches dead and mossy. Ferns — Aspidium in tufts and rosettes — the fertile erect, the others spreading.

Two huckleberries, red and black, eight to ten feet high with pea-green foliage. Salal (*Gaultheria shallon*), eight feet high with fine glossy leaves, slender leaning stems, pale-pink waxy bells, narrow-throated. Rubus, also eight feet high with black leaves, and dwarf cornel with red berries, and the skunk cabbage.[1] Two species of maple, and birch waving in the gray sky solemnly.

The growth of trees on this glacial soil, with a foot or more of humus and brown decaying timber of man's generations, is manifestly rapid. A Menzies spruce thirteen feet in diameter showed only forty-eight rings. One five feet in diameter measured two hundred and forty feet in height. I saw a Douglas spruce ten feet in diameter and an arborvitæ twelve feet in diameter in the Vancouver woods.

[1] *Lysichiton kamtschatcense.* — W. L. J.

Seattle, Washington.
July 20.

The people about the towns and mills and mines and railroads
are fairly drunk with the joy of material action and achievement.
Just hear the hiss of steel saws making boards enough for a house
every minute of the night and day, and the ring of the axe in the
woods. The chips are flying in a perpetual storm, thick as flakes
of snow. Our mines disgorge a thousand tons a day. Railroads
go everywhere to all the world, etc., etc. Our churches, schools,
are also regarded as material wealth. . . .

Steamers on the Sound, forming lovely pictures. . . . Few
realize the civilization going on here, the fierce exulting action
displayed in building, blasting, digging. To Easterners, Washington is still a far wild West — a hazy nebulous expanse of
geography, lurid with tales of Hudson Bay and Indian adven-
turers. But here are towns with no lack of shiny, well-tailored,
back-slanting dudes with noses in the air. Pavements, fountains,
French fashions, business blocks, and residences overburdened
with style, wide-curling cornices, jagged and spiring as the snouts
of Alaska glaciers.

En route by steamer
to Port Blakely.
Undated.

The morning white, with a subdued sun-splendor; headlands
dim in the soft haze, as if distant. The water half-calm, with
dimpling swells ashy black and deep black. The rigging of the
ship reflects in the smooth but waving ripples like wriggling
snakes. The shores gray, being beds of sandstone fringed with
bushes and ferns.

In general views no scar or mark of man is visible, excepting
the tell-tale smoke. Enough of tall trees are left of Thuja, hemlock, and Douglas spruce to make a seemingly dense forest. In
the woods, however, all the best merchantable timber has been
cut for a distance of ten to twenty miles from the shores. In the
lowest bottoms Thuja is the main tree; Douglas spruce stands

on the drier grounds, with quite a sprinkling of Mertensia spruce
or hemlock. Some Menzies spruce or Sitka pine is here also.

The underbrush is made up of salal, red huckleberry, spiræa —
the white twenty feet long, and the red six feet — hazel, willow,
alder, the wild rose, delicate in foliage, Rubus of two species,
Linnæa, honeysuckle, Symphoricarpos, Stellaria, dwarf cornel,
saxifrage, and yew.

Snoqualmie Falls.
August 2 (?).

The fall resembles Nevada, has the same twist at the top, and
the same plunge to a deep pool a hundred yards in diameter.
That the water is deep is shown by the fine radiating beaten foam
and mist — purple rose of exquisite fineness of tone, and the
heavy waves lashing the rocks.

The Douglas and Menzies spruces are both large and small, but
small in front above and below the fall. The slope on the right
side is beautiful with a soft blanket of grass and ferns and Mimulus, near the bottom. Rubus, Vaccinium, yarrow, strawberry,
Smilax, bosses of Hypnum, Jungermannia, and Marchantia
abound, also maple, Thuja, dogwood, salal, Linum, and rose.
And a grand old forest stretches along the banks far down the
valley of the river, forming a wilderness of Menzies spruce and
maples a hundred feet high, mingled with fern beds and beds of
yellow-leaved small maple, its foliage floating on the river....
The cones of this deep forest are the finest I ever saw.

Camp at Yelm Prairie.
August 8.

We set out from Seattle this morning for Rainier via Yelm
Prairie. Camped by the railway track. A fine sunny day, with
glorious views of Rainier. Now in the sunset the foreground
trees are black, bounding the yellow prairie. Beyond lies a deep-
blue mid-ground of low hills and ridges that merge into the green-
blue upsweeping base of Rainier, that blends finely with the pink
and white and lovely gray of the cones. All about is the purple
of the horizon against the gold of sunset....

Here we were visited by my Peter Van Trump,[1] who caught
the mountain fever and wished to join the party. He had been
twice on the summit and guided General Stevens in his ascent.
We got animals from Longmire,[2] a tall, wiry, enterprising pioneer
who hewed his way through the woods and settled here at Yelm
Prairie, raised cattle, prospected with an Indian as guide, hunted,
and claimed springs. He says: 'Drink at these springs and they
will do you good. Every one's got medicine in 'em — a doctor
said so — no matter what ails you.'

August 9.

We started at 10 A.M. with our huge, savage packs. A gypsy
outfit would look tame and proper in comparison, what with our
coffee-pot, capacity twenty quarts, alpenstocks, blankets, and
grub — a ton.

We objected that the animals could not possibly endure the
journey, and that our guide Joe was a mere boy, but were assured
that the toughness of Washington mules and cayuses was immeasurable, and that Joe was more than a man in woodcraft and
knew the mysterious diamond hitch. With Van Trump as a vol-
unteer we felt safe, but were determined to cut loose and walk to
Rainier should the gypsy cavalcade fail by the way.

There was poison and sickness in every pot. The canned goods
were at first called fresh, then in sickness were estimated to be
ten years old, and at length, in agony of dyspepsia, their age was
measured by centuries and antedated the forests primeval.

On we crawled mountainward, and, with so noble a mark, un-
daunted, though drowsy and loitering, save when at intervals
under the sharp spice of yellow jaundice we all awoke to newness
of life.

The woods arise in shaggy majesty, every light giving tints of
exquisite softness to all the wilderness. Trees ancient-looking

[1] Mr. Van Trump's given name is said to have been Philemon. His letters to
John Muir were signed P. B. Van Trump.

[2] James Longmire was one of the leaders of the first party to cross the Cascade
Mountains with a wagon-train, was an early settler in the Yelm region near Olympia,
and one of the early members of the Washington Territorial Legislature.

abound in damp gullies and on stream-banks, forming the forest primeval. . . .

We usually camp on some meadow-edge made by beaver-dams or a filled glacial lake. The animals, turned loose, roll and rub to ease the itching of their backs caused by saddles, then away they go to water and grass, vanishing among the tall grass and willows, as if belonging to the wilderness like the bears and deer. Then for a while the campfire ascends, the men walk to and fro cutting boughs for beds, cooking, gazing, sauntering, or rest outstretched on blankets, with fun and chat.

Here are true Gothic temples with tree-shafts pointed and aspiring. They are singularly impressive in the solemn woods when the boles are illumined by the campfire and all the distracting details retire into the dark. . . .

Walking Alone in the Woods. When one is alone at night in the depths of these woods, the stillness is at once awful and sublime. Every leaf seems to speak. One gets close to Nature, and the love of beauty grows as it cannot in the distractions of a camp. The sense of utter loneliness is heightened by the invisibility of bird or beast that dwells here. . . . But it is not in the deeps of the woods that people are soothed into perfect rest, nor in mountain valleys, however beautifully bounded by lofty walls. One feels submerged and ever seeks the free expanse. Nor yet on lofty summits, islands of the sky, but on the tranquil uplands where exhilarating air and a free far outlook are combined with the loveliest of the flora. In that zone below the ice and snow and above the darkling woods, where the sunshine sleeps on alpine gardens and the young rivers flow rejoicing from the glacial caves, and the groves of eriogonums are open to the light — perfect quietude is there, and freedom from every curable care. . . .

Jubilant winds and waters sound in grand harmonious symphonies — wild music flowing on forever from a thousand thousand sources, winds in hollows of the glaciers glinting on crystal angles, winds on crags, in trees, among the elastic needles sweep-

ing soft and low with silken rhythm, and winds murmuring
through the grasses, ringing the bells of Bryanthus. . . .

In all excursions, when danger is realized, thought is quickened,
common care buried, and pictures of wild, immortal beauty are
pressed into the memory, to dwell forever.

In climbing where the danger is great, all attention has to be
given the ground step by step, leaving nothing for beauty by the
way. But this care, so keenly and narrowly concentrated, is not
without advantages. One is thoroughly aroused. Compared
with the alertness of the senses and corresponding precision and
power of the muscles on such occasions, one may be said to sleep
all the rest of the year. The mind and body remain awake for
some time after the dangerous ground is passed, so that arriving
on the summit with the grand outlook — all the world spread
below — one is able to see it better, and brings to the feast a far
keener vision, and reaps richer harvests than would have been
possible ere the presence of danger summoned him to life. Danger increasing is met with increasing power, and when thus suc-
cessfully met, produces an exalted exhilaration joined with an
increase of power over every muscle far beyond the experience
possible in flat lowlands, where hidden dangers destroy without
calling forth any strength to resist or enjoy. But woe to the
climber, however ambitious, who has had the misfortune to
indulge in tobacco and beer. He is easily nerve-shaken and
daunted, though not naturally wanting in courage, when con-
fronted by frowning ice-cliffs and terrible precipices, with shifting,
crumbling narrow seams.

Camp at Indian Henry's.

Indian Henry is a mild-looking, smallish man with three wives,
three fields, and horses, oats, wheat, and vegetables. . . .

We crossed the Nisqually River, white, with glacial mud, two
hundred yards wide between wooded bluffs. . . . A grouse attacked
Keith.

August 10.

Through glorious woods. We cross branches of the Nisqually, one of them clear, the Mishacol River.[1] The timber is growing better. We camp in a glorious grove on an island. Looking up from my bed I seem to be in a tent of ten posts, with a ceiling of fringing tree leaves and boughs.

August 11.

Grand woods with wintergreen abundant in the fine valley of the Sukatasse.... The trail follows the river up this valley to Soda Springs. We cross three forks, some boisterous and roaring, one bridged with narrow lengthwise stringers.

A fine place for cloud studies. We watch the great masses gather and drift and roll about the crags and glaciers, or hover seemingly motionless in deep-dyed flocks in the sunshine, ever calling back the attention by their marvelous beauty of color and magnificence of motion.

Soda Springs.
August 12.

A fine place. From the springs the mountain is in plain view, distant ten miles.... We bathe in a spring, and sleep in a garret.

[1] This is now called the Mishal River.

Chapter VII : 1890-1894

'Everybody needs beauty as well as bread, places to
play in and pray in where Nature may heal and cheer
and give strength to body and soul alike.'

JOHN MUIR and his wife had definitely decided that, whatever
happened to the ranch and the harvests thereof, he must be free
to go on with mountain-climbing, writing, and his work for the
public good. So, with a greater feeling of release than he had
experienced for a decade, he sailed away in mid-June, 1890, for
Alaska, to undertake the long-planned exploration of Muir
Glacier and its tributaries.

On this excursion into the Northland he wrote two journals,
one describing the voyage along the coast and the events of camping at Glacier Bay, the other largely filled with his observations of the sea. The former has been published in Chapter XVII
of 'Travels in Alaska'; the latter is presented here. The journal
story of his sled-trip of ten days upon Muir Glacier is set forth
in Chapter XVIII of 'Travels in Alaska.' But on odds and ends
of paper — probably taken along to kindle fires under his tin
teacup — he jotted down experiences and thoughts not mentioned in the more objectively written narrative. These fragments, tucked away by him in envelopes and labeled 'Alaska
sled-trip, 1890,' are here published for the first time.

Home again, Mr. Muir resumed his efforts to save the forests,
and to bring about the recession to the Federal Government of
the Yosemite Valley, which, in the midst of the Yosemite National Park, was still controlled and mismanaged by a State

commission of political appointees. To work with John Muir a group of public-spirited men, including William E. Colby, Warren Olney, Sr., Dr. Willis Linn Jepson, and Dr. Joseph Le Conte, organized themselves in 1892 into the Sierra Club, which has ever since been dedicated to the task of enlisting 'the support and co-operation of the people and the Government in preserving the forests and other natural features of the Sierra Nevada.'

In 1894, Muir's first book, 'The Mountains of California,' was published, giving great impetus throughout the nation to the fight to save the beauty of the mountains.

- - - - -

1. *A Voyage to Alaska* [1]

June, 1890.

On the fourteenth of June, 1890, I set forth from San Francisco on my fourth excursion to the grand wilderness of Alaska, full of eager, faithful hope, for every excursion that I have made in all my rambling life has been fruitful and delightful, from the smallest indefinite saunter an hour or two in length to the noblest summer's flight with steady aim like a crusader bound for the Holy Land or a bird to its northern home following the flight of the seasons.

All the wild world is beautiful, and it matters but little where we go, to highlands or lowlands, woods or plains, on the sea or land or down among the crystals of waves or high in a balloon in the sky; through all the climates, hot or cold, storms and calms, everywhere and always we are in God's eternal beauty and love. So universally true is this, the spot where we chance to be always seems the best, and it requires a distinct effort of the will to get oneself in motion for a change of place. On the plains where they are rejoicing in the beauty of wildness it is not easy to leave them and take flight to the mountains, while amid the glories of the

[1] See Muir's *Travels in Alaska*, pp. 329–33.

mountains it is hard to get down to the smooth spacious levels, however richly they may be mantled with flowers and light. Oftentimes on some beautiful glacier meadow of the Sierra I have been so charmed that I would have been willing to be tethered to a stake there forever like a horse at pasture, with only bread to eat and without hopes of seeing a single human being, content to revel in the perfect beauty about me through all the seasons. So also in the glorious exuberance of the tropics we would abide forever, were we not driven and pulled away by the same mysterious forces that send birds and animals on their travels. While, on the other hand, the polar regions are so enchantingly beautiful, we would gladly stay there also to enjoy the huge round days of summer, sunful and nightless like heaven, the mysterious shining ocean of ice, the boundless tundras covered with bloom and throbbing with warm, glad life in endless abundance, and the one huge silent night of winter with its stars and snow-crystals and auroras, one mass of celestial radiance shining in space serene and silent. Easy, then, it is to believe our world is a star, for it looks like one as bright and as silent as any of its neighbors.

Sauntering in any wilderness is delightful, through woods, rocks, bogs, plains and deserts, and green shaggy meadows, and over fields of snow and the crisp crystal prairies of the glaciers, drifting like thistledown responsive to every breeze of influence, however fine, that chances to touch us. Most of one's shorter walks are of this kind, adrift on currents gentle and invisible, bearing us we know not whither, but in all my long excursions there was some main object in view — a mountain, lake, belt of woods, a canyon or glacier — towards which my steps were bent by a course direct or wavering. When we are with Nature we are awake, and we discover many interesting things and reach many a mark we were not aiming at; some new flower or bird or waterfall comes to our eyes, and we gladly step aside to study it; or some tree of surpassing beauty attracts our attention, or some grove, though the species may be well known, or we come upon a specimen that has been riven and scattered by lightning stroke, or bent into an arch by snow, or one or many over which an avalanche has passed. Or we come upon the wild inhabitants of the region — a bear at breakfast beneath the nut-bearing trees or in the thickets of berry bushes, or deer feeding among the chaparral, or squirrels and marmots at work or play. Birds, too, come forward and sing for us and display their pretty housekeeping. All these and a thousand other attractions enrich our walks beyond the attainment of the main object, and make our paths unconsciously crooked and charming. It is as if Nature were saying: 'The way is long and rough and the poor fellow is weary and lonesome. Birds, sing him a song; Squirrels, show him your pretty ways; Flowers, beguile the steep ascent with your beauty; sparkle and bloom and shine, ye Lakes and Streams; and wave and chant and shimmer in the sunlight, all ye Pines and Firs, that the wanderer faint not by the way.'

And thus we find in the fields of Nature no place that is blank or barren; every spot on land or sea is covered with harvests, and these harvests are always ripe and ready to be gathered, and no toiler is ever underpaid. Not in these fields, God's wilds, will you ever hear the sad moan of disappointment, 'All is vanity.' No, we are overpaid a thousand times for all our toil, and a single day in so divine an atmosphere of beauty and love would be well worth living for, and at its close, should death come, without any hope of another life, we could still say, 'Thank you, God, for the glorious gift!' and pass on. Indeed, some of the days I have spent alone in the depths of the wilderness have shown me that immortal life beyond the grave is not essential to perfect happiness, for these diverse days were so complete there was no sense of time in them, they had no definite beginning or ending, and formed a kind of terrestrial immortality. After days like these we are ready for any fate — pain, grief, death or oblivion — with grateful heart for the glorious gift as long as hearts shall endure. In the meantime, our indebtedness is growing ever more. The sun shines and the stars, and new beauty meets us at every step in all our wanderings.

In this Alaska excursion my eye was fixed on the glaciers of

Glacier Bay. I had visited them in a canoe twice before this, in the years 1879 and 1880, before they were known to the world, or the bay into which they discharge was on the charts. But up to the time of this last visit I had seen but little of their sources, and I was eager to make my way back into the mountains where their countless tributaries take their rise, to see what I might learn. I wished also to measure the speed of the flow of the largest one, the Muir, and count the icebergs discharged in a given time, and to study the fossil forests, many interesting vestiges of which are displayed around the shores of the bay and along the banks of the glaciers. Since I made my first excursions and called attention to the wild scenery of Alaska through the newspapers, a bright and lively stream of tourist travel has been developed through the midst of the more accessible portion of it, and fortunately the most accessible is also the most interesting portion of Alaska as to the grandeur and novelty of the scenery.

From San Francisco you can go all the way by steamer; or, if afraid of seasickness, you may go by rail to Tacoma on Puget Sound and there take the regular excursion steamer to Alaska, and as all the way from this point is through inland waters, there is no seasickness, for with the exception of an hour or two's sailing in passing two open sounds the water is as smooth and free from heaving as landlocked harbors are.

I sailed from San Francisco on the steamer *City of Pueblo*, choosing the sea for the sake of the coolness and freedom from dust. It was a fine clear day, as almost all the California days are at this time of year, and as our fine ship passed through the Golden Gate and held proudly on her course up the coast, I was exhilarated with the idea of the magnificent icy region where my summer work lay, and that while merely gazing into the distance the blessings of the trip already began to come in. How fresh and influential the landscape seemed as we glided over the crisp blue water brightened by the downpouring sunbeams, the great ocean stretching away beneath the horizon how many thousand miles on our left hand, the edge of the continent on the other showing a range of smooth-browed hills tanned to a yellowish brown in the sunshine; the seeds of their flowers ripe already and the leaves dry as hay; redwoods filling the hollows and crowning the farther summits, but the trees of no great size owing to the repressing influence of the trade winds! Along the shore the ground rises in jagged bluffs of no great height but severely precipitous and wreathed with foam along their bases. In stormy weather the heavy swells driven by the gale thunder in grand style all along these rocky battlements for a thousand miles, sending sheets of foam and scud over the tops of the cliffs and back into the evergreen woods, making a show of white dashing water incomparably more sublime than that of the grandest waterfalls. Nearly all the way up to Victoria we have the coast in near clear view, but nowhere do we see anything that would hint the presence of the great sequoia belt which stretches unbrokenly along the coast to the Oregon boundary — perhaps the most interesting belt of woods in the world. Only the ragged wind-shorn edge is in sight, but a few miles back out of the direct sea-blast you enter the heaviest, tallest forests to be found on the globe, and the most valuable for lumber. Trees three hundred feet high and twelve feet in diameter are common, and trees of this size with clear shafts stand close together as if mere saplings in a grove.

The conifers along the Oregon coast, chiefly spruce and fir, approach the shore more closely, being evidently better able to endure the buffeting of the gales and salt scud and foam than the redwood.

Sailing is a fine mode of motion; contemplating the scenery we take no heed of the ship, and scarce realize its existence while we are borne through the air as if on wings. We seem to feel the woods as we pass them, and hear their solemn chant as they rock gently and finger the winds with their flat plumy branches. Our rambles come to mind, and we seem to smell the spicy fragrance wafted from miles of plumes and cones, and mingled with the sweet breath of many a mint and lily garden on the banks of the cool, shady streams. We fancy we hear the birds making thin streaks of melody here and there in the calm depths beneath the lofty ceiling of leaves and back of the solemn and eternal song of the sea on the rocks. The sea speaks to the land all round the world, shouting aloud to every island or mere rock, and the land also responds, speaking in a thousand voices with choirs of cascades and falls, to its thousand waves. While looking at the waves coming in endless ranks from the illimitable distance, we are led to think of the paling coral islands and far beyond them, until we conceive the world as a whole, round and watery and bright, bravely spinning through space as a quiet, steady star while so many fine shows are being made on its surface.

It is interesting to note among the passengers the play of quickened action in the minds of those who, brought up in the shadows of city business, have been sleeping all their lives. They gaze at the hills of the coast with curious wonder as if never before had they seen a hill. Objects seen every day are scarce seen at all; clocks strike without being heard, as one may even hear the discharge of a cannon so often in the same tone and volume of sound that it is no longer heard. So much need is there for change of scene, new points of view. How many notice so glorious a phenomenon as the rising of the sun over a familiar landscape? All that is necessary to make any landscape visible and therefore impressive is to regard it from a new point of view, or from the old one with our heads upside down. Then we behold a new heaven and earth and are born again, as if we had gone on a pilgrimage to some far-off holy land and had become new creatures with bodies inverted; the scales fall from our eyes, and in like manner are we made to see when we go on excursions into fields and pastures new, whether superior to the old or not.

How barren and desert-like the ocean looks to a landsman, a waste of waters without sign of life! But sooner or later some whale rises suddenly alongside the ship to breathe and then plunges in grand excess of strength from sight again, or a school of jolly porpoises come racing on like a herd of buffaloes, half in the waves, half in the air, plump, fat, long-winded, and full of merry boisterous life, their pulses beating warmly, and reading us a fine lesson of ocean life. At night, too, the whole mass of the sea, near and far, glows with silvery fire. This is by far the most marvelous manifestation of life, it seems to me, to be found on either land or sea. So far from barren, every chink and pore of the waves seems to burn with life; and between these little animalculæ, little children of light, and the big blunt whales, a hundred feet long, rising like hills above the waters, how wonderful a multitude find a place with food in abundance! Over the heads of this busy throng of water people, our fellow mortals — though we know so little about them — and beneath the birds, the fishes of the air, we make our way, riding our heaving, lumbering horse, which, as it speeds on in the darkness unwearied, its iron heart throbbing in passionate strength, seems also to be one of us, a fellow mortal endowed with hot life.

Our first day out was tranquil and the Pacific Ocean well deserved its name, but on the fifteenth, we encountered a stiff head wind, almost a gale, which soon raised a famous crop of waves that tossed the good ship in grand style and sent most of the fair-weather tourists to their bunks. It was a beautiful sight, the throng of water-hills rising and falling in measured rhythm, each large wave ruffled and dimpled with a multitude of smaller waves, while the tops of the tallest were torn off and drawn out in a mane of spray, among the masses of which the downpouring sunshine was dissolved into all the colors of the rainbow, the irised light flashing and vanishing amid the ever-varying streamers as the waves heaved past. We took the inside passage between the shore and Race Rocks. These rocks are a long range of surf-beaten islets on which many a fine ship has been smashed. The floor of the sea about them must be strewn with a thousand objects — bones, coins, anchors, cables, guns, weapons of many kinds from many countries, savage and civilized, cups, jewelry, etc., like the spoils of a battlefield, telling how many a story could we read them aright. But the sun shines, the world is still young, living beauty mantles the earth anew every day, and we may find better fields for either work or play than this wild graveyard of the sea.

The waves from the deep Pacific, driven by the gale, broke in a

grand display of foam on these bald, hardy islets, leaping over
them, at a height of a hundred feet perhaps, in magnificent curving sheets, jagged-edged and flame-shaped, draping the rocks
with graceful folds of foam-lace from top to bottom, through the
meshes of which the black rock showed in striking contrast,
and brought the white lacework ever wasting, ever renewed, into
relief. I gazed enchanted as long as they were in sight, watching
the exultant, triumphant gestures of the tireless breaking waves,
and the explosive upspringing and gentle overarching of the
white, purple-tinged foam and spray, sifted with sunshine and
fashioned by the wind. How calm and peaceful and graceful they
were, combined with tremendous displays of power! — a truly
glorious show, however common, and a glorious song.

Hereabouts we are so near the land the farms and villages are
in plain view on a grand fertile-looking slope lying at the foot
of the timbered coast range of hills and mountains. The trees
seem fresh and stately, as if in perfect health, but suggest little
of the marvelous exuberance of forest growth that covers the
region lying a few miles farther back.

To the north of the mouth of the Columbia, Long Beach
makes quite a show of raw new houses, but along all the coast
from San Francisco to British Columbia, an extent of six or seven
hundred miles, the country seems about as wild as it was in the
days of Columbus. The old ocean thunders against an almost
solid and continuous line of rocky bluffs showing ragged edges of
woods back of them, while the feeble bits of civilization here
and there make scarce any appreciable mark.

As we entered the magnificent mountain-walled strait of Juan
de Fuca our seasick passengers began to walk again and reappear
on deck. Most of them lay down in their coffin-like bunks as
soon as the waves arose, and thus lost the best of the breezy ocean
scenery. We arrived at Victoria early on the morning of the
seventeenth, in a bland gentle rain, the kind that keeps the
forests fresh and develops the wonderful luxuriance of vegetation
for which this and the adjacent regions are noted. In these few
days we have reached another climate and country and people.

Victoria is an English town, English in dress, in walk, gesture, accent, ways of doing business, and in the style of their houses. In their gardens you will find the old favorite flowers, such as the honeysuckle and tulip, in great abundance, and flourishing in the showery weather if possible better than in the old home; the English currant and gooseberry also, and the favorite apples and pears. It is a pleasant thing to see the pride and privacy of the home transplanted to this far northern young wilderness. The merchant, the day's business done, loves to retire to his residence, well back from the noise of the street and shut in by hedges and fences, making a small private world for himself and family, and where in the old style he can entertain his friends, but where the stranger may not enter, and where the Queen dare not.

Victoria is situated on the southeast end of Vancouver Island. It has grown to its present fair dimensions from the beginning of a Hudson's Bay fort and factory. The Hudson's Bay buildings, old blockhouses, are still standing, and their business of fur-gathering from the wilderness still goes on, though in greatly diminished volume. A vivid picture of wild life rises into view at sight of the thousands of skins piled in their warehouses — wild men and wild animals and wild places, hunting, trapping, fighting, killing and being killed; Nature's hairy people ruthlessly pursued and destroyed for the sake of their clothing.

Vancouver, two hundred and eighty miles long, is the largest of the wonderful group of mountainous rocky tree-clad islands into which the margin of the continent has been carved to the northward of here up to the latitude of sixty degrees, a distance of about a thousand miles, perhaps the most beautiful group of islands and inland waters in the world of the same or nearly equal size. Only on the coasts of Norway and Japan is there any near approach to the Alexander Archipelago in the number and beauty of the islands, or in deep-sea channels. We lay in the Victoria harbor only a few hours to discharge freight, then sailed for Port Townsend and arrived there early in the afternoon.

The sail to Port Townsend is very interesting on account of the beauty and grandeur of the scenery, especially of the Olympic

Mountains, which rise to eight thousand feet above the blue waters, with picturesquely sculptured summits and long withdrawing slopes heavily clad with spruce and fir. Here we met the steamer *Queen*, to which the Alaska passengers were transferred, and we were soon on our way through the islands to the regions of ice.

I had pleasant company on the San Francisco steamer, and regretted having to change. The most interesting man I met on the trip, and the best talker, was an old weather-beaten Scandinavian, a sea-captain and owner of several vessels, well acquainted with the Pacific coast and the Sandwich Islands from long experience and a thousand fights with fogs and storms. He was a passenger on the *Pueblo* on his way to Port Blakely on Puget Sound, where he was having a brig built. He was a bluff, hearty, sturdy specimen of his race, descended no doubt from the Norse pirates, or sea kings as they were sometimes called, and so suggestive of the sea one fancied in talking with him his very words savored of the brine and raised the wind and waves. He was evidently a man of keen courage, keen-eyed and a stubborn skeptic, but the kindest soul on board nevertheless. As a pugnacious iconoclast he was the best-natured growler I ever met. He was particularly severe on priests and the clergy in general, declaring again and again they were good only for people who dwelt in towns and had nothing to do. Sailors had to face real work and real danger every day. All their life was real, and left no space for nonsense and mystery and other worlds. To make one's way through the waves of this one was work enough for him, he said, and took all his attention, etc., etc., plunging on full sail with or without encouragement. He refused to believe even in glaciers — a truly sad state of mind — and I told him he must make haste and repent, for one who believed neither in God nor glaciers must necessarily be the most wicked and dangerous of men, and that unless he reformed he should be sunk in the sea or at least made to ride at anchor in some lonely place where he would be beyond the reach of harm-doing. I charged him with sometimes giving his opinion on things he knew little or nothing about, land things in particular, and then challenged him to say what a glacier was. 'A glacier,' he replied, 'is a great big mountain all covered with ice.' 'Well, then,' said I, 'if a glacier is a big mountain covered with ice, a river would be a big mountain covered with water,' and this he knew would be a poor definition of a river. 'Well, then, what is a glacier according to you?' he inquired in a puzzled tone, and I was glad to get him nonplussed for once ere I told him that a glacier was a river of ice that took its rise on the mountains from the snow and flowed down into the lowlands to the sea like rivers of water, notwithstanding its apparent flinty rigidity. And I at length awakened his curiosity and interest as I described some of the Alaska glaciers, and he admitted he would like to see them; notwithstanding he had heretofore replied to all my invitations to go sightseeing that he had 'seen the elephant, and nobody could fool or bulldoze him.' He was a striking example of coarseness and fineness, like a plant bearing tender delicate flowers buried in thorns and rough awkward leaves.

At Port Townsend I was joined by Mr. Loomis, a young lawyer from Seattle, whose acquaintance I made on an excursion to Mount Rainier, and who had agreed to meet me here and accompany me to Alaska. He had gathered a lot of provisions, a tent, blankets, etc., for the trip and got them aboard. In the evening we sailed for the north, touching again at Victoria and staying a few hours, so that we had time to go up town and gain another view of the charming spring vegetation by the wayside and in the gardens. The wild rose was in all its glory, growing among rank ferns and great thickets of white and purple azaleas. This is the largest of all the wild roses I know of, and it is here very abundant. The garden honeysuckle was also in full bloom, and gave forth delicious fragrance, making many a poor man's cottage delightful, as well as those of the wealthy with ample breadths of laden verandas. The orchards were also in bloom, and altogether the place was charming that June morning after the rain, when the birds were singing and every leaf and petal of the region was giving forth a fresh sweet smell. How strange amid this fine culture and softness of beauty and fruitfulness did the rocks, smoothed by the glaciers and rising here and there above the rank vegetation, appear as they told of the time so lately gone when all the region lay in the darkness beneath an all-embracing mantle of ice! In the harbor, which is simply a glacial excavation, the rocks around the shores all have the smooth moutonnéed form so characteristic of ice-action, and so fresh are their surfaces the swash of the waves has not yet obliterated the fine striated polish. Not even to the extent of the thickness of a sheet of writing-paper have these wave-lapped rocks been denuded in all the centuries since the ice-mantle was melted and they first were exposed to the light of day. Even in the streets of the town resisting glaciated bosses are exposed, whose fine scoring and striation have not been effaced by the wear of travel over them; and in the orchards fruitful boughs shade the edges of glacier-pavements and drop apples and peaches upon unwasted rocks, whose smooth surfaces advertise the tremendous action of down-grinding glaciers in characters that must be plain to the dullest observer. Nowhere, as far as I have seen, are the benevolent influences of the glacial period made manifest in plainer terms and brought forward in more striking and close contrast with the action of man than here. Yesterday a landscape of ice, bloodless, lifeless, no Strait of Georgia or Fuca, no trace of Puget Sound, no Vancouver Island in existence or any other of all the archipelago, all these features unborn as yet, smothered in a mantle of crawling ice. Today apples and wheat and people, forests, green meadows, and exuberance of warm life; all the scenery, all the beauty, and the life newborn. No tale of faerie is so marvelous or so exciting to the imagination as the plain story of the works and ways of snow-flowers banded together as glaciers and marching forth from their encampments on the mountains to develop the buried beauty of landscapes, make fields of soil for food for man and beast and for gardens, make basins for lakes, channels for the streams, bays and sounds, and embroidery of inland channels for the margin of the seas and the continents, fashion the comparatively featureless shores to countless islands,

etc., etc., etc. Then, their appointed work done, in the fullness
of time vanishing from the earth as if they had never been.
We left Victoria at 10.30 A.M., June 18, and with nearly two
hundred tourists sailed away to the northwestward up the
Strait of Georgia, through the midst of the clustering islands.

2. *Alaska Fragments,*[1] *June–July,* 1890

In Glacier Bay, Alaska.
Late June, 1890.

In the early afternoon a dense haze fills the sky. The sun seen
through this becomes a globe of glorious ruby, and its glare on
the sea looks as if the water had been strewn with a crystalline
ruby-dust, or with fine mineral spiculæ of vermilion bordering on
crimson. The ruby-dust seems to possess body, and while it
glows does not in the least dazzle, as if the brilliancy came from
the heart of it rather than the surface. In these intense white-
skied days the heavens are covered by a universal filmy fleece,
and the light comes as if filtered through milk.

The blue of distant hills and mountains when observed in clear
sunshine is subtle and luminous to a degree that surpasses admiration. They do a wonderful thing in the way of color, lifting
themselves all the long summer days in solemn and glorious
beauty.

The cool silver sheen of the waters in tempered sunlight.

Icebergs tossing and drifting in the bay, grating against one
another, moved by the winds on their varying faces, like sails
set at every angle. Crystal to crystal in keen clashing contact.
Small waves lapping and rippling among their blue hollows — in
and out, drenched with rain, drenched with sunshine, dividing
the light into iris glints and stars. Ever wasting, sailing off
solemnly after the crash and war attending their birth. Running
waves publishing the event far and near among rocks and grass

[1] See Muir's *Travels in Alaska,* chap. XVII and XVIII.

of the shores. Dazzling in the light of noon . . . a crowded collection of this glacial jewelry, beyond description glorious!

Ears unable to hear the still, small voice of Nature must needs hear these — the thunders of the plunging, roaring icebergs.

From here a new world is opened — a world of ice with newmade mountains standing vast and solemn in the blue distance round about it.

> In camp on moraine, east side
> of Muir Inlet, Glacier Bay.
> *Late June.*

Nature hard at work. How cordial the blast, the beating of the rain, the rush of the dark clouds about the brows of the mountains! — a hearty blessing to the woods and plants that need the cool rain. How the water roars on the heights all around! . . . We can bear it well, but how about the birds and flowers and the small furred . . .?

Impassioned and imperishable vigor of Nature makes one glow with admiration during these storm-times so easily felt and seen, so real to all the senses. How hazy and trivial all selfish pursuits seem at such times, when the whole brave world is in a rush and roar and ecstasy of motion — air and ice and water and the mighty mountains rejoicing in their strength and singing in harmony!

Orchestral harmony of the storm, the wind in fine tune, the whole sky one waterfall.

How gentle much of storms really is, though apt to go unnoticed! Storms are never counted among the resources of a country, yet how far they go towards making brave people. No rush, no corrupting sloth among people who are called to cope with storms with faces set, whether this ministry of beauty be seen or no. . . .

The tender beauty, the delights delicious of storms!

The storm was a grand festival.

On sled trip exploring
Muir Glacier, its bound-
aries and tributaries.
Hemlock Mountain.
July 11–21.

It has been said that trees are imperfect men, and seem to bemoan their imprisonment rooted in the ground. But they never seem so to me. I never saw a discontented tree. They grip the ground as though they liked it, and though fast rooted they travel about as far as we do. They go wandering forth in all directions with every wind, going and coming like ourselves, traveling with us around the sun two million miles a day, and through space heaven knows how fast and far!

The clearest way into the Universe is through a forest wilderness.

The living and dead look well together in woods. Trees receive a most beautiful burial.
Nature takes fallen trees gently to her bosom — at rest from storms. They seem to have been called home out of the sky to sleep now.

Climbing Fountain Peaks of
Muir Glacier. Morning.

All Alaska fresh and sweet. No desert, all verdant save the mountain-tops.
This green leafy Alaska was once as icy, ruled as sternly by frost as the ice-locked, ice-barred lands of desolation about the pole.

This morning I ate my crust above the clouds. . . . The sun rises clear and the snow on the mountains begins to shine. Bits of sky glad and calm, windows opening into some fair far lands that no storm may reach. Bars of pure crimson on a pale liquid green (mellow green), like tropical sunshine among the islands of Florida when the sea is calm. Clouds sharply outlined, almost crusty on the edges, without any thin airy gradations.

A dewdrop in every Cassiope cup and on the end of every bent grass blade, birds feeding their young, singing with full heart in the sweetness and majesty of Nature's love.... In such mornings it is easy to see that the world is a-making. In this celestial day heaven and earth, radiating beauty each to each, beautify each other.

The sweetest of wild garden odors fill the pores of the ice-cold air — as free from taint of decay as the glaciers themselves.

... The wilderness, I believe, is dear to every man though some are afraid of it. People load themselves with unnecessary fears, as if there were nothing in the wilderness but snakes and bears who, like the Devil, are going restlessly about seeking whom they may devour. The few creatures there are really mind their own business, and rather shun humans as their greatest enemies. But men are like children afraid of their mother. Like the man who, going out on a misty morning, saw a monster who proved to be his own brother.

The Daisy.

> Not worlds on worlds, phalanx deep,
> Need we to prove a God is here.
> The daisy fresh from winter's sleep
> Proclaims His power in lines as clear.

This is one of the still, hushed, ripe days when we fancy we might hear the beating of Nature's heart.

Then flows amain the surge of summer's beauty.

Sometimes mountains seem to swim furred and soft, half melted in a sea of color. Not so those mountains that morning defines clear — they alone are colored, glowing from within. ...

Pleasant to see small rills leave the main boisterous on-bounding flood, go gently trickling, cooing caressingly, to a sod of daisies or a clump of bushes, like pilgrims stepping out from the hurrying crowd to do a kindness, then going on again, as if saying: 'Let the great flood go on; it is thinking of the end, but I am for thee!'

Alaska birds are seldom dry even when it is not raining. When they hop from spray to spray of willows in the glens and about the edges of the streams, they are showered with drops lodged in the wrinkles of the leaves, and depending in beadlike rows along the small bent twigs. Every needle, too, of the evergreens is tipped with a drop easily shaken down. Or larger water masses gather and lodge where the clusters of needles converge at their bases, which, though perhaps not shaken down by the smaller finches, are apt to come plashing down on the heads of the larger jays and thrushes. And yet they are seldom seen much bedraggled.

There is love of wild Nature in everybody, an ancient mother-love ever showing itself whether recognized or no, and however covered by cares and duties.

Hearts don't grow old, but shall ever get nearer to the Source Divine.

To the Indian mind all nature was instinct with deity. A spirit was embodied in every mountain, stream, and waterfall.

In the mountains, free, unimpeded, the imagination feeds on objects immense and eternal. Divine influences, however invisible, are showered down on us as thick as snowflakes in a snowstorm.

The shady recesses of the mountains where the young glaciers are nursed.

The mountains are fountains of men as well as of rivers, of glaciers, of fertile soil. The great poets, philosophers, prophets, able men whose thoughts and deeds have moved the world, have come down from the mountains — mountain-dwellers who have grown strong there with the forest trees in Nature's workshops.

A noble way the glaciers have worn for themselves out of the heart of the snowy mountains to the sea. Shaggy greenwoods and berries beneath them, blushing and ripening their... sweet juices close along the edges of the ice. How bright the crimson cranberries seem as they lie dripping on beds of moss!

Great round silver days, purple around the rim.

The scent of Bryanthus and Cassiope in bloom might well drive one wild. These lovely heathworts give a fine taste to the air.

Cassiope bravely blooming so high in the cold sky and abiding the long winters lying beneath the snow, safe and warm like the marmots beneath the sod.

Around that mighty brown expanse the sun shines unclouded on a thousand rolling hills and wide-bosomed hollows and prairies. The wind rustles the tall grasses and makes them wave and bow like pine trees. The groves of willow and aspen rustle and shake their leaves and shed off the light in spangles. Throbbing sparks show too on many a... lake and stream and pebbly rapid sunken in beds of glacial drift; and the birds sing and the reindeer feed in large flocks in boundless freedom.

Wild sheep, also, seen against the sky, with their great horns now thrown bold above their shoulders, now down among the green plots between the rocks. Woolly goats, too, with jet horns and long white hair. The ptarmigan in huge flocks on the barest heights. A boundless field of Nature's own.

How still and silent and smooth, yet how full of life stretching away to the Arctic Sea!

Turn to the mountains now — life too is there. See how the clouds lift and sink, moving their shadows from peak to peak, now in the full blazing light, now in shade.

Ice in every hollow, a thousand waterfalls foaming and chanting, birch and alder and spruce and carpets of bluebells making whole slopes blue, and red roses and violets white and blue.

Silent, inaudible, invisible flow. The very mountains flowing to the sea. The great heart of the hills sending its life down in streams ... among the stems and beneath the leaves of the lilies. ... Mountains die that we may live. ... 'Surely the mountain fadeth away,' said Job.

July 11.

Day Thoughts on a Glacier. To dine with a glacier on a sunny day is a glorious thing and makes common feasts of meat and wine ridiculous. The glacier eats hills and drinks sunbeams.

When I look on a glacier, I see the immeasurable sunbeams pouring faithfully on the outspread oceans, and the streaming, uprising vapors entering cool mountain basins and taking their places in the divinely beautiful six-rayed daisies of snow that go sifting, glinting to their appointed places on the sky-piercing mountains, joining ray to ray, forming glaciers amid the boom and thunder of avalanches, and at last flowing serenely back to the sea.

Air yourself on the ice-prairies, or on breezy mountain-tops.

Nowhere that I know will you find more and sweeter music than on a broad glacier when the sun is shining ... not even on the eastern edge of a wood in the morning in spring where birds are thickest and best love to sing.

In God's wildness lies the hope of the world — the great fresh unblighted, unredeemed wilderness. The galling harness of civilization drops off, and the wounds heal ere we are aware.

The sweetness in almost human tones of the songs of the streams on the glacier is enough of itself to redeem these frozen floods on whose backs they sing from all that is terrible and destructive, and manifest their essential goodness and charity even when we forget in struggling through their dangerous chasms their gentle origin in flowers of snow.

Every crystal seems to be singing and rejoicing in the delicious sunbeams, the whole vast glacier drinking heaven's light at its leisure. . . . It is all a song of the sun.

Sweet bewitching warbling, tapering trills almost inaudible.

Divine beauty and power and goodness shining forth in every feature of the great icy day. Nature is seen at work transforming the whole face of the landscape, working in tremendous action, yet in perfectly poised harmony, disarranging nothing, the gardens growing calm in the midst of it all, the little birds and marmots fed, the streams singing while the mighty glaciers grind the rocks.

Ice-sheet brooding with supreme dominion over the coming landscape, clasping a thousand mountains in its crystal embrace; and to think that all this mighty glacial geological engine shaping the world, bringing out the face-features of the landscape, is made up of tiny and frail frost-flowers, like children on a frolic with hands joined, children of the sea and the sun — small celestial messengers in infinite flocks, radiant, alighting on the mountain-tops out of the mysterious deeps of the azure. What possibilities are hidden in the tiny messengers! What a work is laid on their shoulders! What songs they sing as they gather and go to their predestined work, every one of the infinite multitude building better than they know, fields and orchards and flowers and birds and happy people, lakes and rejoicing rivers breaking forth into glad existence at their approach! Blessed the ear that hears this snow-flower glacial symphony!

A multitude of mountains glowing in bright and lovely color like any flower, their mighty and majestic curves drawn into noble firmness.

Intense silvery whiteness.

The light is not blazing, flaming effulgence like that which fills and fires the clouds in the tropics at sunset. Only an intense glow, the rapt, passionate essence of light like the glow of a countenance under surpassing excitement, like the soul-glow beaming from the pines on the hills after a storm when the setting sun looks out from beneath a melting cloud. Strange supernatural glory of color that seems to belong to heaven.

I am often asked if I am not lonesome on my solitary excursions. It seems so self-evident that one cannot be lonesome where everything is wild and beautiful and busy and steeped with God that the question is hard to answer — seems silly.

Every particle of rock or water or air has God by its side leading it the way it should go. How else would it know where to go or what to do?

> Camping at base of Snow Dome,
> east side of Muir Glacier.
> *July* 14.

The great white mass of the mountain fairly whitens the night as it looms into the stars.

> On sliding down a ravine near
> Snow Dome, and seeing ravens
> wheeling in the sky above.
> *July* 15.

Not yet, you black imps, not yet. Wait awhile, I'm not carrion yet. Go gack to your gray friends, the wolves. I was only sliding for fun. My body flesh is not yet cast away. I shall need it a long time. Go to the wolves for your dinner! You should have known better. Shame on you not to know better — and they did seem to be ashamed.

> Camping on Muir Glacier.
> *July* 16 (?).

Night Thoughts on a Glacier. Sleeping by a waterfall in a lonely canyon is impressive, but more so sleeping by a waterfall on a great glacier far from land. The sounds as the streams fall into great chasms and shafts seem so strangely wild in smothered hollow tones of the underworld.

Solemn loveliness of the night. Vast star-garden of the Universe.

All the landscape save tips of mountains is submerged in one vast inundation of ice. Yet all overbrooded with perfect repose ... majestic peace.

The poet Burns said that when he was laid in his grave he wished to be stretched out at his full length so he might occupy every inch of ground he had a right to. But on this excursion it is not easy to get so much ground for a bed, for my sled is only three feet by eighteen inches. The rest has to be built of stones.

Sleeping by a glacial waterfall, rills on the ice sound so perfectly sweet, one seems to taste as well as hear them.

In this silent, serene wilderness the weary can gain a heart-bath in perfect peace.

No camp on mountain-top or in depths of woods primeval is so impressively solitary as a camp out in the midst of glaciers among the hills and dales of ice where streams pour into crevasses.

Most people are *on* the world, not in it — have no conscious sympathy or relationship to anything about them — undiffused, separate, and rigidly alone like marbles of polished stone, touching but separate.

At night, such a chaste icy freshness came down the glacier on the sweet north wind. At the bottom of the wind, beneath the general solemn growl and roar of eddies in crevasses, I distinctly heard the gentle rustling made by the glint and swirl of the river of air among the small crystals and walls of the icy prairie — the slow-drawn, icy breath of the north.

Camping on top of Quarry Mountain,
seven or eight miles from the front
of Muir Glacier.

July 18.

How hard to realize that every camp of men or beast has this
glorious starry firmament for a roof!. . . . In such places standing
alone on the mountain-top it is easy to realize that whatever
special nests we make — leaves and moss like the marmots and
birds, or tents, or piled stone — we all dwell in a house of one
room — the world with the firmament for its roof — and are
sailing the celestial spaces without leaving any track.

The marvelous silence of this hill-top solitude.

Camping on Muir Glacier.
July 19 (?).

I always enjoyed the hearty society of a snowstorm; glad,
though solemn, when within a mile or two of safe ground where
a storm-nest could be made. Now on this shattered ice I need
my eyes, but the snow, gyrating, whirling, and sifting, the very
incarnation of spasmodic hysterical mirth, fills them, and I am
blinded as if blinded by kisses, delicious in the eye, and sweet.
The fall of snow in the solid darkness of night.

Camping on moraine,
east side of Muir Inlet.
Late July.

Auroras. The most glorious of all the white night lights I ever
beheld. . . . Radiant glory of that midnight sky, as if the foundations were being laid for some fairer world. Beams of light in
subdued splendor without effort, each shining shaft hasting in
strong joy to its appointed place. Let there be light, and light
was! Resplendent star-atoms shining, stars beyond stars. . . .

Heaven and earth are one — part of the vesture of God,[1]
Around all the earth the deep Heaven lies and is part of it.

[1] So arranged by John Muir.

The dark bodeful night
Becomes divine and glows transfigured in light,
Puts on the garment of Eternity
That comes from no earthly sun — a sight to be worshipped.

3. *A Trip to Kings River Yosemite* [1]

Northern boundary, Grant Park.
May 30, 1891.

We set out from Moon's Mills at 9 A.M.... They are cutting large quantities of sequoia from the ridge above the mill; trees from six to twelve feet in diameter are being dragged down a chute. Some fine sequoia stand in the ravine just at the head of the ridge — one grand old dead giant. Sugar pine is also abundant here.

From here one enters the park, still ascending for a mile, then down into a hollow valley where fine groups and scattered specimens of the sequoia grow in the bottom and up the sides as far as I can see. We pass great gray boulders, with green patches of moss and spots of pale gray lichens on them, ... through chaparral of wild cherry and thorn, with Ceanothus leaning against them for fringe; and by great granite masses bulging out of hill-slopes where a few trees grow on the cracks. Once these masses were strong, but they are now weathering into boulders.

Dead spars, trunks, and branches strew the grove, the sequoia branches being much smashed in falling; some huge giants are burned — noble ruins. One hollow tree has been made into a hotel and dance house!... Twenty years ago I saw five mills in or near the sequoia belt, all of which were cutting more or less of the 'big-tree' lumber. Now the number of mills along the belt in the basins of the Kings, Kaweah, and Tule Rivers is doubled, and the capacity more than doubled. As if fearing restriction or interference of some kind, the mill companies seem to be giving par-

[1] See Muir's 'Rival of the Yosemite,' *Century Magazine*, vol. 43, November, 1891, pp. 77–97. By permission of the D. Appleton-Century Company, certain paragraphs from the article describing rocks and streams, together with passages dealing with devastation, have been substituted for the faded journal version.

ticular attention to the destruction of the sequoia forests, getting the trees made into lumber and money before steps can be taken to save any of them. Trees which are mere saplings when compared with mature specimens are being cut down, as well as the giants up to at least eighteen feet in diameter. Scaffolds are built around the great brown shafts above the swell of the base, and destroyers armed with long saws, axes, and wedges chip and gnaw and batter them down with damnable industry. The logs found to be too large are blasted to manageable dimensions with powder. It seems incredible that Government should have abandoned so much of the forest cover of the mountains to destruction. As well sell the rain-clouds, and the fountain-snow and the rivers, to be cut up and carried away, if that were possible.

I passed through Round Meadow and its superb fir woods, then over the ridge and down into Bearskin Meadow, a lovely emerald plat of violets, larkspur, Parnassia, sedge in bloom, Potentilla, ivesias, Ranunculus, and life-everlasting. This meadow is surrounded by yellow pines, mostly very tall; and by silver fir at the upper end.... Sugar pine cones are an inch long now.... A stream comes trawling through in fine voice.

> Tornado Meadows,
> Altitude about 7000 feet.
> *May* 31. Morning.

A grand summer snowstorm — the silver fir, pines, cedar, and sequoia laden, the snow still falling at 6.30 A.M. .... Quail are calling, and I see an arctic bluebird. Many fine sequoias fill the boulder basin along the trail west of the creek. ...

> At Fox Camp, lower end of
> Kings River Yosemite.
> *June* 1.

At the foot of South Fork Kings River Valley we find ourselves in a spacious park, planted with stately groves of sugar pine, yellow pine, silver fir, incense cedar, and Kellogg oak. The floor is but little ruffled with underbrush, but myriads of small flowers spread a thin purple and yellow veil over the gray ground of the open sunny spaces and the brown needles and burrs beneath the groves. The walls lean well back and support a fine growth of trees, especially on the south side, interrupted here and there by sheer masses a thousand to fifteen hundred feet high, which are thrust forward out of the long forested slopes like dormer windows. Three miles up the valley on the south side we come to the Roaring Falls and Cascades. They are on a large stream called Roaring River, whose tributaries radiate far and wide and high through a magnificent basin into the recesses of a long curving sweep of snow-laden mountains to the eastward on the Kings-Kaweah Divide. But though the waters of Roaring River from their fountains to the valley have an average descent of nearly five hundred feet per mile, the fall they make in getting down into the valley is insignificant in height as compared with the similarly situated Bridal Veil of the old Yosemite. The height of the fall does not greatly exceed its width. There is one thundering plunge into a pool beneath a glorious mass of rainbow spray, then a boisterous rush with widely divided current down a boulder delta to the main river in the middle of the valley. But it is the series of wild cascades above the fall which most deserves attention. For miles back from the brow of the fall the strong, glad stream, four or five times as large as the Bridal Veil Creek, comes down a narrow canyon or gorge, speeding from form to form with admirable exuberance of beauty and power, a multitude of small sweet voices blending with its thunder-tones as if eager to assist in telling the glory of its fountains. On the east side of the fall the Cathedral Rocks spring aloft with imposing majesty. They are remarkably like the group of the same name in the Merced Yosemite, and similarly situated, though somewhat higher.

Next to Cathedral Rocks is the group called the Seven Gables, massive and solid at the base, but elaborately sculptured along the top and a considerable distance down the front into pointed arches, the highest of which is about three thousand feet above the valley.

Beyond the Gable Group, and separated slightly from it by the beautiful Avalanche Canyon and Cascades, stands the bold, majestic mass of the Grand Sentinel, thirty-three hundred feet high, with a vertical front presented to the valley, like the front of the Yosemite Half Dome.

Projecting out into the valley from the base of this sheer front is the Lower Sentinel, twenty-four hundred feet high; and in front of it and on either side are the West and East Sentinels, about the same height, forming altogether the boldest and most massively sculptured group in the valley.

Then follow in close succession the Sentinel Cascade, a lacelike strip of water, two thousand feet long; the South Tower, twenty-five hundred feet high; the Bear Cascade, longer and broader than that of the Sentinel; Cave Dome, thirty-two hundred feet high; the Leaning Dome, thirty-five hundred feet; and the Sphinx, four thousand feet, terminating in a curious sphinxlike figure. It is the highest rock on the south wall, and one of the most remarkable in the Sierra; while the whole series from Cathedral Rocks to the Leaning Dome at the head of the valley is the highest, most elaborately sculptured, and the most beautiful series of wall-rocks of the same extent that I have yet seen in any yosemite in the range.

Turning our attention now to the north wall, near the foot of the valley a grand and impressive rock presents itself, which with others of like structure and style of architecture is called the Palisades. Measured from the immediate brink of the vertical portion of the front, it is about two thousand feet high, and is gashed from top to base by vertical planes, making it look like a mass of huge slabs set on edge. Its position is relatively the same as that of El Capitan in Yosemite, but neither in bulk nor in sublime boldness of attitude can it be regarded as a rival of that great rock.

The next notable group that catches the eye in going up the valley is the Hermit Towers, and next to these the Three Hermits, forming together an exceedingly picturesque series of complicated structure, slightly separated by the steep and narrow Hermit Canyon. The Hermits stand out beyond the general line of the wall, and in form and position remind one of the Three Brothers of the Yosemite Valley.

East of the Hermits a stream about the size of Yosemite Creek enters the valley, forming the Booming Cascades. It draws its waters from the southern slopes of Mount Hutchings and Mount Kellogg, eleven thousand and twelve thousand feet high, on the divide between the Middle and the South Forks of the river. In Avalanche Canyon, directly opposite the Booming Cascades, is another brave, bouncing chain of cascades, and these two sing and roar to each other across the valley in hearty accord. But though on both sides of the valley, and up the head canyons, water is ever falling in glorious abundance and from immense heights, we look in vain for a stream shaken loose and free in the air to complete the glory of this grand yosemite. Nevertheless when we trace these cascading streams through their picturesque canyons, and behold the beauty they show forth as they go plunging in short round-browed falls from pool to pool, laving and plashing their sun-beaten foam-bells, gliding outspread in smooth shining plumes or rich ruffled lacework fold over fold; dashing down rough places in wild ragged aprons; dancing in upbulging bosses of spray, the quiet brave ouzel helping them to sing, and ferns, lilies, and tough-rooted bushes shading and brightening their banks — when we thus draw near and learn to know these cascade falls, which thus keep in touch with the rocks and plants and birds, then we admire them even more than those which leave their channels and fly down through the air.

Above the Booming Cascades, and opposite the Grand Sentinel, stands the North Dome, three thousand four hundred and fifty feet high. It is set on a long bare granite ridge, with a vertical front, like the Washington Column in Yosemite. . . . Above the Dome the ridge still rises in a finely drawn curve, until it reaches its culminating point in the Pyramid, a lofty symmetrical rock nearly six thousand feet above the floor of the valley.

A short distance west of the Dome is Lion Rock, a striking mass as seen from a favorable standpoint, but lower than the main rocks of the wall, being only about two thousand feet high.

Beyond the Lion, and opposite the East Sentinel, a stream called Copper Creek comes chanting down into the valley. It takes its rise in a cluster of beautiful lakes that lie on top of the divide between the South and Middle Forks of Kings River. The broad spacious basin it drains abounds in beautiful groves of spruce and fir and small meadows and gardens, where the bear and deer love to feed. Alas! it has been sadly trampled by flocks of sheep, making it mean and disordered. But keep out sheep and prevent fires, and Nature will soon heal all scars either on ground or tree, and wreathe all in beauty again.

From Copper Creek to the head of the valley the precipitous portion of the north wall is comparatively low. The most notable features are the North Tower, a square, boldly sculptured outstanding mass two thousand feet in height, and the Dome Arches, heavily glaciated, and offering telling sections of domed and folded structure. At the head of the valley, in a position corresponding to that of the Half Dome in Yosemite, looms the great Glacier Monument, the broadest, loftiest, and most sublimely beautiful of all these wonderful rocks. It is about a mile in height, and has five ornamental summits and an indescribable variety of sculptured forms projecting or countersunk on its majestic front, all balanced and combined into one symmetrical mountain mass.

June 6.

The Kings River Yosemite has fewer and less spacious meadows than that of the Merced but more extensive parks, a hundred times more sugar pine, fewer groves of oak. The dry, gravelly parks are the same in both, strewn with brown needles and cones, planted with gilias, Spraguea, mint, and fragrant lupines. 'This is like the Garden of Eden,' said Fox, the hunter; 'none such in Yosemite.' A brier rose with fragrant leaves blooms delightfully about the upper meads. Instead of Mirror Lake is Bear Meadow. Tall lilies grow there.

Many taluses have been shaken down by some earthquake. The boulders are huge, two of them one hundred feet long, rocks enough for a row of pyramids set base to base the whole length of the valley. No azalea, no Douglas spruce are here, but the Fremont pine is abundant, while in one comparatively short climb of five thousand feet up Avalanche Canyon you come to noble groves of Patton spruce, [1] mountain pine, and flexilis pine, offering grand specimens of each. The bark of the mountain pine is red and thick, furrowed and cross furrowed. The *Pinus flexilis* has a four-foot diameter, and is picturesque. There are also fine junipers.

Between and about the main mountain masses of the walls are a thousand gardens and hollows and flowery ledges and glens, dear to the heart of the lover of wildness.

A boy came in with glad subdued excitement, saying: 'Here are some trout for you and two squirrels; one of them is shot so badly he is little good for eating. I can't stay long, for it is now three o'clock and Uncle and I have to go four miles up the valley to take care of a bear I killed an hour ago.' Then he gave particulars of the killing of this, his first bear. He was going along sauntering and hunting, looking for a bear that I had told him lived in the upper end of the valley, and whose fresh tracks I had seen in a little mead under the King Mountain. The ground was open, sunny, sparsely planted with yellow and sugar pine, and the bear, quietly feeding on lupines and grass, was coming towards him. He dodged back of a pine, and, leaning his rifle against it, took aim, but waited until the bear came within fifty yards, when he fired. Wounded, the bear ran past him, pursued by his dog, when he fired again and brought him down. It was a black bear of medium size, weighing about four hundred pounds, sixty-six inches long from root of tail to nose, fat and sleek. The black bear are common in the south Sierra, and common in the middle and north. They are of firm build with masses of muscle, and are able to live on anything. Few grizzlies are here.

Fox made a bear-trap this spring, waiting to bait it with sheep. He told us of a large bear that lived 'holed up' in a north canyon; had seen his immense track. When I went up this canyon I saw the fresh track in the meadow, the grass beaten down, and on

[1] Mountain hemlock.

returning after noon saw another fresh track in the grass, the leaves still wet, which I traced into the brush, where I saw his footprint on the muddy bank of a stream. It was eight inches wide across the toes. I heard and smelled him, and, turning out into the meadow, went round his lair.

June 7.

Calm, clear, not a cloud. . . . Birds in open sunny spaces, doves cooing, feeding in flocks of hundreds, flycatchers on the wing dashing from treetops. The river full to top bank from snow in sun of the last three days. Trout leaping, overhanging branches bending to the current and springing back dripping, as if by strokes of the life-giving river in play. . . .

In spite of sore feet and weariness and the spiky brush leveled against you like bayonets and the rough rocks in battle array, you climb in a kind of natural ecstasy as if lifted by the very spirit of the mountains, which through all the toil seems ever to call, 'Come up higher'; and when the glorious summits are gained, the weariness all vanishes in a moment as the vast landscapes of white mountains are beheld reposing in the sky, every peak with its broad flowing folds of white, glowing in God's sunshine, serene and silent, devout, like a human being. This is the true transportation.

King Mountain, the grandest of them all, hides, and one must go to its feet to see it and feel it, like a grand character that seems insignificant until known. By no work of words, however great, may the tremendous impressiveness of these mountain sculptures be made manifest. At night in half darkness, in the morning as they catch the sun-rays on their outstanding bastions and spurs and domes, or in full flood of noon radiance, or in storm, or cloud-capped and swathed, they grow in bulk and beauty and grand God-like repose.

June 8.

A calm morning. A breeze up the valley as usual begins to blow about 9 A.M., and continues until dark; then the calm and starless sky stretches across from wall to wall. A few white sailing clouds every day appear at nine and vanish before sunset.

At noon I set out with Robinson, the artist, for the top of the Second Sentinel, which is a majestic rock twenty-five hundred feet high, with gable front facing the valley and level roof stretching back to the intersection with the vast vertical front of the Grand Sentinel, the summit of which rises nearly a thousand feet higher and whose face, along which we climbed, is about three thousand feet wide, and almost absolutely vertical. . . .

The talus up which we climbed to the foot of the Grand Sentinel is composed of the largest blocks I ever saw — one over a hundred feet cubic. Magnificent sugar pines grow on it, six feet in diameter — a wonderful growth for such soil. Also maples and a few white oaks and black, and Brewer's pentstemon in flower with hummingbirds attending them, also honeysuckle and rock ferns.

June 9. 5 P.M.

Looking down the valley from the east end of Sentinel Meadow. The meadow is carpeted with Carex, brown and yellow, green-edged with brier. On beyond are yellow willows, fine-grained in texture of foliage, and a dark green line of rushes, then cottonwood and alder, with dark green bars of shadow lingering in the mellow sunlight. A few tall towers of pine; and the majestic mass of Seven Gables, their sculpture bold, brought out in light and shade, violet blue in shadows, pinky yellow gray and white in the sunlight. The shoulder of the West Sentinel has a talus gray and white, the gray blocks dating from some ancient earthquake, the white from the Inyo earthquake. . . .

Fox's Camp. *June* 11.

We started this morning for the Lowlands. Calm and the river a little lower, although still high. Ere we had gone half a mile we were wrecked in the current, and came near to losing our pack mule and all our baggage, including Robinson's sketches. At the lower end of the Sentinel Meadow is a wretched tangle of crinkled cottonwoods living and dead, mixed and jammed with logs, branches and brush, and rocks. Through this mess the trail, such as it is, runs. This morning the river overflowed its banks and the ground of Black Meadow was submerged in muddy water to a depth of several feet, and as the trail at one place runs within six inches of the main current, which sets against it, the bank being precipitous to a depth of ten feet, it is at all times a dangerous place, and was extremely so this morning, as the location of the brink could not be determined. As the mule with his precious burden was floundering through this jam of rubbish and roots and snags and stumps, he toppled over into the flood — his head and part of the pack only being above water. Mr. Lake was ahead riding a horse; the artist and I were following afoot. Mr. Lake, dismounting, made haste to seize the head of the mule by the halter, while he made frantic efforts to keep his head above water and to resist the current, groaning in distress like a terror-stricken human being. Lake shouted lustily for assistance, calling, 'Bring a rope! Bring a rope!' We plashed through the fixed and floating débris and dragged a rope from the saddle, which I made fast to his head; then, taking a turn round a tree, we — the artist and I — pulled on the rope; but as the bank was sheer, we made no progress towards getting him out. Lake, the meanwhile, was trying to remove the pack. In a few minutes in his frantic struggle, bracing against the bank, the animal threw back his head and broke the rope, and instantly was swept away by the flood, mule and baggage drifting in the roaring current, helpless. I confess I never expected to see hide or hair of him again as he vanished round a point of willows, and rapidly summed up in my mind the extent of the loss: the artist's oil and pencil sketches, won at cost of much toil and industry, his blankets, etc., the hundred-dollar mule, and my bearskin from Alaska in which I had braved the cold nights on the glaciers.

A short distance below the place of accident a large pine had been felled as a bridge, and as the wreck drifted near it the mule made desperate efforts to swim towards the shore, but, drawn against the great log, which was nearly submerged, was pressed under by the current. The pack stayed him, however, though his head was under water beneath the log, and in a moment, while being steadily sucked down deeper, Lake made out to seize his head and get it above water. Then, with one more struggle for life, and with Lake's assistance, he was dragged alongside the log until he reached bottom. The pack removed, the frightened, trembling wretch once more walked free in the woods. Then we built a fire, wrung our blankets, and spread them out to dry. But troubles come not singly. The sky was speedily overcast, storm-clouds spread their gray and black wings over the heights, and then came rain; so we were compelled to pack our wet traps and go on to the crossing of Roaring River, getting wet to the skin on the way.

The clouds came down, enveloping the grand brows and battlements of the valley one by one, swathing them softly, a headland standing out here and there adding greatly to their impressiveness and apparent height. Then all settled into a gray mass of rain and cloud and misty vapor. We had intended to attempt the crossing of Roaring River, which is divided into three streams, on a long talus, but now our packer had enough of water and would try no more hazardous work of the kind. Accordingly, all was again unpacked here and carried over three log bridges, and a large fire built, while I hastened on four miles farther to send up Mr. Fox with an animal to bring the baggage down to the lower end of the valley, where on account of the storm we were glad to stay for the night. Here at 3 P.M. we arrived, wet, weary, and excited, but not injured.

June 12.

About to set out again on our way home. Lovely morning, the very spirit of dewy mountain freshness after the storm. White-gray bosses, clouds and detached wreaths, wispy, drifting, creeping like ghosts about the fronts of the cliffs....

Horse Corral.
Altitude 8100 feet.
June 13.

Calm; a few low crawling white and gray clouds, the sun in broad bars streaming across the green sedgy glade.... Many grand sequoias along Boulder Creek, young and old, particularly fine and perfect on the right hand as we go up the trail nearly to the summit of the basin. Also sequoias in three other valleys or ravines west of Boulder Creek show their grand domes rising on headlands fronting the river....

On the mountain-top the sun streams on the majestic array of snowy peaks poised in the calm blue. Like a sun, every godlike mountain seems to radiate beams of beauty that reach and touch us with mysterious influence like loving hands; and we are awed with the sense of infinite space.

Chapter VIII : 1895-1898

'Wildness is a necessity. Mountain peaks and reserva-
tions are useful, not only as fountains of timber and
irrigating rivers, but as Fountains of Life!'

DURING the latter nineties John Muir carried on widespread
activities. In addition to a part-time modified supervision of the
ranch, he wrote numerous articles for magazines, most of them
pleading for the preservation of the trees, and for the creation of
more parks as playgrounds of the people. Moreover, he was
active in the fight for the recession of the Yosemite Valley that
dragged its weary length through all these years. During this
period he made frequent jaunts to the Sierra, two short excursions
to Alaska, and an Eastern journey in which he visited New England, Canada, New York, and several of the Southern States.
The southern portion of the trip was made with the botanists
Charles S. Sargent and William E. Canby.

The policies of President Cleveland and President McKinley
in setting aside National Parks and millions of acres of forest land
precipitated in Washington a mighty battle between special privilege and those who fought for the people's rights. A National
Forest Commission was appointed to investigate waste and
fraud, of which Charles S. Sargent was Chairman. John Muir
accompanied the Commission through Oregon, California, and
Arizona as unofficial guide and advisor. He later had much influence in promoting the national forest policy recommendations of
the Commission. For his labor in the cause of conservation he
was awarded in 1896 a master's degree by Harvard, and in 1897
an LL.D. by the University of Wisconsin.

- - - - -

1. *At Home on the Ranch*

Alhambra Valley,
Contra Costa County, California.
January 19, 1895.

Rain, wind, black weather with tedious monotony, but it matters not as far as my fields are concerned. In writing scenery I am in Alaska and the mind goes there with marvelous vividness. I see the mountains and the glaciers flowing down their gorges and broad shell-shaped hollows.

Sunday, *January* 20.

Dark. The rain-clouds in wondrous depth and fruitfulness. It seems marvelous that so much rain can be stored in the sky.

January 23.

Working on Glacier Bay. . . . A charming day after the rain and darkness for so long, I can't remember how long. . . . This is Helen's birthday and she is greatly excited about it, for it marks, she says, the end of her babyhood. She is nine, and says she will not longer answer to the name of Baby. If people ask for Baby in this house now, they must go and look for a baby. The bright day, she says, seems to have been sent just for her. She celebrated the day on the hills. She climbs well and is in perfect health — an unspeakable blessing after the extreme delicacy of her earliest years. Robins are flying in large flocks, driven down from the hills and mountains by the storm. Frogs are singing lustily, and there is always a crowd of handsome zonotrichias[1] about the house hedge. The cats capture many of them.

January 24.

This is another lovely day, cloudless, calm; slight frost this morning. Went to walk with Helen up the West Hill. She is a

[1] Sparrows of the genus *Zonotrichia*, which includes the white-crowned sparrow (with its subspecies Gambel's sparrow and Nuttall's sparrow), the golden-crowned sparrow, and the white-throated sparrow. — F. H. A.

wonderful climber, and vividly sees and enjoys the mossy rocks
and ferns, etc. The view of the bay was charming — mirror-
calm, shaded slightly by the gentle breeze in streaks. The colors
of the hills far and near are fresh and beautiful. Had a fine view
of the Sierra — solid white from the summits to within two
thousand feet of the plain. The snow yellowish as seen from here.
... All trains are stalled. Wish I could get off into it on snow-
shoes. But this literary work will hold me fast for a long time.

January 29.

I caught a small screech owl (*Scops asio* [1]) this morning — a
handsome horned fellow with gold eyes. He flew about my room
and bumped his head against the window, and soon seemed to
regret he had so easily allowed himself to be captured. Sitting
on a doortop of an outhouse blinking dreamily, he quietly allowed
me to take him in my hand.

January 31.

The brightest, balmiest, best day of all the happy new year, a
true child of the sun. Hard at work on Alaska.... Scot pulling
out Mission vines, Jim pruning, Joe plowing, larks singing. Snow
still lingering on Diablo. Frost in the morning.

February 1.

A balmy day. Sunshine and lark song in glorious measure. A
petition [2] is being circulated in favor of preservation of larks from
the ruthless slaughter of gunners. Larks are as characteristic of
California weather as sunbeams. As well shoot the sun out of
the sky.

February 5.

Still another golden day — all are alike now.... Scot making
fair headway on the vine snags. A few break off when pulled by

[1] Now called *Otus asio.* — F. H. A.

[2] In an article entitled 'Save the Meadowlarks,' published in the *San Francisco
Call* of March 24, 1895, Mr. Muir said: 'Better far, and more reasonable, it would
be to burn our pianos and violins for firewood, than to cook our divine midgets of
song-larks for food.'

two stout Norman horses. Most come out from two feet under-
ground and give no further trouble. The Mission vine, the first
planted in California, is a good table grape, but a poor wine grape,
and brings a very low price for either table or wine. The padres
ought to have known better — such good judges as they were in
most things relating to the stomach.

February 6.

This day I finished 'The Discovery of Glacier Bay.'[1] Will mail
it tomorrow to the *Century*. It seems strange that a paper that
reads smoothly and may be finished in ten minutes should require
months to write.

February 12.

Rain all forenoon with high wind. Fair most of the afternoon.
Rain again at dark. Streams rising. Farmers afraid so much
water about the roots of the orchard trees will kill them. Robins
flying about in large, restless flocks. Am making slow headway
with my literary task, the hardest of all work to me. It is so difficult to say things that involve thought at once clearly and at-
tractively — to make the meaning stand out through the words
like a fire on a hill so that all must see it without looking for it.

February 15.

The blue herons have come. . . . They are wonderfully regular
in timing their year's affairs. Their clock does not even seem to
be affected by the weather.

March 24.

Fine balmy day. Mount Diablo one mass of purple in the
morning. Nature is always lovely, invincible, glad, whatever is
done and suffered by her creatures. All scars she heals, whether
in rocks or water or sky or hearts.

March 25.

Cloudy, threatening rain. Wanda's birthday — the fourteenth
— happy girl. Heaven bless her always. I dread pain and trouble

[1] Published in the *Century Magazine*, June, 1895, vol. L, pp. 234–47.

in so sweet and good a life. If only death and pain could be abolished! She went over to the old ranch to lunch and play and climb the hills with May and Helen and her mother.

March 26.

Thomas Ross was buried yesterday — a fine enthusiastic Scotchman, singer of songs, poetical, artistic, though only a plumber. I saw him first in Yosemite in 1871, when he was with Keith and Irwin on a sketching tour.

March 28.

Clear and cool. Beautiful silvery haze on Mount Diablo this morning, on it and over it — outlines melting, wonderfully luminous....

April 9.

Another charming day of sunshine — only here and there a wisp of cloud in the heavens as if for ornament only. Busy reading. This I can do always from morn till night and never weary. But composition — the devil seems to keep me from it, though I feel that my day of life is fast speeding away and that I must tell my story to the world.

Had a pleasant visit from a brother Scot from Scotland direct — Mr. Rennie, with a letter from my good friend David Douglas — naebody like a Scotchman.

April 12.

Another lovely day, mostly solid sunshine. Took a fine fragrant walk up the West Hills with Wanda and Helen, who I am glad to see love walking, flowers, trees, and every bird and beast and creeping thing. Buttercup, clover, gilia, Brodiæa, Allium, Dodecatheon, larkspur, and portulacas are in flower. The oaks are in full leaf. A fine fragrant walk, the babies delighted.

April 14.

The fifteenth anniversary of our wedding day, and soon will come my birthday. How the years begin to run! Only in the wilderness is Time's flight hidden. A remarkably chilly day though bright — a good fire in my room.... The Louisiana tanager [1] has been here a week.

April 17

Here comes another bright day — the old pomp over again, seemingly monotonous through small local cares and tasks. While like gophers we drive our trade of soil-stirring almost in the dark, how vast a multitude of interesting events are taking place over this busy, loving, hating world!

Three dozen single Cherokee rose bushes arrived today from Oakland to be planted along the new fence by the roadside ... to bless with their blossoms thousands of passengers to come.

April 18.

Planting roses — a fine business. They require much care for a few months, but when fairly established they bless every eye and want no other attention than admiration. I know of no other investment likely to give such delightful dividends of beauty at so cheap and pleasant a price. What a hedge I shall have — fifty rods long, six feet high, with millions of lovely white flowers, three to four inches in diameter, the flowers lasting more than a month and the leaves forever! On their glossy evergreen surface no dust lies.

April 20.

Lovely balmy morning. What a stir there is in every living leaf — every cell a busy factory — and how enormously fertile are the kingdoms of insects! The air is all swirling and thrilling with singing wings — especially of the Ephemera. They rejoice and play in large assemblies and keep time in a wonderful way, going round and round in giddy whirls and spirals, mostly from left to right, with the sun, into which they are geared in some mysterious way.

[1] Now called the Western tanager. — F. H. A.

April 21.

My birthday — I am told the fifty-seventh, and yet I feel only a boy. Must make haste and get my work done ere the night falls. Made an excursion with the babes to Mount Wanda.

April 22.

The flood of warm sunshine pours over California with increasing warmth and creaminess. The hill vegetation has mostly gone to seed and leaves are fading. A couple of Parkman's house wrens are building a nest in the woodpile — merry singers and sprightly workers, bits of bright, unclouded health.

April 25.

Vines beginning to make a fine show of hopeful shoots. Busy staking, and thinning apricots and peaches. How the time flies, and how little of my real work I accomplish in the midst of all this ranch work!... How grand would be a home in a hollow sequoia!

May 24.

Poor Pedro, a wondrous wise and busy and affectionate little gray terrier, was poisoned the other day by eating a bit of strychnined bread in Mr. Westlake's packing shed.... I loved Pedro — so wise and affectionate. He belonged to a neighbor and I saw him nearly every day. A rough wiry-haired fellow, homely, but of such is, I hope, the Kingdom of Heaven.

May 27.

A wild rainstorm mixed with hail this morning that fairly roared through the valley with rapid brown streams rushing down the hills. Wind very high. The severest storm I ever saw in summer. Hay soaked, fruit knocked off, with intervals of sad cloudy hours. The rain lasted all day. Hard on the wee birds. Noticed the mother wren feeding her young in the woodpile.

May 30.

Very high wind from the north, drying the wet hay in cocks and windrows. The little speckled wrens are now very busy feeding their young. They must carry several hundred insects to the nest every day. They can chatter and complain with a large beetle or fly in their bills. This is the Parkman's house wren, a sweet, cheery singer and vigorous scolder of small dogs and cats. I have seen our best cat watching this busy pair. He cannot reach them now, but I am troubled with the prospect of the little ones when they first leave the nest.

June 3.

Very warm and calm. Busy getting in hay. The wrens seem to have led off their young today or yesterday. I hope no accident befell. Many cats lay in their way....

June 5.

Very warm. North wind singing and scorching every man, woman, child, animal, and plant. Horizon dirty and dusty. The night hot....

June 14.

Went to the City. Saw Keith and ate lunch with him.... Saw and talked with Thos. Magee,[1] who told of a minister who did not believe in the age I gave of the sequoias.... This minister may know much of heaven — he knows little of trees....

June 30.

Walked to Swett's[2] to see little Ruth Parkhurst — a sweet, beautiful motherless bairn.

[1] Mr. Thomas Magee of San Francisco was John Muir's companion as far as Fort Wrangell on the journey to Alaska in 1880.

[2] In 1881, John Swett bought a ranch adjoining the Muir-Strenzel holdings in the Alhambra Valley. Ruth Parkhurst was the granddaughter of Mr. and Mrs. Swett.

July 4.

Passed the glorious Fourth at home, occupied as usual. The bunting and gunpowder have no charms for me, only something to escape from. The children burned a few firecrackers — made a show of patriotism with two small flags on the front step, then ate ice-cream and walked with me over to the old ranch. . . . The day has been strangely dark and cold with black clouds and light sprinkling rain, as if Nature frowned on the affair.

The herons are all on the wing — a few still lingering on the old nests. The wrens are feeding a second brood of young. In the evening, Mr. Coleman brought a lot of rockets and candles, which made a handsome show and greatly delighted the children.

July 5.

The moon is shining this evening in glorious majesty. Tom[1] came home with a lizard today, which we took from him and saved its life. A few minutes later he came up with a gopher. . . .

July 11.

The wrens are very busy feeding their growing young in the woodpile. They bring a worm, beetle, or fly about once a minute or two minutes. The little babes will soon fly; they keep up a steady pee-peep all day. The time of greatest danger is just on leaving their cunning shelter. The cats oftentimes look and listen and snuff the air about the nest.

2. *A Mountain Ramble*

Raymond, California.
July 30, 1895.

Breakfast good enough. Recognized Mr. L. and went into the kitchen to speak to Mrs. L. — rosy, fat, in an atmosphere of greasy smells. . . .

Raymond is a dusty, dreary, sunburned station. . . . Yellow poverty-grass, a few thistles, a few Douglas oaks, shallow shadeless

---

[1] 'Our best cat,' Tom, a gray tabby, followed John Muir about the ranch, and sat on the floor by his master's chair at mealtime, receiving tidbits.

pools, doves drinking about them, here and there a willow or poplar. Jack rabbits abundant, mines, the trees wide apart, half desert.

At Grub Gulch a knoll is crowned with a group of noble oaks — a good cause for a town. From here we soon begin to rise, and the grand mountain-ridges with the spires of yellow pine thronging the long slopes come to view, and one's heart burns, and we feel how true it is that going to the mountains is going home....

In Tuolumne Canyon.
August 7.

Camped at the head of the Grand Cascade. A great white day — not a cloud. Yesterday morning was all bright, no clouds nor thunder — save thunder of the rejoicing waters. The walking below Deer Grove was tedious with a heavy pack, yet I stood it well. The night was windy, gusty, and cold; I slept but little. The moon was a fine study as it swept across the sky, silvering the trees it passed. The effect was strange in relieving a small portion of tree the size of the disc while all the rest was in dark. Looked like two droll maskers dancing on the moon herself....

August 8.

A white morning. Camped at the mouth of Ouzel Creek, two hundred yards below the foot of the second Grand Cascade.... The roughest kind of a scramble up and down Ouzel Creek. Fine sugar pines in the gorge. I saw an ouzel....

Much of the way from Ouzel Creek for three or four miles is pretty hard scrambling. The river runs in a gorge lined with huge earthquake boulders, all the small ones having been carried off by the river in floods. First there is a charming yosemite valley a mile or two long, three-quarters of a mile wide, with magnificent sugar pine and yellow pine, Libocedrus, and *Abies concolor*, with ordinary vegetation, such as Rubus, manzanita, and Ceanothus, etc. Then comes the gorge, mostly V-shaped, made through an interrupting bar of granite. I met a rattlesnake while scrambling near the river, the handsomest I ever saw, black and brown in diamonds. It was quite small and good-natured; did not take the precaution to coil, just gazed at me, and, as I passed, went beneath some driftwood. I did not take it for a Crotalus until I saw its half-dozen rattles.

I started from camp this morning at seven-thirty. At eleven I am writing these notes in one of the shadiest, handsomest groves on the river-brink, among silver fir, Douglas spruce, and Libocedrus. All of them are tall and slender for their age, some of the Libocedrus only a foot in diameter while being over a hundred feet high, so also the silver firs.

I found ripe raspberries yesterday and this morning; and saw a green water snake in Ouzel Creek, also an ouzel or two, and four ducks, gliding as if in pure enjoyment down the tossing, boiling stream. I have been following a bear trail today. The bears are mostly away sheep-hunting, though the manzanita berries are nearly ripe. Saw a flock of young, half-grown mountain quail yesterday. The old ones were greatly concerned about them. They flew well when I scattered them. The cock is a much handsomer and larger bird than the hen.

Yosemite Flat.

The flat is grandly forested with sugar and yellow pine, fir, Douglas spruce, and Libocedrus; ferns, azalea, and cherry. Glacier Cliff forms the north wall of it, Dome Cliff the south, while the river in a long nearly straight reach pours with moderate current — all one silver shining plume.... I saw where a bear was eating raspberries and thimbleberries (*Rubus nutkanus*); also, a mile above, saw where one had dug up and devoured a wasp colony. A few survivors were still lingering about the desolate nest. A good bear trail runs through all the difficult passages of the canyon; it is always the best route. I have been following it nearly all day.

Camp at the head of Muir
Gorge, Tuolumne Canyon.

I saw two rattlesnakes today, neither of which threatened me. One was of a species new to me; I did not at first take it for a rattler. It was rather slender, beautifully colored, black and white in triangular pattern. The other was a common dull-gray fellow, thick and muscular, that I came upon suddenly. Towards night in pushing through brush and rocks, I threw my bundle ahead of me. It fell plump on the rattler. He, highly indignant, crawled away a few feet and coiled and thundered. I cautiously withdrew my bundle and passed him a few feet away, he eyeing me suspiciously in wonder and striking at me. It was a rough place, but it was getting late and I was then at the head of the Muir Gorge, and had to camp. I decided to camp on top of a large boulder a snake could not climb except by a log that leaned against it. This I managed to remove. The boulder was about twelve by fifteen feet, sloping like a house roof, but a slight hollow in the middle enabled me to keep from falling. I built a row of stones that barely would lie, for bed and fire, and here I cooked and even slept hard, waking only four times to mend the fire.

The gorge shows no sign of water action above ten or twelve feet; it is about fifteen feet at bottom in width. The walls are sheer, two thousand feet high, glacial action showing at the top and a good way down. The river still runs in a ragged, jagged gorge below the gorge proper for half a mile, falling free in one place twenty to thirty feet and making a descent of two hundred feet or more all together. Below this a mile or two the canyon wall rises to the great height of three thousand feet or more. I call it Three Domes. The altitude of the river where best seen is fifty-five hundred feet above the sea. The braided cascade comes roaring into the river in wild array a few hundred yards below the lower opening of the main gorge and directly into the shattered zigzag gorge, which is a continuation of the main gorge but which you can approach and scramble alongside of, not over fifty to seventy-five feet above the toiling, boiling river....

> Boulder Camp near Yosemite Pass.
> *August 9.*

The gorge begins a short distance below Yosemite Pass; a fine stream comes down this pass; there is snow in it still near the foot.

A fine fall and stretch of cascades come in over a cliff two thousand feet high, just below the gorge on the north side. It is torn into shreds and braided again in a wonderful manner; I heard its roaring more than a mile away....

A little below the gorge, as I was anxiously going to the river to drink after climbing a six-hundred-foot rib, I came suddenly upon another big rattler this morning; my foot was within five inches of his head ere I discovered him and bounced back. He was evidently hiding for birds that came to drink, lying in folds one above another between the water-washed boulders....

Below the gorge the canyon walls stand back for four miles or so, forming a charming and grandly sculptured and planted yosemite valley. The Three Domes form its south wall, three to four thousand feet high; the north wall is wild but less massive....

At the foot of the valley the river goes roaring over a granite dam; first it flows rapid and white, then it calms to emerald in a very deep pool, then away it goes ever faster in a glorious plumy, lacy, upleaping, exulting cascade into a larger emerald pool.... From that the river glides on again at the common rapid gait....

Camp at head of
Hetch Hetchy Valley.
August 12.

Spent the morning washing clothes and myself in the river, which flows in a smooth stately current about a hundred feet wide, with charming reaches and good fording.... Fine groves of Kellogg oak. This, I think, is the most beautiful of all our oaks; the branches so clean and shapely and so intricately woven and interlaced and interarched, the foliage so bright and clean and green and handsomely lobed and pointed and held out horizontal, not drooping; even the slender twigs, so numerous and unsketchably mingled and crossed, never droop, but hold themselves out with leaves on top and bend only with weight. Lying beneath the trees in bright weather, one may realize the tender and intricate beauty of this tree....

Camp in Hetch Hetchy.
August 13.

Left camp at 6.30 A.M. to go to Crocker's. On the trail I met Mr. T. P. Lukens of Pasadena with a complete outfit. I told him I had intended remaining longer in the valley but was driven out by want of food, having been compelled to throw away most of what I started down the canyon with. He said he had abundance and proposed my returning to the valley for a few days. This I was quite willing to do....

We camped about the middle of Hetch Hetchy and soon had a portable table set and covered with new dishes of every description, and had bacon, potatoes, tomatoes, and bread. I felt hungry only after I had eaten well. Potatoes seemed specially good to me. I had only a handful of crackers left and tea when found, which I emptied by the side of the trail. Left my cup also. My outfit was a contrast to Mr. Lukens's....

On Tioga Road. *August 21.*

Yesterday in riding through the forests from Crocker's we met a bright, joyous bevy of short-skirted girls whose faces were radiant with pleasure, as with long strong steps they came through the lofty forest arches, pictures of health and enjoyment, every step telling the exhilaration of the spicy mountain air. They were followed by a large spring wagon laden with camp equipments, drawn by a pair of horses driven by a woman. They had camped at Wade Meadows, where milk was abundant, the owner of which told us they danced and sang most of the night. Girl graduates, said to be sweet, are not so sweet as girl mountaineers. Never before have the mountains seen so many young people camping in their hospitable, life-giving gardens and glens. It is a hopeful beginning. May their tribe increase until, like Switzerland, all the mountains echo with their happy voices.

Lake Tenaya. *August 24.*

A lovely mountain morning, still, clear, white; long lance-rays falling between the arrowy spires of the two-leaved pines and more than a dozen species of grasses, sedges, and rushes that now raise and spread and poise their green and purple panicles unvexed by the feet of hoofed locusts.[1] Only the dainty deer are here to dabble the ground. . . .

Ascending Mount Conners.
August 26.

A fine crisp morning. A gentle snowstorm. Fine to see the snow once more, the loom of a peak in storm. The gentians have closed their bells, the blue daisies their rays, and all are covered with crystals of frozen dew. Viewed through a lens in the sunlight they are glorious — keen fine needle rays making a fringe for every petal and leaf. And how divine a jewelry appears on every grass panicle! Snow falling through the spires of trees, on meads and on the lake. . . .

Yosemite Valley. August 31.

In the afternoon took a walk in search of my 'lost cabin,' as it is now called. Discovered it after a tangly search in the angle formed by the Tenaya Creek with the river about two hundred yards above the apex, and about fifteen or twenty yards from the bank of Tenaya Creek. It is just opposite Royal Arches, completely hidden like a bird's nest in a charming luxuriant growth of alder, Balm of Gilead, pine, fir, and Libocedrus with an undergrowth of *Rubus nutkanus*, dogwood, azalea, tall grasses, sedges, and the finest vaselike clumps of ferns (Asplenium) and high mosses.

The cabin is about fourteen by sixteen feet with a sharp roof, intended to shed snow, walls about six feet high built of Libocedrus logs, floor of the same hewn smooth, also rafters and shingles intended to last long. Here I longed to live in winter after being driven down from the heights by snow like other mountaineers, but the best-laid schemes of all of us gang aft agley. The fireplace was never built. I intended to sleep upstairs

---

[1] Muir's name for the devastating flocks of domestic sheep. — F. H. A.

in the garret, cedar-lined and snug, and to write. The roof is
mostly gone, someone has been camping in it, a bunk in a corner,
a whisky bottle, cans, etc., ferns, and bushes over the floor.

September 1.

I walked to Royal Arches, thence to Hutchings's log cottage.
It is badly dilapidated, used as a hay-barn and shed for implements. Like all the valley [1] excepting the fenced hay meadows,
it is frowsy and forlorn. The apple trees with weeds up to the
waist are still growing well unpruned, wild as oaks, yet laden
with good fruit, some large and well flavored. Also a few hardy
pears and peaches and cherry. The ground squirrels are destroying fruit, eating only seeds after the first meal in the morning.
Many apples lie on the ground. Little or nothing is done with
the Lamon orchards — all weeds, hay, and tangled strawberry,
raspberry, and blackberry.

September 2.

Bridal Veil meads are badly trampled at Black Springs. . . .
Met Mr. Lambert, who has been camped there a month or more,
collecting and studying moths and butterflies; a rare man who
lived alone at Soda Springs seventeen years and fenced one
hundred and sixty acres, carrying rails on his back. Now he is
likely to lose his mountain home by failing to get title before
the park is organized.

Visited the graveyard. Mr. Hutchings's plot, where Floy and
his second wife lie, has the best headstone of unhewn granite.
Lamon's monument is a granite shaft. Albert May has a monument of marble, erected by Mr. Black, for whom he worked long
as a carpenter, and with whom I passed a winter at Black's
Hotel. The graveyard like all else is sadly neglected, and has no
fence. It is strewn with leaves of Kellogg oak and dead grass
and ferns. Fire is likely to burn the wooden fences of graves and
kill the few shrubs. Galen Clark has selected his rest-place here,

[1] The Yosemite Valley in the midst of the Yosemite National Park was still under
the management of a State Park Commission.

fenced and planted it, and keeps it watered. Among other plants are a dozen or more young sequoias, growing well.

Night at Wawona.
Undated.

A whole universe of radiant beauty and limpid purity to revel in.

The thin shining veil of the northern lights waving, streaming, and fading. . . .

Far in this deep heart of the forest the soothing odor of balsam boughs piled deep for woodland beds, genuine perfume of the Land of Nod.

The countless trees of the forest aspiring in boundless, irrepressible, triumphant exuberance of beauty, every shining needle thrilling and singing in the wind like a harp string . . . all brooded by the great solemn night. . . .

3. *Thoughts Upon National Parks*

Yosemite Park is a place of rest, a refuge from the roar and dust and weary, nervous, wasting work of the lowlands, in which one gains the advantages of both solitude and society. Nowhere will you find more company of a soothing peace-be-still kind. Your animal fellow beings, so seldom regarded in civilization, and every rock-brow and mountain, stream, and lake, and every plant soon come to be regarded as brothers; even one learns to like the storms and clouds and tireless winds. . . . This one noble park is big enough and rich enough for a whole life of study and æsthetic enjoyment. It is good for everybody, no matter how benumbed with care, encrusted with a mail of business habits like a tree with bark. None can escape its charms. Its natural beauty cleanses and warms like fire, and you will be willing to stay forever in one place like a tree.

Government protection should be thrown around every wild grove and forest on the mountains, as it is around every private orchard, and the trees in public parks. To say nothing of their value as fountains of timber, they are worth infinitely more than all the gardens and parks of towns.

We believe our forests under rational management will yield a perennial supply of timber for every right use without further diminishing their area, and what is left now of the forest lands which after being surveyed is found to be unfit for agriculture should immediately be withdrawn from private entry and kept for the good of the people for all time. Nearly all our forests in the West are on mountains and cover and protect the fountains of the rivers. They are being more and more deeply invaded and, of course, fires are multiplied; five to ten times as much lumber is burned as is used, to say nothing of the waste of lowlands by destructive floods. As sheep advance, flowers, vegetation, grass, soil, plenty, and poetry vanish.

Had not the Sierra forests grown at a high altitude and thus been rendered difficult of access, they would all have been felled ere this. Meanwhile the redwood of the Coast and the Douglas spruce of Washington and Oregon were more available, though distant. It was cheaper to go up the coast a thousand miles than up the mountains fifty. At Puget Sound, the trees pressed close to the shores as if courting their fate, offering themselves to the axe, while the redwoods filled the river valleys, opening into bays forming good harbors for ships.

Fire. Until lately our forests, the noblest on the face of the earth, and the farms in the lowlands were threatened with utter destruction, mostly by fire and sheep ravages and general vandalism. . . .
Notices were posted on rocks and trees along trails, forbidding trespass, etc., but in the absence of enforcing power, they were of no avail; tens of thousands of sheep were driven into the pastures on the headwaters of the streams as if specially allowed and invited, and, with the fires that followed them, desolation

went on as before, as if a premium had been offered for the most successful efforts looking towards complete destruction. . . .

In the main forest belt of California, fires seldom or never sweep from tree to tree in broad all-enveloping sheets, as they do in the dense Rocky Mountain forests, and those of the East and of Washington and Oregon, where the trees stand close together and are mostly smaller and are consumed like grass or grain. Here the fires creep from tree to tree, nibbling their way on the needle-strewn ground, attacking the giant trees at the base, killing the young, and consuming the fertilizing humus and leaves.

The best service in forest protection — almost the only efficient service — is that rendered by the military. For many years they have guarded the great Yellowstone Park, and now they are guarding the Yosemite. They found it a desert as far as underbrush grass and flowers were concerned, but in two years the skin of the mountains is healthy again. Blessings on Uncle Sam's soldiers. They have done the job well, and every pine tree is waving its arms for joy.

Instead of bands of sheep one may meet new bands of bright girls and boys threading the forest aisles, stepping briskly along, the health and exhilaration of the mountains in every eye. In the exciting times of gold-digging scenery was unnoticed. Many old Californians even, living within a day or two's journey of Yosemite, never saw it; but now in the springtime children go to the park with enthusiasm as if called by the blast of a trumpet.

The park is the poor man's refuge. Few are altogether blind and deaf to the sweet looks and voices of nature. Everybody at heart loves God's beauty because God made everybody.

After I had lived many years in the mountains, I spent my first winter in San Francisco, writing up notes. I used to run out on short excursions to Mount Tamalpais, or the hills across the bay, for rest and exercise, and I always brought back a lot of flowers — as many as I could carry — and it was most touching to see the quick natural enthusiasm in the hearts of the ragged, neglected, defrauded, dirty little wretches of the Tar Flat waterfront of the city I used to pass through on my way home. As soon as they caught sight of my wild bouquet, they quit their pitiful attempts at amusement in the miserable dirty streets and ran after me begging a flower. 'Please, Mister, give me a flower — give me a flower, Mister,' in a humble begging tone as if expecting to be refused. And when I stopped and distributed the treasures, giving each a lily or daisy or calachortus, anemone, gilia, flowering dogwood, spray of flowering ceanothus, manzanita, or a branch of redwood, the dirty faces fairly glowed with enthusiasm while they gazed at them and fondled them reverently as if looking into the faces of angels from heaven. It was a hopeful sign, and made me say: 'No matter into what depths of degradation humanity may sink, I will never despair while the lowest love the pure and the beautiful and know it when they see it.'

Again, how often we see the love of nature in human hearts, made manifest by some lowly plant carefully tended in a can or cracked pitcher in the homes of the poor in cities. The window of my city study overlooked the backyard and porch of a hard-working German family. The eldest child, a girl perhaps about sixteen years old, was housekeeper, doing the washing, scrubbing, cooking. Every morning I saw her getting the boys ready for school, blacking shoes, washing faces, brushing and combing the hair of the youngest. But on a little shelf of the back porch she had a row of plants in cans and damaged crockery — geranium, tulip, a small rose bush. This was her garden, which she carefully and lovingly tended though working so hard, stopping in the hurry of housekeeping duties to touch tenderly and look into the faces of her humble plant friends. A fine study — nature in a human heart, in which the scenery surpassed that of the mountains.

> A pitcher of mignonette
> In a tenement's highest casement.
> Queer sort of a flowerpot — yet

That pitcher of mignonette
Is a garden in heaven set
To a little sick child in the basement.[1]

Flowers, then, are veritable angels in the house or outdoors. It is this love of flowers, Nature's plant people, in the hearts of rich and poor alike, that makes parks in town, and fine gardens on estates, and the one-flower-pot gardens in garrets of the poor. The foundation, therefore, of the great National Parks in the mountains is seen to be a safe one.

4. *At Home on the Ranch*

Alhambra Valley
Contra Costa County, California.
January 7, 1896.

Frosty — a fine glow of barred crimson and red clouds on the brow of the morning....

A small hawk sat on the tail of the windmill today while the wheel revolved freely, without being disturbed by the motion or by the squeaking axle, which required grease.

February 3.

John Reid [2] says one of the herons of the nesting-village of the sycamores in the vineyard came today and took a general survey of the premises.

February 13.

Warm, summery. A peach tree is now in bloom. There are thirty-one heron nests among the sycamores on Alhambra Creek, thirteen on one tree.... The overflow from the regular heronry consists of about a dozen nests on large oaks on the hills to the westward.

[1] So arranged by John Muir.

[2] John Reid married John Muir's sister Margaret. He and David Muir, brother of the naturalist, moved with their families from Wisconsin and lived for several years on the Alhambra Valley Ranch, helping with the work of supervision.

February 15.

Wrote to Sidney Smith declining invitation to lecture before the San Rafael Literary Club. Said that after ten years' silence bravely sustained I had lectured four times this fall, and then escaped with fear and trembling to the shades rural of Contra Costa County, vowing henceforth eternal silence.

February 16.

Walked on the hills with Helen, up Wanda and down the Helen hill. The buckeye will soon be in full leaf. A few of the oaks also beginning to burst buds. Soaproots and Castilleia in flower. Tules on Suisun swamps are burning, sending up huge black columns of smoke like mushrooms with stalks one thousand feet high. The Sierra hidden with smoke.

March 27.

Took long walks over the West Hills and far awa' with Helen. The two white oaks are not yet in full leaf. Some are just opening their buds, but most are well clad with soft, downy newborn leaves. . . .

Saw and caught a small gray lizard that was shedding its skin. The old skin gray, the new nearly black. It first sheds about the eyes, giving it a curious look.

May 8.

Walked over the hills with Wanda and Helen. How the wind did surge and hiss and rustle and shout in the rocks and trees and grass! How the tall grain waved, the billows sweeping onward in endless succession with racing enthusiasm, and how the wild oats danced and rippled and clapped their spikelets like happy hands in a passion of joy! . . .

5. *With the National Forestry Commission* [1]

Ashland, Oregon.
August 27, 1896.

Met Sargent and Abbott at Ashland, and we immediately set out for Crater Lake, we three and the driver. The grades were steep and our horses feeble — one spotted roan with the colic and nervous debility, the other grass-soft and balky — and the spring wagon shackly but tough. Abbott wanted to turn back ... but the team driver said it would soon be all right. Ash on the stream side, also alder and oak, the Kellogg and the white oak, with maple, grapevines, clematis, and glossy dark-green smilax climbing thirty feet up the alders. It was soon dark, and we saw the Douglas and yellow pines and the Murray pine in the starlight. Our astonished horses and driver ran point-blank against a clean-shafted *Pinus ponderosa.*... When we arrived at Hunt's we found them gone to bed, but we drove into a cow corral and I built a fire. The wife arose and good-naturedly gave us an eleven o'clock supper. 'I'm going to double you fellows up,' said she. Tough!

August 29.

Camped six miles north of Klamath on a pumice plain. Firewood was scarce; Sargent and I made a fire between two young contorta pines. Chat and Jersey mosquitoes.

Crater Lake.
August 30.

The lake walls of thirty to ninety degrees slope descended to the shore, where the slope averages thirty-five degrees.... Crater Island is a fine symmetrical volcano and comparatively recent.

The sky in the evening was clouded, but we started for the

---

[1] Members of the National Forest Commission were Charles S. Sargent, Director of the Arnold Arboretum; William H. Brewer of Yale; Arnold Hague of the United States Geological Survey; General Henry L. Abbott of the United States Engineer Corps; Alexander Agassiz, marine biologist, member of the United States Coast Survey; and Gifford Pinchot, practical forester.

island. Halfway over it began to thunder and whitecaps broke into our overloaded boat. We turned back to the shore at the nearest wooded point, and built a fire to dry our drenched clothing. Pinchot and I went a hundred feet up a ridge and made a fire on a flat rock. Arnold Hague and the boatman and Sargent stayed down on the shore. After the rain, it was too late for the island, so we rowed back to the foot of the trail and climbed up to camp; rather tired but none the worse — rather better for the exercise. . . . Heavy rain during the night. All slept in the tent except Pinchot.

August 31.

A wet morning, drizzly, large drops from the hemlocks overhead. Mr. Diller put his head in the tent and talked until we got up. Then we went out to the lake. It was still full of mist, the trees gradually vanishing in gloom, producing a weird effect. We had glimpses of the farther shore, the rim laden with glacial detritus. Started off in the cold drizzle. . . . Found fire desolation nearly everywhere. . . .

On the way to Grant's Pass.
September 1.

We stopped at Gordon's after driving about forty-five miles from the lake — a good woodland ranch on the broad, fine clear Rogue River which we followed nearly all day. Ranches here are made by girdling the yellow and sugar pines where the land is good and irrigation easy. Gordon is a good-looking gray Scotchman, sixty-odd years old. His father is still running a ranch at ninety years. I slept in an alfalfa mow with Abbott and Pinchot; Hague and Sargent stayed in the house, which is a rough camp like most others. . . .

September 4.

Cool morning, calm, delightful in this spacious Yosemite-like spot in the mountains five hundred feet above the sea. A glorious drive twenty-five miles to Crescent City, California, through

Douglas, grandis, and Lawson pines,[1] Thuja, arbutus, sequoia, and Myrica. As we neared the coast, sequoia suddenly vanished and *Picea sitchensis* came in its stead. ... The sudden appearance of sunlight ahead showed the line of demarcation between the sitchensis and sequoia.

A broad sandy plain from the mountains to the sea with woeful charred ruins of a forest — mostly Sitka spruce. ...

Arrived at Crescent City. Examined the mill, and went out on a logging train a few miles and saw the work of ruin going on. It takes three-quarters of a day for two men to cut down a tree eight feet in diameter. Numerous trees here are ten feet in diameter, a good many are fifteen, and a very few twenty. I measured a fallen giant twenty feet at the base, ten feet at two hundred feet, the bark and sap gone.

September 5.

We had a grand drive through a redwood forest four miles along the sand. Being close to the waves, it was smooth as a pavement. Heavy fog. Then up through the grand sequoia. ...
On top of the ridge, back from the sea thirteen hundred feet, the heavens opened lake-like bits of lovely blue through the diaphanous mist. All about us towered the wondrous columns of redwood feathered with saplings, and maple and hemlock, the ground covered with fine rhododendron and red huckleberry. Then the sun, sending long shafts of radiance down through the columns, fell in luminous patches on the boles and flat plumes of hemlock and on the green mossy, ferny ground, and the huge trunks prostrate.

September 6.

Drove to Arcata to take a steamer for Eureka. Had another grand drive through sequoias with a few Douglas and grandis and Lawson pines. Lunched at Trinidad, a dull dead lumber camp, the bay spacious and picturesque, very rocky, the rocks rising in jagged peaks, waves breaking finely over them. The

---

[1] Douglas spruce, white fir, and Lawson cypress. — F. H. A.

wharf is under a cliff.... The redwood has all been cut hereabouts, and is a desolate rugged expanse of black stumps, some
few growing again. We went to a dismal hotel, but could get no
lunch until the regular hour, and as we were in haste to catch
the steamer for Eureka, we lunched on crackers and cheese
bought at a store. Just out of Trinidad the road skirts the forest
along the face of bluffs, the breakers roaring below, and in a little
cove we saw Indians camped, busily engaged in drying smelt.
The rocks and sand were covered with large patches of them,
and the Indians seemed happy and well fed. Gradually we
climbed into the woods, but saw little good sequoia — mostly
Picea sitchensis — and passed many small ranches where the tall
spruces were girdled and the branches cut off about a foot
from the trunk, making curious ruins....

September 7.

From 'Pepper Flat' we took stage for Ukiah, and had a wild
drive through woods along precipices in the dark; changed horses
at midnight. One horse was 'green,' had been hitched up only
twice before, although nine years old. Arrived at Dyerville at
2.30 A.M., and went to bed for three hours' sleep....

September 8.

Ride thirty miles in a buggy with the Scotia saloon-keeper to
Harris, where we are to take stage again. Dry and drier are the
ridges as we approach and pass the county boundary between
Humboldt and Mendocino. Charming oaks, chestnut, and grand
picturesque groves of California *chrysolepis*[1] on top of Rattlesnake Hill. Near the foot of the hill at Bell Springs is a fine
woods of California oak, one of the finest growths I ever saw,
though now much hacked and the ground swept by sheep....
One horse lame and falling. We changed horses three times during the night.... Weird effect of lamplight on the bushes and
trees lining and overhanging the road, oak *Garryana* perhaps the
most beautiful of all. Young sequoia beautiful also. Most of the

[1] *Quercus chrysolepis*, live-oak.

way from twenty-five miles southeast of Scotia to Ukiah is on top of mountain waves or around the sides, chiefly bare of trees and swept with sheep; fine oaks here and there.

September 9.

Rode all night, horribly sleepy, but enjoyed the sudden illumination of the bushes by the wayside — of the Ceanothus, manzanita, oaks, and tassels of yellow pine ... but the oaks all the way to Ukiah were the glory of the vegetation — *agrifolia*, *chrysolepis, Californica, lobata,* and perhaps *Garryana.*[1] ...

Arrived at Willit's about six o'clock in the morning, giddy and tired out with thumping. There was no side to my seat, and I was wrapped in mail-bags — curious neckties. Had breakfast and started for Ukiah; thence to San Francisco on the train.

September 10.

Went home on the nine o'clock train. Saw my wife and babies, changed clothing, repacked satchel ready to rejoin party next day.

September 11.

Start for Santa Lucia Mountains to see the *Abies bracteata.* Arrive at King's City and are fitted out with a team and two-seated carriage for Newhall's ranch, General Abbott and I in the livery carriage, Sargent with Mr. Newhall in his private buggy.

September 12.

Started out in a spring wagon to see the San Miguelito Ranch of thirty-five thousand acres and had a delightful ride through oak parks and Sabine pines. Remarkable rocks, Yosemite-like, on the Nacimiento River. The San Miguelito Ranch is mostly in wide valleys north and south, is surrounded with low smooth hills, fringed with *Pinus sabiniana* in fine effect on the skyline, and airy, feathery Douglas, lobata, and agrifolia

[1] Unquestionably *Quercus garryana.* — W. L. J.

oaks on the sides, with Manzanita, Rhamnus, Ceanothus, Adenostoma, sycamore, willow, and *Populus Fremontii* and fine oak parks on a grand scale. . . .

September 13.

We start well mounted for the *Abies bracteata* groves up the mountains at the head of the San Miguel Canyon, about nine miles distant from the ranch-house. At about eighteen hundred feet we come to a few vigorous *Pinus tuberculata* and at twenty-five hundred feet we find our first *Pinus Coulteri*, a rough, shaggy, knotty, sprawly-limbed tree with huge cones. At the head of the Miguel Canyon it is the predominating species, a large sturdy rough tree two to four feet high, sending out the wildest, roughest, longest arms imaginable, turning up at the ends with bristly brush, a foot or more long, of long stiff needles, and with one or a pair of the heaviest of all cones about ten inches long and dripping with amber resin — handsome, somewhat like the shaggiest of the Jeffrey pine, but gray like the Sabine pine, some limbs forty feet long bending nearly to the ground, similar to the sugar pine in the throwing of its limbs, but always turned up at the tip instead of down as in the sugar pine. 'A brute of a tree, a rough navvy Irishman,' says Sargent.

The *bracteata* is the roughest of all the Abies, most like the *Abies subalpina*, irregular in length, silvery below, having cones somewhat like the Douglas spruce, with long bristly curved bracts one inch long or more, the cones being two and a half to three inches long. They all grow near water, and are evidently doomed to speedy extinction in the wild state. They are now growing in Europe.

September 15.

Returned to San Francisco, and Sargent and I set out for Tomales Point. From San Rafael we took a team and surrey for the country club near Point Reyes and arrive at dusk. Before breakfast we discover the muricata pine we were in search of. It is a broad upper-spreading tree, with branches in wide flattish layers in foliage bluish green. The cones stay on forever. . . .

San Bernardino Reserve.
September 22.

From Los Angeles we came to San Bernardino and were welcomed by Colonel Wood, who showed us every attention, gave
us good rooms in his cedar-lined cottage, and showed maps and
statistics of rainfall and drainage basins. The Colonel was the
prime mover in the establishment of this San Bernardino Reserve.
He is alive to the value of forest protection; no fear of this reserve
being destroyed.

September 23.

Colonel Wood drove us up to a mountain-top where we had a
fine view of the valley of San Bernardino, and also of Mojave
Desert, the long black slopes leading out into it covered with
Adenostoma and perhaps juniper — extremely picturesque. The
yellow and sugar pines seem to have been pressed down from
above with snow and ice; therefore many are crown-topped like a
palm, and some have limbs only on one side, as if exposed to a
blast.

Arizona, en route to the
Grand Canyon of the Colorado.
September 28, 1896.

Set out from Flagstaff for the Grand Canyon. Much to my
surprise, we had an easy and delightful ride of about twenty hours
through a beautiful forest, mostly yellow and flexilis pines, with
the Douglas spruce and some of the *Abies concolor.* . . .

The Grand Canyon of the Colorado.

At 6.15 P.M. I ran up to the verge of the canyon and had my
first memorable and overwhelming view in the light and shade
of the setting sun. It is the most tremendous expression of
erosion and the most ornate and complicated I ever saw. Man
seeks the finest marble for sculptures; Nature takes cinders,
ashes, sediments, and makes all divine in fineness of beauty —

turrets, towers, pyramids, battlemented castles, rising in glowing beauty from the depths of this canyon of canyons noiselessly hewn from the smooth mass of the featureless plateau.

September 29.

We all set out for views along the brink of the canyon through the queer extensive forest of nut pine and cedar. Pinchot and I afoot traced the rim and enjoyed endless changing views; standing with our heads down brought out the colors — reds, grays, ashy greens of varied limestones and sandstones, lavender, and tones nameless and numberless. The light and shade of clouds fell on the wondrous city of structures and side streets, and storms swept softly out over the rim, gently dimming here and there the separate buildings with the silken brush of rain. . . .

An hour before sunset we chose a fine camp among the little pines and cedars, collected wood — cedar as incense to the gods — and camped for the night. The glorious fires brought out the beauty of the grass and sage tufts, glorifying even these, and transforming the little gray nut pines and the cedars with their many gray branches and stems to a divine beauty. A charming sunset. We chatted until midnight, then dozed on juniper leaves, rising to replenish the fire. At four-thirty in the morning we started for the hotel, where we washed and waited for breakfast.

September 30.

We went to the bottom of the canyon down the Hance Trail. Passed mesquites and Indian huts. The river flowed strong, broad, brown, and mud-laden. Much Clematis on the way, and charming Abronia near the foot of the trail. Had Mr. and Mrs. Hague and an Italian naval officer for companions. Sargent and Pinchot went to Flagstaff today. We start tomorrow.

1897.

In Retrospect. One day last summer, when Professor Sargent, General Abbott, and myself were driving through the great redwood forests of the Pacific Coast, we came suddenly on a scene of most desperate and startling disorder that seemed strangely out of place in the solemn woods. At the foot of a shallow dell where a little stream crossed the road we discovered three Indians, a white man, and a bay horse, all in wild commotion, especially the horse. He was a fine-looking, well-bred animal, but laboring under furious excitement, snorting, shrieking, groaning, standing up on his hind legs and beating the air, then dropping on his knees as if trying to stand on his head, butting his owner like a goat, butting the bank on one side of the road and the gnarled base of a giant tree on the other. His eyes had a terrible stare, while he groaned and foamed and froth flew from his distended nostrils.

The Indians scrambled up the bank into the bushes out of danger. We stopped a little way back. The white man holding to the halter was dodging and dancing to keep from being crushed, was swearing in despair. We all knew something about horses, but none before had seen anything to match this. Professor Sargent was a great rider. General Abbott had led cavalry through the Western wilderness in wartime, and I used to be considered a good trainer of wild mustangs, but we were all puzzled and had no idea what could be the matter with the animal. He seemed to be torn and tossed with devils. The fit, it seemed, had come on when he was quietly drinking at the roadside. At length his master, now at his wits' end, noticing that he kept striking and rubbing his head, at length discovered a yellow jacket in his ear, which thoroughly cleared up the mystery.

6. *Rambling Through the Southern States* [1]

The Adams House, Boston, Mass.
September 17, 1898.

Sargent called for me and took me to Wood's Hole on Cape Cod. Fine woods all the way. Many fine residences of rich

---

[1] John Muir's companions on portions of his journey through the South were Charles S. Sargent, Director of the Arnold Arboretum, and William H. Canby, botanist.

people seeking summer coolness; magnificent asters and golden-rods, deep glacial bays, heavy drift, the extreme end of the Cape sandy. Met Mrs. Sargent and the fine girls and manly boys, just getting ready for Harvard. . . .

September 19.

Spend the day at Sargent's, sleeping, sauntering, reading. Met Walter H. Page;[1] had good chat.

North Carolina. *September* 24.

On Roan High Bluff, over sixty-two hundred feet above sea-level. Gray granite rock joints weathering, lovely slopes feathered with coloring trees descending in fine lines. . . . A marvelous view of mountains, fold beyond fold, ridge beyond ridge. The Big Smoky mountains all timbered save a few bald heads; these grassy and covered by tall *Rhododendron catawbiense*. We came up here by surrey from Cranberry, eighteen miles, through a glorious mountain forest; streams singing everywhere, cornfields among the rocks, weeds and girdled trees and fallen trunks and stumps on steepest hillsides, and good corn. Easy to get a living here, too easy; yet one sees miserable overworked young mothers, and pale barefooted girls. . . .

Huntsville, Tennessee
October 7.

The place is overrun with soldiers, many of them drunk, a hundred and forty or so allowed out of camp every day; the streets filthy in the business part; many of the beautiful old ante-bellum residences with noble trees, especially white oak, maple, and magnolia; the finest of each of these I ever saw. . . . Boxwood walks, statuary in neglected gardens, with weeds up to my shoulders. . . . Soldiers at games of chance getting fleeced easily, infinite beastly drinking.

[1] Editor of the *Atlantic Monthly*; during the World War, United States Ambassador to Great Britain. — F. H. A.

At Canby Home,
Wilmington, Delaware.
October 12.

Arrived at Wilmington a little after dark, and drove to the fine neat home of Canby and had a good supper. Went to the Canby herbarium and then home to bed.

Had a fine sleep. Miss Canby is a good housekeeper — refreshing to see a clean town and clean home after the squalor of the South. . . . Saw a fine aboriginal forest along the estuary of the river — a wild place so near civilization, wild on account of the sandy sterility of the soil, where people live by scratching out meagre crops from sand, and by 'crabbin', froggin', and fishin'.' . . . We got tired and hired a horse and wagon from a farmer; searched in vain for Vaccinium and made haste back to the station for the train, to discover that the train ran only every other day. . . .

Boston, Massachusetts
October 24.

Spent the day in Boston and Cambridge with Page and Mifflin.[1] Dined with Page, Bradford Torrey,[2] Allen and Gibbs.[3] In the afternoon Mr. Allen took me to The Riverside Press, a grand bookmaking establishment.

October 25.

At 3 P.M. went with Page to his home in Cambridge. Mrs. Page took me to call on Mrs. Asa Gray; had a pleasant time talking over old times and a later pleasant evening with the Pages. The streets are in fine color, now at their best, every street in mellow light, an avenue lined with elm and maples; Mrs. Gray's residence with a botanic garden about it.

---

[1] George H. Mifflin, then head of the publishing house of Houghton, Mifflin & Co.

[2] Nature essayist (1843–1912), whose books cover observations extending from New England to Florida and west to California. — F. H. A.

[3] Francis H. Allen and Herbert R. Gibbs, members of the editorial staff of Houghton, Mifflin & Co., and the *Atlantic Monthly.* — F. H. A.

New York. *October 26.*

Set out at noon for New York with Mr. Sargent, his daughter Mollie, who is to sail tomorrow for Europe,... and Mr. Stratton.

October 27.

After lunch, drove through Central Park with Sargent and Stratton, a fine wilderness for a town, heartily appreciated by rich and poor. I notice the curious monkey-like gravity of expression worn by drivers and footmen of the rich. Two great stiff, formal, brilliantly buttoned monkey men and two docked massive horses to one little weakly woman or girl rolling in a big capacious carriage. Squirrels, gray, very numerous and tame, come close... frisking, eating nuts, playing, fed by the passers — charming effect....

October 28.

R. U. Johnson [1] is well and as funny as ever. He marched me through the wildest, maddest parts of the town last night, pretending he was taking me to jail for vagrancy — stopping now and then to ask little ragged boys the way to the police station, took me out into the middle of the streets among the whirl of cars and pretended he was afraid he would be run over, to frighten me — showed me the moon and minutely told me how to know it among the million electric lights, etc.; and the way he fooled with Tesla was too funny for anything.

Tyringham, Mass. *October 29.*

Breakfast with Sargent and go to Fourbrook Farm with Gilder [2] by the 9 A.M. train. Pass through charming scenery along the Housatonic River in the famous Berkshire Hills; the colors of the foliage fading, yet grand in the mass — brown, red with a good deal of yellow; all the hills feathered to the top.

[1] Quoted from a letter to Wanda Muir, October 28, 1898. Robert Underwood Johnson was associate editor of the *Century Magazine* and co-worker with John Muir in the National Park and Forest Reserve movement.

[2] Richard Watson Gilder, editor of the *Century Magazine*.

The meadows and lowlands are boulder-strewn, and the few openings have been laboriously cleared of stones that have been made into fences. The grass is close and lush even among the stones. Find Mrs. Gilder waiting for us at Lee, and drive with her four miles to Fourbrook Farm. As we step into the big house, an old-fashioned frame farm cottage, Mrs. Gilder turns, shakes hands with me, and with a kind smile says, 'Let me welcome you.' ... I am given a nice restful old-fashioned room, sweet and fresh.... The two little girls make a fine love-stir in the home.

After lunch Gilder and I take a stroll along the stream, full of trout; swift, curving, surging from side to side with many small rapids between banks where grow birch, alder, and huge willows in clumps. We walk through a fine meadow with a wonderful richness of grass that will produce more wool and mutton, milk and beef than almost any other equal area I know of; yet the country is full of abandoned farms.... The inhabitants are going to towns to work in factories and stores, seeking fortunes — they know not how. They look down on labor. Health, manhood are all given away for the sake of something beyond their reach, and which even if attained would be found far less desirable by any sane person than the home farm life they despise....

October 30.

I start out with Gilder on another long walk. First we go back of the house up the valley, and walk to an extensive moutonnéed mass of gray lichened gneiss or granite rock, with acres of Kalmia, two to six feet high; and a few tufts of grass or a tree — oak, birch, pine, maple, chestnut — here and there. Views up and down the main valley are very fine; many small streams and small bogs over all the hills. The hill-trends in general show that the ice-sheet flowed southeasterly; the sheer broken cliff-faces on the moutonnéed rocks tell the same, and on the quartz veins weathered out to a height of two or three inches a few spots are polished and striated, and these striæ also indicate the southeasterly flow of this portion of the bottom of the glacial sheet. The loftiest moutonnéed hill mass hereabouts, that stands out in the middle of the valley five or six hundred feet high, tells the same story....

October 31.

The girls start off in a comical little cart drawn by a comical little donkey, with a handful of green backs to buy cats this morning. I visit a rake factory; I could make rakes at half the cost of those made here.... After lunch we all start for New York. In the evening we go to Johnson's for dinner and meet here Hobson,[1] a quiet, keen, handsome fellow with firmness and daring shown in his face. He is writing his story at Johnson's house.

Spoke with Choate [2] an hour or two on the way up.

Wing-and-Wing on the Hudson
November 1.

Spent half a day in offices of the *Century*. Went to lunch with Johnson and all the Century Company officers at the Players. Met Professor Osborn [3] and went with him to Wing-and-Wing. Received hearty welcome by Mrs. Osborn.

November 2.

A bright, crisp day, slightly hazy, the first of Indian summer. Fine tulip-trees and oaks, white, black, and scarlet, and maples — a fine, rich row along the road where I walked in the morning, yellow-green tinged with red, fine in tone after the rain. Hickory leaves brown, withered, falling fast over the lawn, sailing, glinting, some fast, others slow, meeting in the air, parting, alighting far apart or together. Colors mostly fading but many maples in glory, some solid red, also oaks. Red and gray squirrels abundant,

[1] Richmond Pearson Hobson, hero of the Spanish-American War.

[2] In a letter to his daughter Wanda, John Muir said: 'On the way here, on the car, I was introduced to Joseph Choate, the great lawyer, and on Sunday Mr. Gilder and I drove over to his fine residence at Stockbridge to dinner, and I had a long talk with him about forests as well as glaciers.' See Badè's *Life and Letters of John Muir*, vol. II, p. 312.

[3] Professor Henry Fairfield Osborn, American paleontologist.

the latter shorter than California gray, heavier; the red smaller, slenderer than Douglas; both glide along the leafy highways from tree to tree, up, down, jumping across from limb to limb with wondrous speed and ease, unhesitating.... Charming views of the Hudson bends from Osborn heights. Glacial lines on the hills and scoring clear — all lines glacial.... I enjoy a clean, fluffy, light bed,... charming dainty embroidery on the bureau cover, and coffee in bed.

November 3.

I walked with Mrs. Osborn, selecting trees for transplanting.... A merry fairy girl is Josephine — bewitching, witty, beautiful. Perry, a manly boy fourteen years old; his brother a hopeful naturalist who has many pets, ring doves, rats, etc. Virginia is at boarding-school....

November 4.

We all go to West Point in a carriage to see a football game on the parade ground between the Princeton and West Point teams; result a tie, five to five — tremendous excitement....

November 5.

Yesterday Professor Osborn delivered a lecture on paleontology at the church. It was well illustrated with lantern slides. Several visitors, including an Episcopal clergyman, sadly, comfortably hidebound....

New York. *November 6.*

To New York with Professor Osborn.... Lunched at the Players. Talked with Tesla from nine to twelve-thirty.... He talked of his invention for blowing up ships.... He wants us to dine with him and spend an evening at the Waldorf — a swallow-tail affair, I fear....

November 7.

Election day; lunched with the Gilders, Johnson, Hobson, and

Mrs. Van Rensselaer;[1] the latter handsome and a good writer. After lunch, Mrs. Van Rensselaer took me to see New York from the top of the World Building and from a cab through Wall Street. Then to her home; chatted; told Stickeen story to her friend; was driven to Johnson's to dinner, then downtown to the election jam; was almost killed.

Wilmington, Delaware.
November 8.

Start for Wilmington, meet Sargent and Canby,[2] go to hotel for dinner, then drive in a carriage ... through a park along the Brandywine, through many fine wild oak woods — scarlet oak (*Quercus coccinea*) gloriously vivid, blending with the purple and bronze and brown of the white and black oaks, and the yellow of maples. The red oak (*Quercus rubra*) leaves are always yellow and brown — only the wood is red. Had magnificent views of Chesapeake Bay. Canby took great pleasure in showing off his fine Delaware....

November 9.

Saw the old Johns Hopkins house and fine meadows with the noble Taxodium like the Sequoia. Here also oak is king, the glory of the park and woods in general.

Washington, D.C.
November 10.

We drive in a carriage to the Soldiers' Home in a very cold wind. Heavy overcoat and robes required. Good smooth streets, charming oak woods, scarlet oak vivid red hot — every leaf. We

---

[1] Mariana Griswold Van Rensselaer (Mrs. Stephen Van Rensselaer), art critic and author.

[2] In a letter to C. S. Sargent of October 28, 1897, John Muir described Canby as 'bright, time-defying, plant-loving, plant-plucking Canby. I enjoyed his companionship ever so much,' he said. 'The way he compelled the startled plant people to tell their names and pushed them into their places is a warning and a judgment and an inspiration to the lazy — a capital, kicking, indomitable, untirable fellow, with a knowing eye for everything from Cuscuta up and down. May his tribe increase!'

drive to a park which is still in the rough, a good stream, a branch of the Potomac, running in wild ravines through it. The Zoo has many animals, good specimens of buffalo, elk, kangaroo, llama, mountain sheep, deer, antelope, coons in forks of dead trees in pairs, keeping each other warm; many birds — pelican, swan, geese, etc. The beavers have built dams. In the afternoon we drive to the Capitol. Fine grounds, the ginkgo a peculiar pale, pure, pleasant tulip yellow; acres of marble. The Library mighty gaudy in fresco but tomblike, sepulchral in blue, vivid marble outside and in, overdecorated.

St. Augustine, Florida.
November 11.

Open country, curiously level.... Liatris, goldenrod, and a few daisy-like Compositæ among the general growth of dwarf palmetto; young long-leaf pine, electric, each tassel standing out in a rich cluster, light between the needles except in the dark center.... We drove around town until dark, saw the ancient city gates, the old castle, a fine cathedral, and huge hotels. We stopped at the Alcazar, of Moorish architecture, of coral concrete, with cool arched colonnade cloisters, a central square with fountains, electric lights — a strange, bewitching effect; the whole evidently the work of a good architect.... Why Moorish architecture instead of American?

Miami. *November* 13.

Began botanizing in a vacant lot before breakfast; many fine plants in flower; saw a beautiful blue Clitoria, with a fine thread stem, twining around smilax. Drove through a dark forest to Cocoanut Grove, the tree limbs laden with Tillandsia, both the long-thread kind and the pineapple kind; the live-oak is especially laden on its long arms. Saw live-oaks small with grayish foliage; the Ficus, or banyan tree, with its strange mass of roots grasping, choking other trees, sending down innumerable roots mixed with fine grapevines.... The pawpaw is a queer muggins, with its palmlike crown of large-lobed figlike leaves and yellow spangling racemes of flowers and green fruit three or four inches long by two inches in diameter, yellow when ripe, drooped close about the stem beneath the leaves and flowers. The jumbo-limbo is a rather ugly irregular-trunked tree with a mean dingy-red bark, a species of Rhus with handsome glossy leaves, abundant in these woods and very poisonous.

Called at Kirk Monroe's place near Cocoanut Grove; he has been here ten years, sometimes staying all summer; lives on Mellin's Food and writes two or three books a year; has a nice palmy home, a fine spring, bananas, oranges, lemons, and has a charming wife, Scotch, born in Glasgow. She made eggnog and gave me the first glass as the talker....

> Going down the
> Coast to Key West.
> *November* 14.

A low green coast, round-headed trees, no pines or oaks, mostly jumbo-limbo, Ficus, Sideroxylon, and mangrove. The colors of the water are varied and charming — pale Nemophila blue and lovely green in ever-changing cloud-like patches, some purple, some black, with glistening brightness on the sides of small waves; no heavy swells, those being kept out by a barrier reef low under water marked by a row of lighthouses. The green and pale blue of water are from particles of coral — the whole land being the work of coral. Fine cumulus clouds taking the place of mountains over the low green water and land.... Arrived Key West.

> Key West. *November* 16.

Drove around town, a miserable place of squalid shanties, mostly with sickly cats, dogs, and pale, weak-looking Cuban cigar-makers. Saw a wonderful Ficus, or banyan tree, with ten feet or more of compound trunk from descending roots, bracing, reinforcing the most complicated trunk imaginable, with leaves a shiny bright green and thrifty-looking. No wind can overthrow such a tree.... I went aboard the steamer *Miami* and sailed about 1 A.M.; the negroes singing as they wheeled away the freight, a

merry set barefooted in the rain on the sloppy deck, kept us from
sleeping.

November 17.

Sailing on the deep blue waters of the Gulf outside of all the
keys and shoals, the ship *Miami* is rolling, but few if any are
sick. A fine series of lighthouses, said to be well kept, all along
the coast.

Palatka. *November* 19.

Went to a show where a lady sang, played guitar, and danced.
Walked along the beautiful river-banks, where grow the largest
hickory trees and thriftiest I ever saw; the leaves, dark green,
are said to stay on till January. They are almost evergreen and
have widespreading branches — a noble tree. The largest
Magnolia grandiflora stands a hundred feet high and is four or
five feet in diameter. It has much character in port and mode of
branching. The sunshine glances and glows on the glossy dark
leaves, throwing off spangles.... I started for Cedar Keys, got
on the train, then waited over an hour for the arrival of another
train. A crowd of negroes in Sunday dress, attracted by some
prisoners on the train, stood about the cars — boys, girls, and
comic babies. Close by was a grove of pines, Cuban or long-leaf
or both, and I wished to get cones. The ground was wet and
swampy, so I offered a negro barefoot lad a nickel if he would
bring me a cone. Instantly the bog was as full of boys as of
frogs, pushing, plashing, grabbing cones. The crowd, wondering
what the excitement was about, turned their attention away
from the train. 'W'at's the mattah, w'at's in dem crazy boys
a-hoppin' like frogs in the ma'sh?' etc. I offered another nickel
for a green cone from the top of a tree sixty or seventy feet high,
and there began a comic scramble in which climbing power was
tested. Two climbed opposite sides of the same tree, and one,
feeling in danger of falling behind, locked his legs around those of
his rival and tried who could hold on longest. Another got tired
twenty feet up or so and had to descend, resting at times, leaning

back and looking up 'like a woodpecker,' as I heard some of his companions say. It was a merry show and illustrated how money made the boys go, cheered by onlookers who roared with laughter. 'Go it Sammy!' — 'Go it, Abram!' — or 'Wash, see dem boys, clean daug-gone crazy,' etc., etc.

Archer, Florida.
November 20.

Inquired about the Hodgsons [1] at Cedar Keys. Heard Mr. Hodgson and the eldest son were dead, the others were at Archer. Stopped off there and found Mrs. Hodgson in her garden. She did not know me, asked my name, 'Muir, John,' I said. Then, 'The California John Muir?' she almost screamed. Then she introduced me to her friends, telling that thirty-odd years ago I was the finest, handsomest young man one could hope to see, etc. We talked old times, glad to hear that in all the long icy and sunny years intervening we had not forgotten; only my sickness had been forgotten.... She said she knew the Muir Glacier must have been named for me.

November 22.

Much of the way up from Archer to Live-Oak the forests are utterly ruined by fire; black ghastly stumps and poles of every length cover the sandy ground with charred ruins.... Many factories or mines of phosphate are hereabouts, and I saw one kaline factory. Towards the beautiful Suwanee River the country rapidly improves. Where the railroad crosses it is a beautiful stream with fine leafy, ferny, viny banks about a hundred yards wide; and pines, magnolia, cypress, oak, laurel or bay tree, Nyssa, and the Liquidambar on the banks or near....

Mobile, Alabama.
November 23.

Found Dr. Mohr,[2] and went out riding along Mobile Bay,

---

[1] John Muir, while in Cedar Keys in 1867, worked in Mr. Hodgson's sawmill. Mrs. Hodgson nursed him through a well-nigh fatal attack of malarial fever.

[2] Dr. Charles T. Mohr, Botanist of the Forestry Division of the United States Department of Agriculture.

through a fine forest of magnolia trees, eighty to a hundred feet high, two to nearly four feet in diameter — grand old giants, and the tupelo (*Nyssa multiflora*) about the same size, prettily colored; oak and laurel, and a very fine live-oak.

November 24.

Dr. Mohr took me out four or five miles on car to a low hill; saw fine old woods, mostly planted; live-oak, hickory, and a juniper forty feet high, thirty-nine feet wide, the trunk eight feet in diameter. It is a graceful, feathery, airy tree, like a tent under its drooping boughs, with a long hopeful shoot at the top of light yellow....

New Orleans, Louisiana.
November 25.

Bought a ticket home; then went for three or four hours to Dr. J. H. Mellichamp,[1] a charming old South Carolinian. He had read my book and is a good botanist....

Texas. *November 27.*

Nearly all western Texas is one garden of yucca lilies, grass, Compositæ, and sage that must be fine sight in flower.... Going to bed last night I saw a fine snowy mountain-range to the south, strangely fascinating, the only mountains with snow on the trip. Many striking residual masses at the summit and this side of it. A noble view; I felt like jumping out of the window to go to them.

Southeastern California.
November 29.

Remarkably broad gate-like San Gorgonio Pass, much washing at the eastern end. A mountain-range on each side with ragged spurs. We went through by moonlight. The Indio and Palm stations are handsome with large filifera palms[2]; some at Indio are nearly three feet in diameter and twenty feet high, though young

---

[1] Botanist, resident of Bluffton, S.C. — W. L. J.
[2] *Washingtonia filifera.* — W. L. J.

and full of fruit. A very wide basin full of fine gray sediments, evidently slowly accumulated from the surrounding picturesque crumbling barren mountains. No trees in basin save those planted, scarce a sage or a salt-bush seen at long intervals. . . . Glad to see the trees on the Tehachapi Mountains.

Chapter IX : 1899

Silent, inaudible, invisible flow. The very mountains
flowing to the sea. The great heart of the hills sending its life down in streams. Mountains die that we
may live.

AN OUTSTANDING event of 1899 in American scientific circles was
the Harriman Expedition to Alaska. Financed by Edward H.
Harriman, the railroad magnate, the party that cruised north
in the steamship *George W. Elder*, under command of Captain
Peter A. Doran, included Mr. Harriman and his family, a group
of his friends who wanted to hunt big game, a staff of artists and
photographers, and twenty-three eminent scientists. John Muir
was a member of the party.

June and July were delightfully and profitably spent cruising
along the coast and among the fiords, bays, and islands as far
north as Plover Bay, Siberia, and Port Clarence on the American
coast, gathering valuable data as to flora, fauna, climate, geology,
and glaciers.

The presence of women and children on shipboard contributed
to a pleasant social life, and the distinguished scientists themselves seem to have had their hours of jolly relaxation. John
Muir and John Burroughs, friends of many years, added much to
the amusement of their fellow passengers by their usual exchange
of whimsical, growling banter. In this instance, Mr. Muir appears
to have had the last word in an hilarious doggerel poem written
after his return home, but characteristic of the impromptu
jingles composed and recited on shipboard.

-    -   -   -   -

I. *Cruising with the Harriman-Alaska Expedition,* 1899

On the train,
en route to Portland.
May 26.

Left San Francisco at 7 P.M.... Fortunate in having Charles Keeler,[1] a charming companion....

May 27.

Along the way are myriads of small oaks, blue and lobata. Going up the canyon of the Sacramento, we see the giant saxifrage in flower in purple fringes on the stream-sides. The most showy and beautiful bloom is that of the blue Ceanothus, in large beds and belts — a lovely bright tone of blue.... All go out to drink soda. People look at what they are told to look at or at things that have been named. Nameless things, however fine, go unnoticed. Shasta is cloud-capped. The prostrate Ceanothus is in flower at the base of Shasta and a beautiful Phlox in beds six inches to a foot or two in the sandy, gravelly ground in the open woods.

Portland, Oregon.
May 28.

Down the Willamette; rainy, the whole country sloppy, the grain rather short. Hops spindling, the buildings slight and shabby, bread-getting too easy for thrift. We go to the Portland Hotel to wait for the Harriman train....

May 29.

Met Judge George. Had a long talk on forest protection, found him lukewarm. Mr. Steel uncertain on the same subject. Told him forest protection was the right side and he had better get on record on that side as soon as possible. He promised to do what he could against sheep pasture in the Rainier Park and also

---

[1] Charles A. Keeler, California poet and ornithologist, and Director of the Museum of the California Academy of Sciences, San Francisco, California. Mr. Keeler was Muir's cabin-mate throughout the cruise.

in the Cascade Reservation. Met Hawkins, fat and easy, who said he did not like to fight like Quixote on the sheep question or any other. . . .

May 30.

The party arrived at noon. The Mazamas[1] went to the station to welcome them; Keeler and I also, and met many of the club. John Burroughs grabbed me in the crowd. 'This is John Muir,' etc. All rode up to the Portland Hotel. Saw Brewer[2] on the car. A hearty salute and introduction to Mrs. Harriman. We started at 3 P.M. on the steamer for Kalama. The river very full, the banks submerged, a dense growth of maple on either side. Got aboard the fine special train and arrived at Seattle about midnight. Everything fine. . . .

Seattle. *May 31.*

Sauntered about town most of the day with John Burroughs. Registered at the Rainier Grand, lunched there, and was interviewed by a newspaper man who wrote a queer notice of 'Two Famous Men.' Left on the *George W. Elder* at 5.45 P.M.

June 1.

At Victoria early in the morning. Stayed two hours. Saw fine specimens of *Sequoia gigantea.*[3] Wild roses just coming into bloom — the largest of wild roses, fragrant like the brier after rain. The sunset opposite Nanaimo glorious. To the east the water was a rose lavender, the sky at the horizon blue, eight or ten degrees above a red purple. In the west gold and purple on horizontal bars of cloud, shading off into lilac. Islands dark purple.

June 2.

We went ashore near the north end of Vancouver Island. Found magnificent woods — *Thuja gigantea, Cupressus nootka-*

---

[1] A mountain-climbing club.
[2] William H. Brewer, Sheffield Scientific School, Yale University, New Haven, Connecticut.
[3] Cultivated trees in the city park. — W. L. J.

tensis, Abies amabilis, Mertensia hemlock, and *Pinus monticola,* also the devil's-club (*Echinopanax horridum*), just opening its leaves. The Douglas spruce is abundant. One seven feet in diameter. A fine river at the head of the small bay.

June 3.

At Lowe Inlet on Princess Royal Island. Visited the cannery. Went ashore from the wharf and climbed in extremely rough woods. No Douglas spruce, but Thuja, *Cupressus nootkatensis,* and *Abies amabilis* abundant. Yew seen by Fernow.[1]

Annette Island. *June 4.*

In the morning we visited Duncan[2] at Metlakahtla. Heard his story of separation from the Episcopalian Church. He did not wish to follow closely or insist on ritual. Wished to recognize all Christians as one, quarreled with the bishop, appealed to the Canadian Premier at Ottawa, wrote a letter of explanation to England, was suspended, and all Church property was confiscated. The Indians were threatened with arrest if they took aught from the gardens of their own houses. The Indians had no rights save to that given in charity by Queen Victoria, so said the Judge at Victoria. They threatened war, decided to leave the country, got Annette Island from our Government on condition of swearing allegiance to the United States.

Found *Pinus contorta* dwarf trees in bright abundant bloom; also trees thirty feet high. Staminate flowers are of a brighter red-purple than in the Sierra, and the pistillate smaller but with brilliant jets of crimson three-sixteenths of an inch in diameter by one-fourth of an inch long, cheer the Alaska bogs in company with *Andromeda polifolia,* Coptis, the pure white *Rubus chamæmorus,* huckleberry, *Kalmia glauca,* and buckbean of two species — one found in bogs and pools, with fringed petals, the other with round leaves. Also *Menyanthes trifoliata. Pinus contorta* quite abundant

[1] Professor B. E. Fernow, of the School of Forestry, Cornell University, Ithaca, New York.

[2] William Duncan, a Scotch missionary who brought a group of Indians under his care on Annette Island to a high state of industrial civilization.

in boggy ground when not choked by Menzies spruce, Mertensia hemlock, *Thuja gigantea,* and *Cupressus nootkatensis,* the last two of which are here common. Andromeda is pale pink, with lovely bells, the huckleberry showy.

June 5.

Arrived at Wrangell last night. Got up at five this morning, went ashore, and walked in the outskirts of the woods.... *Picea sitchensis* in flower, yellow and red, ... one inch long, very handsome, the buds just opening, the brown caps of scales coming off, the needles yellow like the flowers, exceedingly rich, growing all around and turned up like *Abies magnifica.* I measured a log of Picea at the mill, one hundred feet long, and three feet, eight inches in diameter inside the bark, straight as an arrow, its age one hundred and sixty-eight years. *Linnæa borealis* is abundant at Wrangell. Also Streptopus and Menziesia. We had fine views of the mountains after passing through Wrangell Narrows. The Devil's Thumb is very striking, inaccessible, nine thousand feet high, the actual pyramid being sixteen hundred feet high. A fine view of Patterson and Baird Glaciers.

Went ashore an hour and a half this evening. Dense woods; deer trails and bear abundant. Found Listera, a delicate thing, and Menziesia abundant, and fine lupine, also the large blue hairy hummingbird [1] and eagles, crows, the Oregon junco, and a warbler.

June 6.

Arrived at Juneau about 9 A.M. Stayed a half-hour. Mailed letter home. Went to Douglas Island and visited the mines. In the stamp mills I could not hear myself shout. [2] At the mine

[1] This looks like one of John Muir's pleasantries. Can he have referred to the kingfisher, whose hovering may have suggested a hummingbird? The only hummingbird found in Alaska appears to be the rufous hummingbird (*Selasphorus rufus*), and that is neither large, nor blue, nor hairy! On the other hand, Mr. Keeler, in *Harriman-Alaska Expedition,* vol. II, p. 209, records a kingfisher at Wrangell. — F. H. A.

[2] The party visited the Treadwell Mines. John Burroughs in referring to the stamp-mill uproar remarked: 'It dwarfs all other rackets I ever heard. Niagara is a soft hum beside it.' See Burroughs's 'Narrative of the Expedition' in *The Harriman-Alaska Expedition,* vol. I, p. 29.

we saw many big blasts go off. A large concern — many small mills in one. The ore is shot down a shaft, loaded into cars, hauled to the mill, put through the crusher, and made into pieces of two to three inches in size. On the north side of the Juneau channel trees grow up to near three thousand feet. On the south — being in the shade — the snow lies long, preventing much tree growth above one thousand feet, and avalanches keep trees off their paths. This side is green with grass and alders, willow, and perhaps birch.

Sailing up Lynn Canal.

Had a fine view of the Auk, or Mendenhall, Glacier. No change apparent from ship since last I saw it. Eagle Glacier is very striking with lofty fountains, the heads of the mountains veiled in filmy clouds greatly enhancing the beauty and grandeur. Lynn Canal seen in clear or cloudy weather is always beautiful. As we sail along the glaciers in rows, canyon doors open and close. Davidson Glacier, at the head of the canal, is much shrunken and crevassed laterally — a grand lesson on glacial trunks.

The traps — ninety of them — were set last eve. Five white-footed brown mice and one toad were caught at Taku Indian village, where I camped nineteen years ago.

June 7.

A typical Alaska morning — cold and rainy. At 10 A.M. we went on a special train twenty miles up to the summit of White Pass, which is twenty-five hundred feet high, one tunnel going on to Fort Selkirk. *Abies balsamea, Pinus contorta* slim as the lodgepole pine and seventy-five feet high, cottonwood and birch nearly as high, also *Picea sitchensis* at the summit. *Tsuga patto-niana* beautifully dwarfed, also *Abies balsamea* smoothly thatching the steep slopes. Flowers of birch just opening. . . .

In the pass we saw water-polished surfaces of a few feet in area on the sheer wall by the stream on the edge of the glacier. It looked somewhat like glacial polish, but was down-curved instead of horizontal, and the striæ ran into each other. Had a good time with Young; [1] we talked over old times.

Glacier Bay. *June 8.*

We reached Glacier Bay about 5 P.M. A party at once started for Howling Valley [2] to hunt. Six packers went ahead, the five or six hunters an hour later. They walked too fast and got tired by 11 P.M. In the morning it rained, but they went ahead over the divide into one or two feet of snow, roped together.

June 9.

Howling Valley being snowy, all turned back, arriving at the ship in late afternoon. Meantime nearly everybody else went ashore over the moraine, and along the shore to the face of the glacier and up Red Knob for a view. On the glacier I led a party of three girls [3] to the smallest island three miles from the front. All were impressed with the vast extent of ice, the lovely blue grottoes, shimmering in the light, like fairyland. . . . The surface of the ice is dirty; the many medial moraines are now broad like a plowed field, now small like wagon-tracks, now rocky like a railroad embankment. . . .

A grand berg discharge. One large berg, being two hundred feet long or more, slid down like a whole bastion among the hundreds of small bergs making acres of floe. The large mass rose, with sublime dignity and deliberation, a hundred and fifty feet, the water like hair streaming from it, a wave twenty feet high combing and dashing up spray a hundred feet against the bergs, and roaring like the ocean. I hurried the party high up the shore — had not the wave been spent among the bergs, it would have risen very far up. The dancing, clashing, clapping,

[1] The Reverend Samuel Hall Young, companion of Muir on the Alaskan journeys of 1879 and 1890.

[2] Howling Valley, so named by John Muir, is about twenty miles back from the front of Muir Glacier.

[3] Four girls accompanied the Harriman Expedition, whom John Muir dubbed 'The Big Four.' They were Mary and Cornelia Harriman, Elizabeth Averell, and Dorothea Draper. See Muir's letter to 'The Big Four' in Badè's *Life and Letters of John Muir*, vol. II, pp. 329-32.

lapping of bergs, big and little, rocking, welcoming the newborn as it slid ahead at the rate of ten miles an hour after rising again and again, were sublime, glorious. All this motion and commotion after a hundred years of slow invisible action in sculpture work, free at last to speak out, cry aloud, and dance to its own thunder....

June 10.

One party landed at Muir Glacier early, a second party on moraine islands near the entrance of the bay, the third near Willoughby Island, and the fourth [1] was landed as high up the bay as possible from the ship, with a rowboat, to explore and examine the Pacific Glacier and others as to changes made in twenty years. A fifth party in a launch included those who simply wanted to see.

Canoe Trip. *June* 11.

A lovely morning. Reflections in the bay. Many ducks and gulls at the breeding place by the Hugh Miller Glacier, calling plaintively, and a plover stood on the shore while the tide was out, eating his dainty breakfast, nicely picked out of the mud. Willows are in flower and Dryas on the new land, and moss dark on the gray granite....

All the day was bright pure sunshine. Everybody sunburned. We saw bergs discharged from the Hugh Miller and Charpentier Glaciers, but they are comparatively feeble.... The outlet of the Hugh Miller near the mouth of Reid Inlet is very beautiful — a mile of mud-flat in front, with a cascading tributary on the west near the mouth. The surface of the main trunk is smooth and white. Fine view from here of the Carroll Glacier, much crevassed like the Muir, two or three miles in front, and broad spreading to the right and left. We were baffled by firmly packed ice in trying to cross to the east side of Reid Inlet, and had to

[1] John Muir headed the fourth party on their three-day exploration and camping trip. His companions were G. K. Gilbert of the United States Geological Survey and Dr. Charles Palache, mineralogist, of Harvard University. See *The Harriman-Alaska Expedition*, vol. I, p. 40.

return and camp at the mouth of the Inlet on the south side on a large moraine deposit. In scrambling down the moraine slope of round boulders and pebbles we started a mud avalanche which poured in a gray oily stream, bearing boulders down with it. . . . Charming willows, with silky leaves and red stamens, lovely red-pink bells, urns of magenta. Rubus, salmon-berry, and huckle-berry. Glorious views of Fairweather. Ptarmigan eggs found at Harriman Glacier. Also snow buntings.[1]

June 12.

We started for the Pacific Glacier up the right (south) side, along a fairly passable mountain, and discovered a new tidewater glacier. Immense melting. The Pacific Glacier has receded four miles, and there is now a large island, two miles long, the face of which only was open when I visited it twenty years ago. Three tidewater glaciers have been made out of one. Reid Inlet is not only three to four miles longer but double in width. The left portion of the Pacific Glacier is dirty and likely to die soon. . . . A yellow warbler sang near the Harriman Glacier.[2] Sunrise back of the spiky crest; the sunbeams took flame and sent off halo rays from each peaklet.

We saw bergs of immense size in the upper portion of the bay. I landed on one yesterday and paced it. It was seven hundred feet long, dirty, and looked like a glacier as we rounded a corner. Saw one or two others a thousand feet long. . . . We camped in a cove well sheltered and found two Indians, father and son, just carrying their precious canoe up beyond tide and bergs and berg waves, taking pains even after sundown on a cloudy eve to sprinkle well the stern and bow to protect it from cracking. The canoe was almost as light as birch bark. They offered us gull eggs obtained from the Hugh Miller rookery. Carefully testing

[1] These were the common snow bunting, not the hyperborean, or McKay's, snow bunting found later on Hall's Island (see page 410). — F. H. A.

[2] This Harriman Glacier, named by John Muir, is one of the three glaciers formerly composing the Grand Pacific Glacier, at the head of Glacier Bay. It is not to be confused with the Harriman Glacier later discovered and named, in Prince William Sound.

them in water, and rejecting those that floated, they boiled them hard and ate half a dozen each. Then they invited us to eat wild celery, which they peeled and ate after dipping in oil, brown like molasses. The petioles were hollow but crisp, and tasted well. The Indians also had two woodchucks. They got off two hours before us this [-next?] morning, unembarrassed with camp wealth of non-essentials....

June 13.

Cool, half clear, everybody sunburned among the icebergs yesterday. We got off at six-twenty, pulled down to Geikie Inlet, found we could not reach the head and be in time to intercept the steamer on her way up to Composite Island to meet us there. Turned back at eleven-thirty. A few minutes afterwards we saw the steamer pass the point of Muir Inlet, and had to chase her all the way to the Island, ten or fifteen miles. Gilbert and Palache, wild to get on the steamer, wigwagged from the top of bergs, fearing they would not be found, and thinking every eye on the ship was searching the berg-encumbered bay for us even three or four hours before the time set for meeting. At length, about 3.30 or 4 P.M. we were taken up by a launch six or eight miles from the steamer. All well. Meantime the other launch was seeking us in Reid Inlet. I lectured in the evening on the trip. The Harriman Glacier name was received with cheers. Mr. Harriman came to thank me for proposing the name.

Sitka. *June* 14.

Just entering Sitka Harbor. The day fine, seems warm after Glacier Bay. The fine woodsy scenery of Peril Straits was enjoyed after the icy splendor and rocky bareness of the bay. Rainy all day. I strolled up town, called on Mrs. Vanderbilt,[1] and visited curiosity shops. We entertained Governor Brady, Captain Goodrich, and Lieutenant Emmons and ladies — a dozen or so —

[1] John Muir, on his first journey to Wrangell, Alaska, in 1879, had found 'a real home' in the house of Mr. Vanderbilt, a merchant, the pleasure of his stay being enhanced by the presence of little Annie Vanderbilt, a 'doctor of divinity two years old.'

with champagne and a merry, chatty time. Had gramophone songs and speeches at table, then smoke and chat. Following that a lecture on the plants of Glacier Bay and fossil forests by Coville.[1] Emmons has a lovely orchid, Calypso, brought him by a Russian from a neighboring island. It is purple, one-flowered, one-leaved, and has a white bulb like Calypso.

June 15.

A lovely morning, then showery. We set off to visit the Hot Springs, twelve or fifteen miles from Sitka, in a cove among islands. Rich vegetation of hemlock, spruce, balsam, yellow cedar, salmon-berry in bloom, dwarf cornel, three vacciniums, a charming purple Dodecatheon. Sauntered about the edge of the dripping woods gathering purple Rubus flowers one inch in diameter, and white-flowered *Rubus chamæmorus*; dwarf cornel and huckleberry are now in flower, and the glossy leaves and charming pink flowers of cranberry, also sedges, seaweed, and tall rye. Magnificent boulders are capped with two or three feet of humus, and a rich growth of Empetrum and vacciniums and seedling hemlock spruce and cornel.

The keeper of the Hot Springs murdered a mother deer and threw her over the ridgepole of his shanty, then caught her pitiful baby fawn and tied it beneath its dead mother.

June 16.

A fine day; did some shopping, then went to the woods and bogs. The spruce is now in beautiful flower — the difference in size considerable. Some pistillate flowers are an inch and a half long, some green, some quite red. They stand upright, as if determined to do so, very hopeful and rich-looking, whole trees being covered with them. The cones are pendant when full grown — I suppose from weight. Found Coptis, *Rubus chamæmorus*, small purple Ribes, and strawberry in flower. Met a lot of people on the street from the Topeka, Professor Moses among

---

[1] Frederick V. Coville, Curator of the National Herbarium, and Botanist of the United States Department of Agriculture.

the rest, and talked with them. . . . Huckleberry, Kalmia, and a species of Cassiope in flower, also Streptopus and Dryas.

June 17.

Sketched in the forenoon and went to the woods in the afternoon. Found hermaphrodite flowers of *Picea sitchensis.* Sailed at 7.30 P.M. Many fear seasickness, but the ocean is about as calm as the harbor.

June 18.

Still calm as a harbor and inland waters, scarce any perceptible swell. Dry but cloudy down to about two or three thousand feet, with occasional glimpses up to four or five thousand. The many glaciers make a wonderful show, beginning with the Brady and ending for the day with the Yakutat. Some ten reach the level of the sea, but, though large and fed by fountains so high and lavishly snow-laden, not one sends off bergs, while the other flank of the range sends no fewer than nine berg-producing glaciers to the sea in Glacier Bay. The west flank should and probably does receive more snow than the east or northeast. No doubt one of the causes of greater waste by melting and evaporation is greater exposure to the sunshine. Another perhaps greater cause is the warm winds from the sea. The temperature of the water today a mile or two off shore was forty-eight degrees.

The glaciers are from one to about four miles wide. They melt in broad bands of dirt covered with moraine belts and hills, tree-clad mostly. Many of them are evidently still more or less underlaid by ice, and in the gradual melting through the centuries are losing their forests by slow sloughing, making round-topped hills, sharp-peaked. Only La Pérouse Glacier touches the sea, at high tide, and is on the north side white and bluffy, with blue spots where bergs and crumbled masses have fallen, the waves melting the ice taluses. This is the only glacier I know which illustrates the very first stage of decadence. The amount of moraine material is enormous. At the north side of La Pérouse, the forests have been uprooted to some extent by the glacier, as in the case of the Brady. The sea far out is green with mud particles.

Yakutat. *June* 19.

The village of Yakutat is made up of a few houses in a row in a side bay at the mouth of the main bay. The ship ran out and up the main bay on the west side to the Malaspina Glacier, and put off hunting parties for bears, and ornithologists, etc. Also left a small party at Yakutat. These landings took till 4 P.M. Meanwhile, as Gilbert, Gannett,[1] Kearney,[2] and I waited to be taken up the bay as far as possible, Mount St. Elias was gradually clearing itself. First the left-hand peak, then the main peak appeared, white, sublime in magnificent cloud-robes, snow and ice laden to the summit, a noble spectacle rising from the Malaspina ice-plain. At 4 P.M. we started up the bay, not hoping to go far with the ship and intending to be put off below Haenke Island, but, finding a strip of fairly open water along the east side, we steamed all the way to the head of Enchantment Bay, and anchored for the night. The scenery was wild and enchanting. All agreed to leave off the 'Dis'[3] from the name. The slopes of the mountains around Enchantment Bay are mostly low and green and yellow and purple with grass, willows, alders, and cottonwood and hundreds of species of charming flowers and mosses; among them a large hairy, bossy-hearted lupine and *Epilobium latifolium*, a charming purple-flowered Rubus in large beds making carpets, with two tri-parted leaves and a large single flower an inch wide, very red in the bud, and when nearly opened growing pale with age as the red berry begins to form; also handsome yellow mellow-flowered Potentilla in big rugs; delightful beds of willow in smooth carpets; purple-flowered saxifrage, and Romanzoffia like Œnothera. Egg Island, joined to the mainland at extreme low water, is the floweriest of all, and we could easily detect the fragrance a half-mile off. Lupines grow here in hearty luxuriant abundance, and a sturdy Composita, Romanzoffia, a large red Castilleia, the purple Rubus, two saxifrages, strawberry,

---

[1] Henry Gannett, Chief Geographer, United States Geological Society.

[2] Thomas H. Kearney, Jr., Assistant Botanist, United States Department of Agriculture.

[3] The bay is named on the maps Disenchantment Bay.

blue violets large and fine, Allosorus, willow carpets and shaggy almost impenetrable thickets covering a large area on top; rushes, grasses, Pedicularis, dwarf Vaccinium, a handsome dwarf Composita dotted over with yellow flowers in cushiony tufts, like the stemless Silene found on the moraine at Nunatak Inlet.

June 20.

All today was spent in Hidden Glacier and Nunatak Glacier Inlets. I sketched Hidden in the morning. Was told I had only two hours ashore while the ship lay waiting, but the surveyors, Gilbert and Gannett,[1] took five hours, much to the Captain's disgust, who wanted to take on water farther down. In the afternoon, we steamed in a launch twelve miles up Nunatak Glacier Inlet. The Misses Harriman and Averell went with us, as did Dall and the artist Dellenbaugh.[2] . . .

Russell Fiord.[3]
June 21.

All day was spent on Hubbard Glacier and charming Egg Island. I had a long walk on the moraine of Hubbard, an immense affair underlaid by ice, much of it raw; with willows, saxifrage, tufty Compositæ, and Epilobium, etc., on the firmer portions. The whole glacier is about four miles wide, about half of which is discharging bergs at a lively rate — mostly small stuff from the shattered spiky front about two hundred and fifty feet high. They kept up a fairly constant cannonade last evening. A short time ago — a century or two — the Hubbard and Dalton ice and the other glaciers above these flowed southward by Enchantment Bay, the way by Yakutat Bay being crowded full by the Malaspina. The shoulders of the tributaries show this. The Hubbard and Dalton cut off the mouth of Enchantment Bay at one

[1] G. K. Gilbert and Henry Gannett surveyed and mapped Hidden Glacier. See *The Harriman-Alaska Expedition*, vol. I, p. 59.

[2] Dr. William H. Dall, Paleontologist of the United States Geological Survey, and Fred S. Dellenbaugh of New York City.

[3] The *George W. Elder* was the first large vessel to go to the head of this fiord. See *The Harriman-Alaska Expedition*, vol. II, p. 261.

end, the terminal moraines at the other forming a lake after the ice was melted out of the lower portion, as is shown by a terrace and by the difference in vegetation. The only glacier at the southern arm of the fiord now reaching tidewater and discharging bergs in Enchantment Bay is the Nunatak, at the head of a fine inlet nearly ten miles long. It sends off but few, and is fast fading. Many others descend nearly to sea-level, but melt a good way back from the bay. The largest of these is the Hidden Glacier, at the head of an inlet two or three miles long and a mud-flat three or four miles long. Immense quantities of stratified moraine material lie at the base of the mountains or load their sides to a height of a thousand feet or more, mostly falling sharply, furrowed and black, giving a curiously fluted appearance to the grand, simple slopes. The rocks are chiefly soft contorted and folded shales, with coal veins or stained with coal. These, so weak in structure, gave way to the action of the ice without offering any controlling influence to determine the forms, and thus, too, the immense quantities of moraine material are accounted for. But it is easy to explain the rapid waste of these glaciers in mountains so high and so snowy. Below the Dalton Glacier on the northwest side of Yakutat Bay are two that come nearly to sea-level, quite small and dirty, a half-mile wide. . . .

June 22.

After enjoying a walk on the Hubbard moraine-clad glacier and reveling in the fragrance and beauty of Egg Island, we camped on the mainland opposite it. Got enough wood for fire and enjoyed the time though slightly rainy. Camped in a bed of big lupines.

June 23.

Arrived at an Indian summer village this afternoon at three, and got on the steamer. Enjoyed a stroll on Haenke Island on the way down. It is as flowery as Egg Island, and the views of the glaciers and mountains are glorious. Indians last eve camped on Egg Island. They had been hunting seals all day. Four men

got fifteen seals, which they skinned at Egg Island the night we
camped on the mainland opposite. We heard them shooting and
heard the seals barking or half howling, in a strange, earnest voice.
We saw many seals also in the open water, swimming in pairs.
Some of them came close alongside, as if blind, perhaps wounded.
Two of the Indians came over and ate breakfast with us, and
were greatly interested in our aluminum utensils. Parties went
ashore; I did not — the stink of decaying seals is awful. The
naturalists set traps up the mountains, and caught shrews and
mice. The hills and mountains of both Yakutat and Enchantment Bays are green with grass and brush, but have few trees
except at the mouth of Yakutat. I saw a high mountain three or
four thousand feet high yellow with moss. Only very sheer or
crumbling fronts are naked.

June 24.

We anchored at Yakutat last evening. The naturalists went
back to examine the traps this morning, and we waited three
hours for them, then sailed for Orca, Prince William Sound. Had
a grand view of the Alps from anchorage, but fog and clouds shut
out all the glorious show today.

Stopped for soundings in the thin fog. A glorious clearing at
3 P.M. Along the north side of Prince William Sound and westward is the richest, most intricate range of glacial sculpture I ever
saw. The sun is shining on fields and strips and patches of snow
and ice as we approach Orca. Clouds, white, streaked with purple
bars, make a grand addition to the noble mountain-range, a
thousand peaks big and little around the horizon. We saw a large
broad low-descending glacier near the mouth of Copper River.

Prince William Sound.

At 8 P.M. we arrive at Orca, anchoring between a small island
and the mainland. The town has a cannery, a post-office, and
a store. Half the salmon catch is already put up, and a large ship
rides at anchor ready to take it away. The mainland slopes are
steep and heavily timbered with Menzies spruce and Mertensia

hemlock, with Patton hemlock forming the timberline at a height of two thousand feet. Some of them are twenty feet high. The other hemlock and the spruce are four to five feet in diameter and one hundred and fifty feet high. Broad light-green, down-swooping spaces, cleared by snow and avalanches, are richly embossed with alder, willow, balsam, and poplar.... Just below the snow are rich beds of Bryanthus and the bristly-leaved Cassiope. The Oregon junco and the yellow-crowned warbler sing nearly alike; gulls, kittiwakes, and fulmar petrels fly in the harbor. Saw a small sooty albatross yesterday following the ship — a graceful bird, seldom flapping his long narrow flat wings.... A marbled murrelet was shot from the launch — a handsome bird, a diver about the size of a pigeon with a short tipped-up tail. Went ashore. A fearful smell, a big greasy cannery, and unutterably dirty, frowsy Chinamen. Men in the business are themselves canned. Mice caught. Was introduced to a Mr. Brownlee. He and the storekeeper had read 'Mountains of California.'

Columbia Glacier.
June 25.

A charming calm Sunday morning, fairly balmy. The slopes with spruce and pine rise sharply above each other in crowded ranks, up rich green avalanche slopes, sunny and peaceful. Went ashore and climbed in the woods along a hearty bouncing stream. The roots of the trees are mossy, interlaced, ferny, and tangled. Devil's-club, Aspidium, and spruce. We sailed in the afternoon from Orca to the Columbia Glacier near Valdes. The glacier occupies a fiord eight miles broad, and short — five or six miles, blocked and interrupted in part by islands, most of which are completely overgrown with spruce and hemlock. The cascading tributaries — many of them flat and broad — make a glorious show as the front is approached. The discharging portion is two miles wide. The left east side is rounded and has a gravelly sandy flat in front and an island. The trees on the point of the island have been uprooted by the advance of the ice; some are leaning like those of Brady Glacier. Comparatively few bergs are now being

given off. The wall is white and has but few apparent streaks of
medial moraine stains — the quietest ice wall of equal size I have
yet seen. Very few bergs encumber the fiord. We approached
within half a mile and found the water about six hundred feet
deep. A very complicated glacier — many of its tributaries must
be interesting in bold, striking features, judging by what is visible
from the approach. A series of three flat cascades on the roof-
like slope of a mountain near the front is very striking and beautiful in detail. The ice in falling is delicately sculptured instead
of being broken up and soiled with moraine material. The
sculpture of the basin of this glacier in general is very rich and
instructive. Here we left Professor Gilbert to study and map
while we went to a copper mine. I went ashore in a boat laden
with campers. After making camp, some started up the mountain
three thousand feet high after 10 P.M. for a view. Harriman went
hunting and climbing on the same mountain with Indian Jim of
Yakutat. He arrived at the ship about midnight. The ladies
came ashore after church services at 10 P.M. and enjoyed a ramble
on the bog. In the meantime, I walked in the woods and a big
bog of peat four to six feet deep. Found a spruce three hundred
and eighty years old, four feet in diameter. Found a fir in flower,
not fertile hereabouts this year, and a P[atton?] hemlock also in
flower; it grows tall here with other species and spruce. Cassiope,
Kalmia, and a delicate cranberry are abundant, also the aster
and Fritillaria and the dark purple Caltha.

June 26.

We sailed for Port Wells Glacier, arrived at a point in a fine
long, straight inlet fifteen miles from the two glaciers, and set out
for the west branch of the glacier in a launch. The view up was
glorious, of four cascading glaciers descending in a jagged white
torrent, strange, silent, and seemingly motionless through the
rich green-furred pastures of the mountain goat. Alder, willow,
beds of Bryanthus, and alpine bloom in general. Two of these
four cascades send off a few bergs from a well-formed front. Mr.
Gannett was busy with map work, the naturalists were shooting

birds in the bushes and along the shore and collecting gull and
tern eggs on the moraine, both kinds dull earth-colored. I saw
Mr. Harriman coming with the other launch. The children [1]
played on the sand and gathered flowers. The main glacier has
glorious tributaries swooping down from the mountains, pure
white, and the sculpture of the basin is extremely rich, sharp,
and deep, with immaculate domes in the sky — God's big tents.

Later we started for the east branch of the glacier, about the
same size as the west, neither sending off a great number of bergs.
The basins, of course, are alike, seeming as if the main trunk were
melted and its seven main tributaries separated. We landed on
a point near a good rich forest, with a fine undergrowth. Found
a new species of Arctostaphylos. Had much bother with the
launch among the bergs. Got back at 7 P.M. and started for the
glacier in the next fiord to the west. When we reached the mouth
of it, Captain Humphrey [2] said he would not take the ship farther,
did not believe in going into every frog pond, etc., and stopped
the ship. But Mr. Harriman ordered full speed ahead, rocks or
no rocks. We saw a fine glacier and went within three hundred
yards or so of the front. Ventured around the point on our left
and saw another glacier; still venturing, we saw another and
another, until the fifth glacier was seen discharging bergs into the
main inlet,[3] which is about fifteen miles long. Harriman, excited, said, 'We will discover a new Northwest Passage.' At
9.30 P.M. I left the ship with Gannett, Captain Kelly, Inverarity,
the photographer, and young Morrill and Indian Jim, to camp
and explore for two days, while the propeller, which had lost a
blade in the ice, was being mended at Orca. We found a paradise
of a camp on a bench covered with Cassiope three miles from the
head of the inlet on the right side, fifty or sixty feet above the sea.

[1] The children of the party, called by Mr. Muir 'The Little Two,' were Carol and
Roland Harriman, the sons of Mr. and Mrs. E. H. Harriman.

[2] Captain O. J. Humphrey, having expert knowledge of coast navigation, joined
the party at Orca.

[3] This glacier, as well as the fiord leading to it, having been discovered by the
Expedition, was named in honor of E. H. Harriman. See *The Harriman-Alaska
Expedition*, vol. I, pp. 71–73.

Timber abundant. We made another supper at eleven, and went to bed at 2 A.M., after a grand exhilarating evening, pitying them on the ship.

Exploring Harriman Fiord,
Prince William Sound.
June 27.

Spent the day walking along the shore nearly to the uppermost glacier. Crossed the dirt glacier crevassed badly, and a wedge glacier sending down a large stream, also a large rocky delta, hard to cross. Lovely Cassiope beds among the logs and rocky benches and surging up against stumps and trees . . . each flower rising on its own stem . . . and nodding, its petals waxy white, some pink separate, sepals and pedicel red, all facing downhill eight or ten to a square inch, each making a letter S. A charming plant mixed with a woody saxifrage, lacerated leaves, Caltha and Menyanthes with green glossy leaves, Veratrum, mountain aster ten feet high, young hemlock, bronze-colored, Vaccinium, Potentilla. The first pure forest of Patton hemlock I ever saw at sea-level. It seems old, storm-beaten, snow-crushed, yet strong and cheery and irrepressibly lovely in form, contriving to be beautiful under every condition of weather and soil or bare rock, bare save the cover blessed Cassiope spreads. The rock, slate, is variably metamorphic, giving rise to very irregular forms under glaciation, little moraine matter being left on the ledges. On such rocks this hardy hemlock often attains a height of ninety or a hundred feet, and a diameter of three feet, two inches. One nine inches in diameter, three hundred and twenty-five years old, was cut by hunters two years ago. Other old mossy stumps are said to have been cut by the Aleuts, so says Indian Jim. Many of the trees seem to have been shattered by lightning. The young trees are curiously bent and kinked by heavy snow, all, old and young, stiffer, less lithe and graceful than in California. The bark of old trees is pale gray, and flaky dark purple inside. Now in flower; the pistillate, five-eighths of an inch long, one-fourth of an inch in diameter, is dark blue; the staminate one-fourth of an inch long, one-eighth of an inch in diameter, is of the same color. Ptarmigan are abundant. Captain Kelly hunted night and day, and found one bear track. No goats were seen, nor any sign of them. Seven miles down the inlet near the shore, this hemlock forest is mixed with Sitka spruce — a shrub or small tree. . . . How far up it extends I don't know. I found one young tree in flower, the pistillate, a glorious purple and red, changing in rich plushy furred shadow like the throat of the brown hummingbird.[1] Never elsewhere have I found spruce flowers so fine — the most superbly beautiful of all the coniferous flowers.

June 28.

Started from the upper camp about noon, three miles from the head glacier, and rowed in the rain about seven miles to another charming camp opposite the third glacier from the head, hoping the steamer would not call for us until next day.

We set our tents at 4 P.M., enjoyed the view of the two fine glaciers[2] on the other side of the channel and examined the superb bosses of Cassiope in bloom. The Indian visited a campground five hundred yards below ours and returned with a chip, a bone, and a few hairs, which he offered as proof that Indians had killed a bear and camped there eight days ago. . . .

At this camp the Sitka spruce form a considerable part of the forest with the P[atton?] hemlock still predominant and taller — a hundred feet. A fine Will[iamson?] established, old, mossy, lichened, and if possible still more superbly embossed with Cassiope, etc.

June 29.

The ship's whistle awoke us. At 6 A.M., while the cook was preparing breakfast in the rain back of a big slate boulder which had fallen from the cliff and stood stuck in the shore ground, the wide end up, we left the slimy rock mess and breakfasted on the

---

[1] The rufous hummingbird. — F. H. A.

[2] Cataract and Surprise Glaciers. See map in *The Harriman-Alaska Expedition*, vol. 1, p. 66.

ship. Then bade farewell to our fine new glaciers and sailed for Cook Inlet, stopping to examine copper mines on the way out, but not the glacier or glaciers in Icy Bay, as we intended, Mr. Harriman saying we had no more ice time. We saw bergs from it as we sailed past its mouth. I must come again. Altogether we have seen seven large glaciers of the first class and three small cascading ones which send off a few bergs and make a wonderful show, coming jagged, foaming, surging down through the midst of green pastures on majestic, simple, broad, and high back-leaning slopes — grand white cataracts of ice in cow and goat pastures. And we saw ten berg glaciers in the Prince of Wales Sound and many broad low descending ones in the first stage of decadence, and many cascading ones separating for one hundred or five hundred feet and then regelating and flowing on as if nothing had happened. These in falling make nearly as thundering a noise as the great glaciers. Montague Island has mountains all about the same size cut into broad regular scallops side by side, the sun shining on them, very many low-descending out of cloud-capped mountains up the coast. Arrived in Kachemac Bay at Homer, a post office on a sand spit.

June 30.

Decided this morning to sail to Kodiak Island instead of to Cook's Inlet, to give the bear-hunters a more promising field. Glorious views of Kachemac Bay; many glaciers; bright weather. Fine views of Iliamna, Redoubt, and other volcanoes, the former smoking and steaming distinctly at times; the glaciers on them snowy to the top. The four peaked mountains have broad smooth solid white slopes, surrounded by sharp lower peaks and peaklets — the most beautiful, icy, and interesting of all the mountains of the Alaska Peninsula. Many glaciers descend to sea-level or nearly so.

Kodiak Island.
July 1.

Arrived at Uyak Bay early this morning after putting off a party of botanists and ornithologists at midnight at Kukak Bay

on the Peninsula. Here the big-game party got off on a launch,
gun-laden for war, thence going up the inlet afoot. They took a
light canvas boat and all the packers. Uyak Bay has a lovely
green slope, white mountains of the Peninsula looming in the distance; no trees, but the greenest of bright green emerald grass
with alder. Started for Kodiak about eight. A whaling steamer
company has a cannery here. The sail is lovely — green slopes
all the way to Kodiak through the northern strait and the narrow
strait between Raspberry Island and Spruce Island, Kodiak and
Afognak Islands. Spruce appeared in patches about halfway,
increasing to the eastward, all inclined to be low and brown with
a dense growth of last year's cones. Only a few of the patches
seemed old. Arrived at Kodiak at 3 P.M. and took a walk with
John Burroughs a few miles into the woods. Saw or heard the
yellow-crowned sparrow,[1] yellow warbler, song sparrow, the
hermit thrush, Townsend sparrow,[2] and the red-poll chickadee.[3]
Found black Fritillaria, curious spotted dark Cypripedium,
purple Arctostaphylos, Anemone, daisy, Geranium, Hydrophyllum, cranberry, Rubus, salmon-berry and crowberry, Castilleia,
dwarf birch, shiny-leaved willow, and Rhododendron.

                                                   Sunday, *July 2*.

A calm misty morning; everybody going shooting, sauntering
as if it were the best day for the ruthless business. I went up the
mountain east of the town with Trelease and Gifford;[4] vegetation
charming; Cypripedium in patches, Geranium large blue, Geum,
Anemone, Trientalis, purple Rubus, Rhododendron, starwort,
dwarf willow, daisy, Ledum, grass, Carex, Phegopteris — two
species, *Phegopteris dryopteris* and *Phegopteris polypodioides*, the
latter exceedingly abundant over all the hills here and, indeed, over
all the damp coast foothills of Alaska from Fort Wrangell to
British Columbia. No green mountains and hills of any country

[1] The golden-crowned sparrow. — F. H. A.
[2] Townsend's fox sparrow. — F. H. A.
[3] The chestnut-backed chickadee. — F. H. A.
[4] Dr. William Trelease, Director of the Missouri Botanical Garden, St. Louis,
Missouri, and R. Swain Gifford, artist, of New York City.

I have seen, not even those of the Emerald Isle, can surpass these of the Alaska Peninsula and the Aleutian Islands, where no trees grow, or at least now grow at the east end of the Peninsula. Grass, ferns, dwarf willow, Empetrum, Vaccinium, Geranium, bluebell, daisy, Rubus (salmon[-berry]), gentian, Wistaria, Primula, *Orchis cypripedis*, violets (blue), and a curious berry-bearing dwarf carpet-making plant like *Salix reticulata*, growing well up towards the summit of the Kodiak green mountains. Good botanists mistook it for Salix.

July 3.

We went on a picnic to Wood Island with Mrs. Harriman a little later in the morning, returning at 3 P.M. Found part of an old forest and got good approximate ages of ten spruces — one nearly five feet in diameter and perhaps three hundred years old. Found Moneses in fine ferny, mossy woods, and some good-sized cottonwoods. Heard of a large birch growing on rivers emptying into the head of Cook Inlet, and saw specimens of the bark in a basket, said to be even better than canoe birch. It was brown and large enough for a canoe, thirty inches wide, fourteen deep, with a four-inch strip at the top. I am promised specimens of the flowers by Mr. R. G. Slifer of Kodiak.

July 4.

Took a walk with Captain Doran and had a pleasant chat. After dinner we had a Fourth-of-July oration by Professor Brewer. It was good — stated principles fought for and won and for the first time in the history of the world put into practice and spreading into all countries. In spite of corruption we are well governed as any. We were the first to fight for humanity, in Cuba. The English were afraid to interfere in Armenia and other places. Charles Keeler read a poem, Ritter [1] and Fernow danced jigs; we sang patriotic songs and the Doxology after the jigs. Then we had boat races, canoe and bidarka races. Two men in a

[1] Professor William E. Ritter, President of the California Academy of Sciences and Professor of Zoölogy in the University of California.

canoe were badly beaten by two men in a three-hatch bidarka.
The boat races were close and exciting. No lecture this evening.
Thanks!... After supper I called on Mrs. Washburn, Mrs. White,
and Mrs. Fisher. Had a pleasant call. Mrs. White is handsome
and Mrs. Washburn appreciative and bright. They have pretty
rooms. Mr. Harriman returned last evening after killing two
bears, mother and child. In color light brown, they are allied to
the grizzly.[1] The hind foot was only eight inches long; some
tracks seen were five inches longer.

A mountaineer came aboard this evening who was familiar with
a great part of the coast from Wrangell to Unalaska; said he was
just like me, liked to go on glaciers and mountains alone. Had
been chased by wolves, had crossed many glaciers, most admired
the two in Lituya Bay because they were so steep and sheer as
they enter the water in the west arm of the bay, and thunder so
loud. He lives in a cabin near the head of Cook Inlet, on the river.
Thinks the birch there is the same as the Eastern canoe birch; it
has white bark. Said he was glad to meet me, had heard of me.
We started about midnight, but got aground after going about a
ship's length.

July 5.

The ship, lying over on a gravel bed, nearly rolled us out of
bed. At 10 A.M. we got afloat.... The day is perfectly lovely,
balmy, not a cloud in the sky. Temperature seventy-three degrees
in the shade. The party of hunters, lavishly armed and attended,
got no game whatever of the kind sought, but did get healthy
exercise and a fine waterfall and flowers.

July 6.

We started for Unga Island of the Shumagin group. The day
has been perfectly bright and cloudless; the sunshine falling in
dazzling brightness on the deep blue water, spangles flashing, and
on the wonderfully peaked and notched mountains of the immediate coast and the broad, solid, white, glacier-laden, volcanic cones

---

[1] The Kodiak bear, *Ursus middendorffi* Merriam. See picture, *The Harriman-
Alaska Expedition,* vol. I, p. 84.

and mountains behind the nearly snowless, richly sculptured and composed ranges of the coast. A truly marvelous range. One of the volcanic mountains was in sight most of the way from Kodiak to Uyak [1] and almost all this day. It must be about ten to twelve thousand feet high, is very broad and laden with immense snow and ice-fields, and was spouting up a white cloud of steam three hundred feet high on the northeast side about three thousand feet below the summit. One of the craters of a raggedly fluted, sturdy, gnarled cone about five thousand feet high was very striking. Another — perhaps Pavloff — is very beautiful in the perfect smoothness and symmetry of its lines. From the east end of the Shumagin group the sunset was glorious. From behind the dark jagged edges of this wonderful range flashed up-leaping billows of gold. A party of naturalists are to stop at Unga instead of going on to Unalaska. Just as the sunset faded, we came to a curiously weathered rock called the Haystacks.

Sand Point, Unga Island.
July 7.

Another lovely sunny morning. Only a few wisps of white and gray clouds wreath the old volcanic mountains, which the passengers are discussing as to whether they are clouds or smoke. The Pavloff volcano with the symmetrical cone to the east is in sight and is smoking or steaming. We are at Sand Point on Unga in a charming cove. A few houses are here, a trading-post, and a deserted, wrecked schooner on the beach. Here the Kamchatka rhododendron is in full bloom, the stems about a quarter of an inch in diameter, prostrate, the year's branchlets erect two to three inches high, one to two flowered, the flowers being one and a quarter to one and a half inches wide and one inch long, richly purple, the petals finely haired on the margin, stamens and pistil dark purple, calyx greenish purple, hairy on the margin of sepals, branchlets and pedicels purple and moderately hairy, leaves about one and a half inches long, five-eighths

---

[1] The party had returned to Uyak Bay to pick up the hunters before proceeding to the Shumagin Group. See *The Harriman-Alaska Expedition*, vol. I, p. 87.

of an inch wide, obovate, rounded on end, fringed with hair, white, an eighth of an inch long, separate buds and branches, stems brown, bark scaly or shreddy, pale, nearly white on the oldest stems, leaves glossy beneath.

All the rock in sight seems to be volcanic. Magnificent views of Pavloff (altitude about eight thousand feet) and the adjacent volcano, about the same height, which is smoking, have been distinctly visible nearly all day. When last seen the latter had a plume of whitish steam and smoke perhaps three hundred feet high, slender like the plume of a mountain quail. Fog began to set in at 4 P.M., shutting off the view of the high mountains until 7.45 P.M., when the summit of Shishaldin appeared above the clouds, seeming immensely high, though only eight thousand nine hundred and fifty feet. Then Isanotski's craggy, unsketchable top broke forth, a sturdy Titan, expressing grim, unmanageable strength in every feature. Both were at length unveiled, the one an exquisite type of calm, smooth, broad-based, immovable symmetry, the other of rough powers uncouth and gigantic. Darkness is falling. Clouds rise and droop, exposing and hiding the majestic mountains. A few years ago the summit of Isanotski was blown off in a violent eruption. Dr. Dall assures me that four years ago it was naked and Shishaldin snow-clad. Now the latter is bare, evidently hot at top, and the other has a crevassed, well-characterized glacier on it. At ten o'clock Shishaldin in the gloom looks gloriously wild and impressive, revealed from finely drawn top to broad base except the portion covered by a bar of clouds slanted across it. A very wide snow-and-ice-field a few miles to the west sends down three or four glacial lakes, perhaps the last of the low-descending glaciers we shall see to the west and north on this trip. To the west of these glaciers rises the fine volcanic cone of Pogromni on the west end of Unimak Island.

Dutch Harbor, Unalaska.
July 8.

Arrived at Dutch Harbor early this morning. A new town has been made by the American Company, which succeeded the

Alaska Commercial Company in the fur-seal business. A busy town since the Klondike discovery. Hills green and flowery, with a large Anemone, Palemonium, Geranium, Cypripedium, Empetrum not now in flower; ferns and grass very rank. Nineteen trees, of Sitka spruce brought from Kodiak a hundred years ago and planted in the hills between Unalaska and Dutch Harbor, nearer the latter, are alive and thriving; the largest two feet one inch in diameter, twenty-five or thirty feet high, broad and branching to the ground and now fairly red with magnificent purple or crimson velvety flowers and young cones one to two inches long. I never saw a more fertile or more magnificent mass of bloom on any spruce. Was told that grass burned every year. A dry day or two, even in spring or summer when the grass is green, on account of the mass of last year's grass below and the stems, alive or dead, of Empetrum, makes fires possible, and constant care is required to protect these trees and others on the adjacent island which were raised from the seeds of those brought from Kodiak. Where the brushy spaces begin, there is the same tendency to running fires. Captain Kelly told me that ten miles from Kodiak he saw a whole mountain-side lately burned over. Fire seems to be the main cause of treelessness here as on the Western prairies, notwithstanding the wetness of climate. The earth is mostly volcanic and covered with rich soil.... We kept John Burroughs on the ship.[1]... Sailed at 2 P.M. for the Seal

[1] Charles Keeler in his prospective book, *Friends Bearing Torches,* gives the following account of how John Burroughs was 'kept' on the ship:

'When we reached Iliuliuk on the island of Unalaska in the Aleutian Islands, we had some hours ashore before continuing our voyage into Behring Sea. Mr. Muir and I had been strolling about on shore and were just returning to the steamer when we saw John Burroughs walking down the gangplank with a grip in his hand.

'"Where are you going with that grip, Johnny?" demanded Muir suspiciously.

'Burroughs tried to give an evasive answer, but on seeing he was caught he confessed. He had found a nice old lady ashore who had fresh eggs for breakfast, and he was going to board there and wait for us while we went up in Behring Sea. He didn't like to go into those tempestuous waters.

'"Why, Johnny!" exclaimed Muir derisively. "Behring Sea in summer is like a mill pond. The best part of our trip is up there — seeing the fur seal on their breeding grounds on the Pribilof Islands, seeing the Eskimo at Port Clarence! Come along! You can't miss it!"

'Mr. Burroughs did not want to go into Behring Sea, but he could not with-

Islands. First we steered for Bogoslof Island, more volcanic rocks upheaved a hundred years ago. Another is close by and joined to it by a reef, thrown up fourteen years ago, and still hot. Many murres nest on the rocks, and many sea-lions have their young on the shore. Two of the handsome little fellows were stolen from their mothers and brought aboard by a party that landed on the island.

St. Paul Island, Pribilof Group.
July 9.

Reached St. Paul Island at 2 P.M. Tried to reach St. George but could not for fog. Most of the passengers went ashore in three boats, piloted by a party who came off to see us. Heavy surf except at landing. We walked over the grassy, flowery, smooth meadow-like ground to the rookery to the northward. Many songbirds were here, especially the Lapland longspur, beautifully colored, and my favorite, the leucosticte, or rosy finch, darker than in the Sierra. Saw thousands of seal on the black lava rocks, one big watchful, defiant male to every ten or twenty or thirty females. It was a rough place to lie on, to say nothing of taking care of the young. The males, with heads erect, long necks, gray hair, fighting, roaring, rushed at the photographers with surprising speed in two or three jumps when they approached too near, shaking their heads and snapping their jaws, showing their teeth through fierce long-bristle whiskers. The general cry of the multitude is like the ba-a-ing of sheep. Females with love in every gesture and look of their great eyes, with heads

stand Mr. Muir's scorn. He weakened a bit and was lost. I carried his satchel back to his room, and personally conducted by Muir and me he returned to the steamer. Later, when we had been aground on a rocky reef off the Pribilofs and had tossed about on the fog-hung, tempestuous waters of Behring Sea, Mr. Burroughs lay in his berth and groaned while I attempted to atone for my share in persuading him to venture into those stormy northern waters by sitting beside him and reading Wordsworth to him. Toward the close of our voyage, while many versifiers in the party were writing rhymes about people and incidents of the trip, it was not to be wondered at that Burroughs got back at his chief abductor by substituting "Muir" for "man" in the old missionary hymn, and speaking of Behring Sea,

"'Where every prospect pleases,
And only Muir is vile.'"

bent down from arching smooth round necks, gazed at their new-born, helpless, jet-black babes — a beautiful sight. Also the play in the waves was a wonderful thing — old scarred veterans fanning themselves with their flappers, sea-lions' pups, grave and long-flappered like fur seals, crying like children and grunting. Very fat, they weigh seventy-five pounds though still babes. The male seals eat nothing for three months.... Saw myriads of pygmy auks.

The ground is covered with a rank grass and is gaily decked with cream-colored poppies and blue polemoniums, violets, Claytonia with low phlox-like white flowers.... Got aboard and started for St. Matthew's in the fog, just after dinner. At 8.15 P.M. the ship grated heavily on a boulder reef and set free lots of kelp. The keel floated and then struck again and again, three times within a few ship's lengths. Everybody was solemnly startled. Land within two miles was seen as the fog lifted. We hoisted the jib, got free in fifteen minutes, and gave the shore rather a wide berth. We steered for St. Matthew's, but gave up the attempt to find it on account of fog and steered for St. Lawrence Island.

July 10.

A grim gray fog all day with a rather rough sea, rainy, the deck sloppy. Heard a lecture on whaling by Captain Humphrey. A large number of murres are flying about the ship as if curious to see it.

Plover Bay, Siberia.
July 11.

A high head wind and rather heavy sea began about 2 A.M. The ship was pitching a good deal, the screw out of water at times, therefore we had to slow down to four to seven miles per hour. A few whitecaps. John Burroughs is sick. At 12 M. the wind abated. We have been steering for Plover Bay instead of St. Lawrence Island on account of the heavy sea. Arrived at 4.30 P.M. and anchored. A boatload of Chukchis came off; the men, with shaven or sheared crowns, seemed familiar and

brought back the long-ago of eighteen years. Then nearly everybody went ashore and sauntered about the little village on the
sand and gravel moraine spit. The weather a clear, crisp cold.
Perhaps about fifty inhabitants live here in a dozen huts covered
with walrus hide. The contact with civilization of the whaler
seamen sort has, of course, spoiled them. In spite of all this they
make a living in this seemingly desolate land of frost and barren
stone.

At eight o'clock we sailed for Indian Point, a larger village
with a mission establishment. We reached it three hours later
and looked at it in the orange and purple light of sunset. A heavy
surf was breaking on the beach; nobody wished to try to land,
so we sailed for Port Clarence. Most of the passengers tried to
sit up to see the sunrise, calling it the midnight sun. Though
only a few degrees below the horizon, it did not come up until
about two-thirty, and, the fog rising, it was not seen until three.
A sleepy lot were they at gong time; few appeared before 9 A.M.

Port Clarence, Alaska.
July 12.

Arrived at Port Clarence at 1 P.M. Had a good view of King
Island, Prince of Wales Cape, and faint views of East Cape and
Diomede Islands. A bright day, the temperature about forty
degrees. Picturesque brown mountains and hills and valleys and
low finely curved slate, moors, and tundra hills along the north
side of Port Clarence sound. We found ten whalers at anchor
back of a low spit. Six of the captains came aboard, smoked, took
a drink, and told stories in the Captain's cabin. I was introduced
and we had a talk on whaling. . . . They said plenty of whales are
about the mouth of the Mackenzie, but did not expect any near
the strait. One said he got twenty-five last year. . . . They outfit
in Port Clarence, coal, etc., and go around Point Barrow the first
of August.

About fifteen Eskimo umiaks came alongside, fifteen to
thirty persons in each, beside babies and dogs, provisions,
skiis, etc. — a merry gypsy crowd. They had fine walrus and

sealskin canoes propelled by oars, paddles, and sails, and
were extremely buoyant and fast sailers. There was lively trad-
ing at big competitive prices — three dollars for a pair of walrus
tusks, four dollars for deer hides, two dollars for boots, one dollar
for a bit of carved ivory. Dr. Merriam [1] got inflated saddle-
back seals old and young. . . . We sent to the north shore to get
water. A large party of us, including the ladies and children, went
ashore at the watering-place. Had a hard surfy landing, and to
get into the boat again the ladies and children had to be carried
through the surf. It was a sight to see the demolishment of dignity and neat propriety. We had stiff rowing against wind, tide,
and heavy waves. The naphtha launch came to our help and towed
us laboriously to the ship. She danced like a feather; so did our
boat, the launch propeller whirling high in the air, as we crossed
every big curling swell. It was difficult getting aboard; the ladies
and children had to climb a rope ladder on the lee side, the boat
dancing and wobbling awfully for landlubbers. We found the
tundra boggy and wet, of course, over hill and dale. The five or
six willows had gone, or were going, to seed, the Spiræa was not
in flower, but there were beautiful purple polygonums, Primula,
Polemonium, small dark-blue gentian, blue violet, saxifrages,
one of them large and showy, drabas, *Vaccinium ovalifolium* and
uliginosum, cranberry, and dwarf birch. One crinkled stream-
side willow, the only bush, looks black from bad storms. It is
tussocky, spongy walking. Mosquitoes, lemmings, and Lapland
longspurs abound. The small long-tailed jaeger gull whips the
larger.

July 13.

We started early for the east end of St. Lawrence Island, passing King Island at 7 A.M. Arrived at 4.30 P.M., the wind high and
rainy, yet two boats put off with botanists, birders, etc. It took
fifty minutes to reach the shore after a hard pull. Got possession
of a few goslings and plants and sight of three polar bears, which

---

[1] Dr. C. Hart Merriam, Chief of the Biological Survey, United States Department of Agriculture.

turned out after a hot pursuit to be swans! In the evening we sailed for St. Matthew Island.... No natives seen on this still, desolate land.

July 14.

Rather dull and wet at times, slight showers and a tendency to fog. Arrived at Hall Island, a mile or two from St. Matthew's, and landed a party, then immediately started for St. Matthew's, where another party was landed, care having been taken to record the course so that in case of fog they could be found, the whistle being sounded as a signal for the parties seeking the ship. Both islands present bold, bluff precipices to the sea in many places, though there are a few broad smooth valleys opening between the bluffy sheer portions, and as far as I can see the interior is smooth and has an overswept glaciated appearance. The capes are bold and jagged mostly. On a considerable part of the shore it would be impossible to land. At the mouths of the streams there are large quantities of driftwood, all undisturbed, as neither of the islands is inhabited. The nearest point on the mainland is two hundred miles distant, which is perhaps the cause of its being uninhabited.

July 15.

We picked up our parties, one last evening from Hall Island, the other this morning from St. Matthew's, and the traps were collected, set mostly for lemmings. None caught. The handsome snowflake[1] was found on both islands in great abundance, pure white with a sprinkling of black dots on extremities of wing and tail feathers. We landed again down towards the south end of the island today, nearly all going ashore in launch and boats. The ground was everywhere flowery, though said to be less so than at the other two landings. Lupine, Astragalus, Oxytropis, charming Primula, Stellaria, mats of pink ball, shrubby Chamæcistus, Silene, Arnica, Tussilago, Anemone, viola, dwarf willow, pink

[1] A name sometimes used for the snow bunting. This was McKay's snow bunting, or the hyperborean snowflake. See the colored plate in *The Harriman-Alaska Expedition*, vol. I, p. 110. — F. H. A.

Andromeda, very fine Claytonia, etc. Great banks of snow indicate violent winds on dusty snow; some banks seem perpetual, have terminal moraines and a sort of glacial motion; some reach sea-level. The country in part is brown and barren with green sticks here and there, no hint of wondrous floweriness. The shore bluffs are rugged and deeply worn, made of all sorts of lava, some of it bedded horizontally. Myriads of puffins and murres, plovers, and sandpipers. A fine pure-white Point Barrow gull [1] was shot, also a snowy owl, and two young blue foxes and one old one, the mother; the pitiful things were laid out on the wet deck. Nests were robbed of their young and eggs, and the parents killed. Many little birds were left to starve.

July 16.

Making a straight course to Unalaska in dull weather. Burroughs is sick. We sighted one of the Pribilofs as we passed, twenty-five miles to the west of us.

Dutch Harbor, Unalaska.
July 17.

Arrived at Unalaska at 4.30 this morning. A charming calm, balmy day in a landlocked, mountain-locked bay. The sweetest fertile sunlight lies on the green slopes unruffled by either rocks or bushes, grassy to the top and around the harbor. Volcanoes to the northward, Makushin seen early, later clouded. The *Corwin* [2] is about to leave for the north. Captain Humphrey left us to go to Bristol Bay on the *Excelsior*. We left at 10 A.M. Beautiful mountains on our right as we go through the pass. A lovely yellow-green colors even the rough pillars of mountain fronts, resulting from degradation. Soon ran into a fog, but a good course is laid and the whistle is blowing. Will have to anchor when we come to the island tangle a hundred miles distant if the

[1] Really an immature glaucous gull (*Larus hyperboreus*). White individuals of this species were once described as *Larus barrovianus*. — F. H. A.

[2] On the steamer *Thomas Corwin* John Muir sailed in 1881, with the Jeanette Relief Expedition, to Behring Sea and Wrangell Land. See Muir's *Cruise of the Corwin.*

fog does not lift. Solid fog. Dropped anchor at 10 P.M. Started again at 3 A.M.

Shumagin Group. *July* 18.

The fog lifted; a fine morning; we had to slow down only once for a belt of fog. Fortunate yesterday, while the fog blotted out all the grand volcanic scenery, that it was already in our minds, having had so clear a view ten days previously on going west. We had delightful sunshine all day sailing among the yellow-green isles past Pavlof to Sand Point, on Popof Island. Arrived in the afternoon. All the party except Kelly, Palache, and Morris,[1] who were on the peninsula mainland, joined us here. Soon we started for the mainland camp thirty miles distant. A launch was lowered and the party was brought aboard at 8.30 P.M. and we started for Kodiak. Professor Ritter gave a general account of the work and fortunes of the party. They found the population of the town to be one man, who kept a store and supervised cod and salmon fisheries. There is no canning; all the fish are salted. They were successful in collections, caught a devil-fish with a ten-foot spread of arms, fifty pounds in weight, also some new things in the warm and low-growing things. He spoke of the wonderful richness and luxuriance of the grass cover and wondered when it would be put to human use. Professor Saunders[2] spoke of the plants of land and water, both rich, and of coal mines, tertiary, etc. Also of the great growth of alder. There is none in Unalaska or Behring Sea, fire probably preventing both bushes and forest trees. Young Kincaid's[3] address on insects ... was one of the very best of the trip. He has genius and will be heard of later, I hope. They found few beetles on the island, probably on account of mice, and few butterflies on account of high winds. They would fly up a few feet, and then some made haste to dive into bushes, having learned caution; the others of

[1] Dr. Lewis Rutherford Morris, physician, New York City.

[2] Dr. Alton Saunders, botanist, South Dakota Experiment Station, Brookings, South Dakota.

[3] Professor Trevor Kincaid, Professor of Zoölogy, University of Washington. Seattle, Washington.

course were blown into the sea. He told of several remarkable cases of mimicry. The yellow-tailed bumblebee is a strong stinger and is let alone to a great extent, but is mimicked so closely by stingless flies as at first to mislead the very elect. All cheered him heartily at the close.

July 19.

Fairly clear. After breakfast we were off Foggy Cape, one hundred and eighty-five miles from Kodiak. Little land was seen. I woke with a headache from the extreme heat of the cabin. The game scout reported a few bear and reindeer tracks, but saw none of the animals. He was left to look out for game but to shoot none, only to be ready to point it out to the game fellows, that they might have the pleasure of making a hole in the animals, shedding the blood, satisfying the savage instincts that should be kept down rather, after civilization has gone far enough for trousers and prayers.

In Charles Keeler's lecture the evening before last he said heredity is conservative, and likened it to a pyramid standing on a good firm base. Most changes along lines of variation perish; only a few that chance to fit best the environment survive.

We had a long talk on book-making, with much twaddle about a grand scientific monument[1] of this trip, etc. John Burroughs will probably write the narrative. Much ado about little.... Game-hunting, the chief aim, has been unsuccessful. The rest of the story will be mere reconnaissance. As we could not reach Kodiak before midnight, we stopped four or five hours at Sturgeon Bay so as to pass through Shelikof Strait, narrow and full of rocks, in daylight, reaching Kodiak at 7 A.M.

July 20.

It was rainy at Sturgeon Bay, but two boatloads went ashore and built a big fire of drift. They waded in wet grass, but for-

[1] The 'monument,' entitled *The Harriman-Alaska Expedition* and edited by C. Hart Merriam, consists of twelve volumes, containing among other papers the 'Narrative' by John Burroughs, 'Pacific Coast Glaciers' by John Muir, and 'Days among Alaska Birds' by Charles Keeler.

tunately shot nothing. I arose at 4 A.M. and saw the ship pass the strait against the swift tide current and boiling rips. Myriads of waterfowl; many whales; here and to the westward along Unimak shores are the favorite sea-otter hunting grounds. I watched the gradual thickening of the forests; no doubt fire had much to do with forestless regions, also the slipping of the snow during years of extra-heavy long-lying snow. The heaviest, oldest forests are not exposed either to fire or snow action.

Were at Kodiak until 5 P.M. A lovely, balmy day; we went walking through glorious, fragrant floweriness — among wild roses in prime, some three and a half inches in diameter, and a blue (Gentian family) plant, Geranium, Polemonium, etc. In the afternoon a large party went to a fox-farm island, of a thousand acres. They also saw salmon being cleaned, salted, and barreled. We took on water and coal and left for Kichamak at 5 P.M.

Kichamak Bay,
branch of Cook Inlet.
July 21.

Arrived in the morning. Put off a party of naturalists, including Ridgway [1] and Kincaid, at Saldovia. At Homer we put off another party (Gilbert, Palache, and Dall), then started for the head of Cook Inlet. But we turned back after going fifty miles or so, discouraged with the shortness of time, and for want of arrangements as to getting a smaller steamer for the shoals and rivers at the head of the bay. We recalled the Homer party at 7 P.M. Palache had killed a Canada grouse with a stone. It ranges north as far as the spruce grows. They found birch, brown or black, thirty feet high, one foot in diameter, and white Campanula. On the stratified deposits (Tertiary) on the west side of Kichamak Bay and Cook Inlet considerable areas were covered with dead forest, said to have been killed by showers of ashes and cinders ... from Iliamna; some say by ordinary forest fires. The

---

[1] Robert Ridgway, Curator of Birds, United States National Museum, Washington, D.C.

volcano, Mr. Woodbridge, the superintendent of the coal mines, told us, flamed up lurid red last Wednesday night, the nineteenth, with flames or red glow two or three hundred feet high. He killed a moose for us on our first visit on the east side of the Bay, but we sailed away just as he reached Homer.

July 22.

At 1 A.M. we got the Saldovia party aboard, Mr. Harriman having to go ashore for them in a launch. Not expecting us until the morning of the twenty-fourth, they were asleep and did not hear the steamer's whistle, though less than a mile distant. Weather was quite rough this morning early. Spray and rain were driven in at our window, wetting our clothing. The sea is now moderate, the barometer rising, the sun shining, the water lovely blue at 10 A.M. as we steer south of Middleton Island. . . .

Yakutat Bay. *July 23*

Foggy and cloudy until about 9 A.M., when the clouds lifted, revealing the west side of the Malaspina Glacier pouring from apparently two main glaciers, the fountains of which could be traced fifteen or twenty miles up to a height of six or eight thousand feet. We were just opposite this portion of the glacier, and at my suggestion the ship was run in closer at right angles to its former course in the direction of Icy Bay or Point, and by going then ten miles nearer we got a good view of nearly all the great glaciers. The west side, draining the slopes of St. Elias direct, is white nearly to the sea, and terminates in bluffs like the ordinary front wall of berg-producing glaciers, though not a berg was visible. Occasionally a mass is undermined by the waves, which at high tide beat on several miles of the front, but as far as I could see or learn, a mud or sandy flat is exposed at low tide everywhere and the glacier wall is being melted back and broken like a rocky coast. The rest of the glacier as seen from the ship is a dirty brown-black-yellow with moraine accumulations. This margin is eight or ten miles wide in some places, in others not more than two or three. Trees of good size are growing on the solidest parts, and of course on the outer parts, making a nearly continuous zone from the beginning of the dirty portion. The widest and apparently the oldest section is near Yakutat, being several miles wide thereabouts. Of course this moraine material on the ice is being slowly re-formed, the more rapidly where streams draining the glacier emerge. More of the surface is covered with willow and alder than with trees. Red alder forms dense tangles and is a serious obstacle to travel, since it grows higher than either the hemlock or spruce. A single individual tree of black cedar was found. Mertensia hemlock is less abundant than spruce. *Salix Barclayi*, five to ten feet high, has large upright catkins and forms tangled thickets that are the favorite resorts of the ptarmigan.... All along the Fairweather Range there is an outlying mass of moraine material, much of it stratified, almost continuous, more or less tree-clad, enclosing lagoons. Some hills of it one thousand feet high or more....

The clouds gradually lifted until nearly all the mountains of the Elias Alps were unveiled in majesty, the moving clouds assuming a thousand forms and hues, dark and flaming, glowing gloriously, red, purple, and yellow. We had fine views of Mount Vancouver and Mount Cook, especially of the former, broad-based, covered with gable peaklets, a very rich sculpture, more impressive than Elias from here (Yakutat). Elias is clear towards sunset, or nearly so, the peak of the summit several times being cut off sharply by clouds. Then it seemed tremendously high. Everybody is happy. Mr. Harriman and bear party went ashore, or rather attempted to go, but returned baffled and disgusted. No bears, no bears, O Lord! No bears shot! What have thy servants done?

July 24.

At 4 A.M. most of the mountains are clear. A huge berg went floating by. We intended to take on ice for water and ballast. Now we are going to Juneau and home. The want of bears is the cause of change of plans. We started at 6 A.M. Soon the clouds lifted as if under our control, just as we arrived opposite the large

Yakutat Glacier, six or eight miles east of Yakutat Bay; by the time I had finished my sketch it was clear, and all the way down the coast to Cape Spencer the Fairweather Range was absolutely cloudless, and glowed in its robes of snow and ice with ineffable beauty and glory of light. The water, too, was beautiful, in shades of green and blue from the vast quantity of glacial mud poured into the sea by the majestic array of glaciers. Mount Saint Elias was clear except the tip top. The wind was at our backs and just enough to balance the ship's motion, making a delightful calm for us as we gazed and sketched and took photos of the glorious show of mountains beheld at just the right distance to show the range from top to base with all their glaciers and fountains as a whole. The mountains from St. Elias are continued along the coast as an unbroken regular range. Mount Augusta, a beautiful gray sharp-fluted mountain, stands well back. Mount Cook looms as a huge house or barn on top, and Vancouver, nearest to Yakutat Bay, feeds the eastmost tributaries of the Malaspina. This is the grandest range, the richest in sculpture, with infinite gables and V-shaped valleys and canyons. That fine mountain seen to the north is perhaps Mount Hubbard. From Yakutat the mountains visible from the coast are pretty regular in height, say seven or eight thousand feet, with glaciers at regular intervals of five or six miles, the largest of which is Yakutat.

As we approached Fairweather there was a gradual increase in height and richness of peaks and peaklets, culminating in Crillon. Our view of Fairweather and company was most impressive as we came within fifteen or twenty miles. From here the general show surpassed even the St. Elias Alps, and perhaps any other view to be had on the coast. When we halted abeam of Fairweather, it was much less impressive, flat, wall-like in feature comparatively. The best way to climb it is from the west shoulder by ascending a glacier to the ridge and following to the top. Only a short distance of step-cutting would be required. The company all declared the view of the range from the northwest, twenty miles or so from Fairweather, and six or eight miles from the shore, with the yellow, mellow cloudless light, was the most glorious of all the trip, surpassing even the scenery of Prince William Sound.

Crillon, as seen from this side, shows a tremendous precipice on the north face, dropping sheer from the sharp pyramid summit to at least eight thousand feet — one of the wildest mountain walls I ever saw. The bottom of it was not visible to us. Both Fairweather and Crillon at top were immaculate. The Crillon Glacier is the only one that presents an ice-wall to the sea. It is undermined by waves at high tide, and bergs fall on the beach. None that I saw reached the sea. Not a berg between Yakutat and Glacier Bays, although four glaciers in Lituya Bay are thundering off bergs, but I saw none outside of that and Prince William Sound, not even in the inlets. But as we entered Cross Sound I saw a berg sailing out into the open ocean, and all the way along the sound and Icy Strait they were sprinkled, all, of course, coming from Glacier Bay. This bay yields more ice than all those to the westward put together, perhaps more than all the rest of the coast beside. Just as we entered Cross Sound, clouds, or fog, began to fall, and after this glorious show the trip seemed done, and everybody, exhilarated by the icy glorious day, began to loosen the tension of nerves by shouting, joking, cutting up generally. We all felt more free and friendly, and reserve went to the winds. After dinner, though nothing stronger than Apollinaris was served, there were college yells, rival yells, and lusty jostling. Then with one accord all went to the upper deck and danced reels, jigs, etc., with utmost abandon and laughter to music that was a mixture of whistle and gong. At a lecture on Somaliland by Elliot,[1] to his astonishment he was cheered and greeted with cries of 'What's the matter with Elliot? He's all right! Who's all right?' etc. Interesting result of ice action.

I got a view of the island-blocked entrance to Taylor Bay, and thought of my wild storm day there on the glacier with Stickeen; also of my stormy tossing on the waves of Cross Sound in my canoe.[2]

---

[1] Daniel G. Elliot, Curator of Zoölogy, Field Columbian Museum, Chicago.
[2] See Muir's *Stickeen* and *Travels in Alaska*, pp. 294–95, 298–311.

July 25.

We arrived at Juneau at four this morning. The town is quiet, the miners all away getting gold or trying to get it, mostly for poor or vicious uses. A wonderful wealth of forest on the steep mountains. At noon we sail to Douglas City for coal, take on sixty tons, and start home in the lovely sunny calm. Many small bergs from Taku Glacier are silent and stranded, like patches of snow lasting into summer. Many stand also at the approaches to Holukam Bay. A charming night, purple and gold, with a full moon, and fish splashing silvery. Everybody sad and glad that the end of the glorious trip is near.

July 26.

We passed the Wrangell Narrows at 3.30 A.M. How rich the sunshine, how warm it is, and how delightful to get back into the still inland waters embosomed in green mountains whose grass is trees!... Marvelous richness of the forest. Landed in the afternoon at Tongass, an abandoned village at the Dixon entrance on the south side. We found apple, Lonicera with black fruit, Menziesia, red huckleberry, salmon-berry yellow and red-ripe, most of the fruit already eaten by man or bird or squirrel. In taking down a totem pole a nest of Douglas squirrels was found; two of the half-grown young were caught and of course made into specimens.

July 27.

A lovely morning, the light dancing on the water as we pass through Grenville Channel.... The trees, the warm yellow of Thuja, cypresses, and hemlock, and the dark bluish green of most of the spruces, the Sitka and Douglas. Cedars along the shore. At eight-thirty we anchored in Safety Cove for a jinks on the top deck; a mighty lively and merry, witty time. A Sioux war dance by Captain Kelly was very effective. The laughter of Captain Doran rang out at jokes such as how to catch jack rabbits and rats: lie back of a log and make a noise like a turnip. After songs and the reading of a poem to Captain Doran by Keeler, it was growing dark and we adjourned to the social hall, where dance, song, and speeches went on in wild glee and abandon. Also the Harriman-Alaska Expedition yell, 'Who are we? Who are we? We are, We are H. A. E.,' and Brown's song to the officers and sailors were very good. One of the sailors has a good voice and is a good actor. He sang songs pathetic and comic, and was wildly cheered. The Captain was made the happiest man on the ship with cheers, resolutions, poems, and songs. 'What's the matter with Captain Doran?' etc. All called downstairs for beer, then Gannett and I retired to the Captain's cabin for smoke and talk. The ship left Safety Cove about midnight. It is a charming cove, with shores finely finished. Many dead spars are there, mostly fire-killed. Fire creeps in dead moss and humus even when wet, and kills trees without making much show. A good many Douglas spruce are here, dark blue-green when young.

July 28.

A cheery breeze; mountains of the mainland are dimly seen through smoke. Many committee meetings are held, but little accomplished.

July 29.

Arrived at Seattle 3.30 A.M. Burroughs and I walked up town to 27th Avenue. Clouds hid the mountains, though the air was unusually clear after a thunderstorm. . . . At 1.00 o'clock all the party lunched at Hotel Butler. I had to submit to an interview for the *Examiner*; the *Post-Intelligencer* got interviews from Gannett, Fernow, Merriam, Grinnell, and others. The results as shown published this morning are very good save those of Gannett and Fernow. After escaping from paper men, went to the Curtiss gallery. With the ladies, Burroughs, Coville, and Mr. Averill,[1] we went to the park by rail, guided by Dr. Young. Later we had a sail on Lake Washington to another park and saw miserable eagles, pea-fowl, and tame elk, extremely gentle, begging food from visitors. Young sequoias of both species are doing

[1] Mr. W. H. Averill of Rochester, New York.

well. Magnificent madroñas fifty feet high, three feet in diameter, are dying from hacking, trampling, and general bad usage, the trunks being covered with advertising cards and plates....
We all went aboard to dinner.

July 31.

We leave Seattle for Portland at 7 P.M. Pass Cape Flattery at sunset. A calm, glassy sea, with a strikingly defined bar of gold laid down by the sun on the smooth water equal in width from end to end. Unlike the wild storminess of dashing spray of former visits. Emerson [1] lectures on India.

August 1.

After a remarkably smooth voyage — even the Columbia bar was almost without breakers — we arrived at Portland. A dredger was at work in the river churning mud, pumping it aboard and letting it pour in heavy streams to be carried off by the current. The river views were fine — picturesque bluffs.... We were met at the dock by the Harrimans, John Burroughs, Gannett, and others, who had gone by rail. I bade them good-bye, and was invited to visit the Harrimans at Arden, New York. Burroughs and Palache accompanied us to the station. We left for home at 7 P.M. Timber along the Columbia is mostly Douglas spruce with many arborvitæ, and a fringe of willow and maple.

California, homeward bound.
August 2.

A fine dry day. The dry woods seem strange and wondrous open. Lovely young sugar pines and madroña. Fine to see the Ceanothus once more, and the chinquapin, a striking tree in the Umpqua and Rogue River valleys.... Shasta is rather smoky. Little snow save on the glaciers.

[1] Professor Benjamin R. Emerson, Professor of Geology, Amherst College, Amherst, Massachusetts.

John of the Mountains

2. The True Story of J. B. and Behring Sea [1]

John B.,
Slabsides, he
Said he could never love Behring Sea.
And said it most doleful,
Rhyming it soulful,
Moaning it,
Groaning it,
Pa-thetic-alee.
'It's big and it's blue,' he said,
'And has whales and a crew,' he said,
'Of sea lions and seals
That plash merry reels
With ravings and wrangles,
Contortions, contangles,
But the waves won't keep level,
They keep only mad revel;
My poor head is aching
And every nerve quaking,
And oh my interior
Grows queerier queerier,
The sea's shaky all over,
And its kelp is not clover,
And its seals and its whales,
And its gulls and its gales,
Care nothing whatever for Slabsides or me.'

So he lay down and howled it
And shivered and growled it,
Flat and limp on a deck chair
With naught but his nose bare,
Bemoaning and sighing,

[1] In a letter of August 20, 1899, to 'The Big Four' Mr. Muir wrote: 'I have a good mind to tell his [Burroughs's] whole Behring story in his own sort of good-natured, gnarly, snarly, jungle, jangle rhyme.' See Badè's *Life and Letters of John Muir*, vol. II, p. 332.

'Come, girls, I am dying.
Quick, tuck in my blue toes,
All ten of 'em's true froze,
And send grim Muir away
To cold storage, I pray,
On his ice peaks and passes
And bergs and crevasses.
Let him laugh in a glacier-crack,
Freeze on a nunatak
For in all this bad land or sea
There's nothing, there's nothing that's too bad for he.
Oh, Cornelia and Mary,
I've no comfort, and nary
A glimpse of my York State.
Soon, soon, I'll be shark-bait!
Dorothea and Bettie, dears,
Can you not give some wetty tears
For the spray the mad winds are throwing,
For the snow the gray clouds are sowing,
On the head of J. B.,
The poor head of J. B.'

But soon a big sea-change came o'er him,
No winds or wobbling waves now bore him;
Repentant, sane, his Slabside days
He spends in changing growls to praise.
And thus he sings from dawn to dark
As glad and blythe's a meadow lark.

Slabsides. *August 15, 1899.*

'Home again from Arctic climes —
Jingle jangle, merry chimes —
Amid my birds and trees and vines
I fondly trace bright memory's lines.
With heart and soul devout I stand
Gazing on wild Alaska-land.

Its mountains clad in snow and ice
Seem calm and pure as Paradise;
Their sculptured domes and peaks and towers,
Enwreathed in purple mist and flowers,
Rise range o'er range in song and rhyme
Triumphant, wild, serene, sublime,
Fountains of strength, of life, of motion,
Guardians alike of land and ocean.
The purple tundras far extending
Seem with the purple heavens blending;
The vast unbounded nightless days
Are filled with their Creator's praise.
Endless waters, endless woods,
Endless gardens, endless floods,
Silvery fiords, and balmy air
Spread endless beauty everywhere.
New lands, new seas, new heavens, new earth,
Day by day we saw their birth.

Where all is beauty, all is love,
Through earth below to heaven above,
Enchanted, wondering, throbbing, blowing,
All the show in one flood flowing,
I scarce can make my memory bring
From out the whole one separate thing,
Unless, perhaps — unless it be
That broad effulgent Behring Sea,
The brightest gem of all the bright North,
Reflecting every beauty right forth.
Its miles and leagues of purple dulses
Still thrill and tingle all my pulses,
Its gracious waves, and breezes mild
Still rock and fan me like a child.
Its flower-embroidered shores and bays,
Its wondrous skies, its nightless days,
Its teeming life, its bloom-clad islands,

Its far blue wavering lines of skylands,
All rise before me now untroubled,
Their blessings, beauties, more than doubled.
Therefore my proudest song shall be
The glories of great Behring Sea.

Ah me! poor doubting doleful sinner
I feared its waves might spoil my dinner.
And had not H.[1] and Muir insisted
In faithless dread I should have missed it.
I thought to stop at calm Dutch Harbor
Boarding with an old Dutch barber,
And, while the ship was gone, go hunting
The gold-crowned sparrow and snowflake bunting,
And thus I might have lost the whole,
My head, the sea, perhaps my soul,
Where blinding clouds and fogs roll down
On Unalaska's icy crown,
While famous wolves were howling longer
To pick my bones to make them stronger.
As now I view it all again
It seems an awful might-have-been.
This dreadful fate, thank Heaven, was passed,
For Muir most kindly held me fast
And made me leave that old Dutch Harbor
Give up the sparrows, give up the barber,
And stay aboard the ship and sail
To Behring Sea with blissful gale.

If on that ship there was one friend
For whom my thanks should have no end,
Both faith and reason I am sure
Will say his name is just John Muir,
A name from which the fogs are clearing
As from the wind-swept face of Behring.

[1] Harriman.

Henceforth where'er my lot is cast,
As long as life and memory last
My grateful prayer to Heaven shall be,
God bless the Harrimans and he
And every wave of Behring Sea.'

JOHN MUIR, 1899.

Chapter X : 1900-1914

'Not like my taking the veil — no solemn abjuration
of the world. I only went out for a walk, and finally
concluded to stay out till sundown, for going out, I
found, was really going in.'

THE early years of the new century were devoted by John Muir
to work in behalf of national forest reserves and parks. This
included the new battle to defeat the rising movement to dam
the Hetch Hetchy and convert it into a water reservoir for the city
of San Francisco. In 1901, was published his book 'Our
National Parks.'

President Theodore Roosevelt, planning in the early summer of
1903 a Western journey, prevailed upon the old mountaineer to
be his guide and companion in a tour of the Yosemite. So in
mid-May they spent three days and three nights together in
the wildness talking 'forest good.' The result of this conference
was the immediate strengthening of the President's conservation
policy.

A few days after this historic trip John Muir set off with his
friends Charles S. Sargent and son Robeson on a world tour.
Returning home a year later, he helped bring to a triumphant
issue, in 1905, the long warfare for the recession of the Yosemite
Valley. But on the heels of this victory came anxiety and sorrow.
The health of his daughter Helen was again precarious, and she
was ordered to seek the dry climate of Arizona. Her father accompanied her, but was soon called home by the illness of his
wife, who died August 5, 1905. Returning to the south, Mr.

Muir, riding with Helen over little-known parts of the State discovered new areas strewn with fossil trees, which were incorporated in the Petrified Forest National Monument set aside by President Roosevelt in December, 1906.

The years yet remaining to John Muir were filled with labor. His efforts to set in order his notes and memories were largely frustrated by public demands upon him, especially by the 'everlasting Hetch Hetchy fight.'

In 1911, he journeyed to South America and Africa, to realize at last that dream of his youth of sailing up the Amazon and visiting the araucaria and baobab trees in their native habitats. It is a noteworthy fact that the loneliness and weariness of age never quenched his joy in the wilderness.

He returned home in 1912, to throw his final energies into the desperate cause of the Hetch Hetchy. Dr. William Frederic Badè went to Washington as a representative of the Sierra Club. Lovers of nature all over America redoubled their efforts to save this threatened portion of the Yosemite National Park, but all to no avail. In the summer of 1913, the Raker Bill was rushed through Congress and the beautiful valley was delivered over to be despoiled.

John Muir, broken in health, but serene and courageous, retired to the labor of preparing his Alaska notes for publication. When he died on Christmas Eve, 1914, in a Los Angeles hospital, some of the proof sheets lay scattered on his bed where he had laid them down.

- - - - -

1. *Trees*

Undated.

Sit down in climbing, and hear the pines sing. So savage and inaccessible at first sight, yet so easy to find one's way. No conscious effort, no dangers to tell about, for all seems a grand, smooth song.

A tree does good like medicine, enriching the air with rosin and balsam, and its beauty, type of steadfast strength, makes one feel immortal.

About 1900.

Sequoia. We are often told that the world is going from bad to worse, sacrificing everything to Mammon. But this righteous uprising in defense of God's trees in the midst of exciting politics and wars is telling a different story, and every sequoia, I fancy, has heard the good news and is waving its branches for joy. The wrongs done to trees, wrongs of every sort, are done in the darkness of ignorance and unbelief, for when light comes, the heart of the people is always right. Forty-seven years ago one of these Calaveras king sequoias was laboriously cut down, that the stump might be had for a dancing floor. Another, one of the finest in the grove, more than three hundred feet high, was skinned alive to a height of one hundred and sixteen feet from the ground and the bark sent to London to show how fine and big that Calaveras tree was — as sensible a scheme as skinning our great men would be to prove their greatness. This grand tree is, of course, dead, a ghastly disfigured ruin, but it still stands erect and holds forth its majestic arms, as if alive and saying, 'Forgive them, they know not what they do.' Now some millmen want to cut all the Calaveras trees into lumber and money. But we have found a better use for them. . . .

Could one of these sequoia kings come to town in all its god-like majesty, so as to be strikingly seen and allowed to plead its own cause, there would never again be any lack of defenders. And the same may be said of all the other sequoia groves and forests of the Sierra with their companions and the noble *Sequoia sempervirens*, or redwood, of the Coast Mountains.

In the noble groves and forests to the southward of the Calaveras the axe and saw have long been busy, and thousands of the finest sequoias have been felled, blasted into manageable dimensions, and sawed into lumber by methods destructive almost beyond belief, while fires have spread still wider and more lamentable ruin. In the course of my explorations twenty-five years ago I found five sawmills located on or near the lower margin of the sequoia belt, all of which were cutting more or less Big Tree lumber, which looks like the redwood of the coast and was sold as redwood. One of the smallest of these mills in the season of 1874 sawed two million feet of sequoia lumber. Since that time other mills have been built among the sequoias, notably the large ones on Kings River and at the head of the Fresno. The destruction of these grand trees is still going on.

On the other hand, the Calaveras Grove for forty years has been faithfully protected by Mr. Sperry,[1] and with the exception of the two trees mentioned above is still in primeval beauty. The Tuolumne and Merced Groves near Yosemite, the Dinky Creek Grove, those of the General Grant National Park and the Sequoia National Park, with several outstanding groves that are nameless on the Kings, Kaweah, and Tule River Basins, and included in the Sierra Forest Reservation, have of late years been partially protected by the Federal Government, while the well-known Mariposa Grove has long been guarded by the State.

For the thousands of acres of sequoia forest outside of the reservation and national parks, and in the hands of lumbermen, no help is in sight. Probably more than three times as many sequoias as are contained in the whole Calaveras Grove have been cut into lumber every year for the last twenty-six years without let or hindrance, and with scarce a word of protest on the part of the public, while at the first whisper of the bonding of the Calaveras Grove to lumbermen almost everybody rose in alarm. This righteous and lively indignation on the part of Californians after the long period of death-like apathy, in which they have witnessed the destruction of other groves unmoved, seems strange, until the rapid growth that right public opinion has made during the last few years is considered, and the peculiar interest that attaches to the Calaveras giants. They were the first dis-

[1] Mr. James L. Sperry, a pioneer of California, owned the Calaveras grove of Big Trees from about 1853 to 1901, when he sold it to Mr. Robert Whitehead on condition that the sequoias should not be cut down.

covered and are best known. Thousands of travelers from every country have come to pay them tribute of admiration and praise, their reputation is world-wide, and the names of great men have long been associated with them — Washington, Humboldt, Torrey and Gray, Sir Joseph Hooker, and others. These kings of the forest, the noblest of a noble race, rightly belong to the world, but as they are in California we cannot escape responsibility as their guardians. Fortunately the American people are equal to this trust or any other that may arise as soon as they see and understand it.

Any fool can destroy trees.[1] They cannot defend themselves or run away. And few destroyers of trees ever plant any; nor can planting avail much towards restoring our grand aboriginal giants. It took more than three thousand years to make some of the oldest of the sequoias, trees that are still standing in perfect strength and beauty, waving and singing in the mighty forests of the Sierra. Through all the eventful centuries since Christ's time — and long before that — God has cared for these trees, saved them from drought, disease, avalanches, and a thousand storms; but he cannot save them from sawmills and fools; this is left to the American people!

2. *Arizona Notes*

Undated.

The shimmering plains of Arizona — the burning of leaves. Eternal verities — basic truths. Radiant days and months, hot, dry, with a tinge of sadness and haze on the horizon.

A great calm day fading like a flower purple in the morning, and glowing noon, purple and crimson in the evening — fading like a flower.

There are no shallow, colorless people in Arizona. Like the rings of wood in trees, the wonder-working climate has been

[1] See *Our National Parks*, pp. 392–93.

absorbed in rings of character. These people meet misfortune with indifference — a characteristic of strenuous life. Small, finicky-minded men, full of mixed odds and ends of curiosities, are rare here, likely to be lost in wide horizons. The cowboys on horseback are superbly statuesque without an apparent desire to pose.

I never breathed air more distinctly, palpably good. It is clean, fresh, and pure as the icy Arctic air. It fairly thrills and quivers, as if one actually felt the beatings of the infinitely small vital electric waves of life and light drenching every cell of flesh and bone, bringing on a complete resurrection after the death of sound sleep. It is easy then to believe in the theory of vibrating light and heat waves.

A Sand Storm. Gleaming particles of quartz, swift on the wind, flying thick as the dust particles of snow-crystals in a frosty blizzard. At first mostly confined to sand-washes of the streams, now on account of pasturage it prevails everywhere, uprooting sage bushes, etc., making drifts on every brow and hill, and over all the plateau.

Don Pedro. Kindness is not taught, not acquired, but is a gift direct from God, a natural force invisible.

He acted unconsciously like a law of nature, as if he couldn't help it any more than the sun could help shining. He did the tenderest, finest, rarest things as naturally as common things, so naturally one felt them rather than saw them done — all alike with apparent unconsciousness. We were drawn into full fellowship with him without knowing how or why, establishing friendship without time for its growth as one instantly feels the warmth of fire. Indeed, a steady mellow sympathy radiated from his presence like heat from glowing coals. Him I regard as one of the noblest, most devoutly revered enrichments of my life, on which distance, and even death, makes no difference.

He referred only once to the Apache War. When walking a little way from his magnificent hacienda, I was admiring the broad, spacious valley with its massive mountain walls and ranges; he pointed to an isolated peak on the west side and said it always reminded him of the wild Apache War, for on its top he had oftentimes seen their signal smoke columns rising to give warning of the approach of troops.

How wide he sowed, and what a harvest he reaped! His whole life long and broad, like his twenty-by-thirty-mile ranch, with its famous flocks and herds, revealed steadfast courage and will. But not these were great, but the man himself. . . . Cold writing is a feeble medium for heart-hot ideas.

3. *Thoughts Written on the Birthday of Robert Burns*

January 25, 1906.

It is surely a fine thing to stop now and then in the throng of our common everyday tasks to contemplate the works and ways of God's great men, sent down from time to time to guide and bless mankind. And it is glorious to know that one of the greatest men who appeared in the last century was a Scotsman, Robert Burns. . . . His lessons of divine love and sympathy to humanity, which he preached in his poems and sent forth white-hot from his heart, have gone ringing and singing around the globe, stirring the heart of every nation and race.

And yet what a hard, sad life he had in his own Scotland, amang his ain folk. 'The largest soul of all the British lands,' said Carlyle, and perhaps no man had so false a reception from his fellowmen. Wae's me that Scotsmen let our best Scotsman starve. And though now he has love and honor beyond bounds, and noble monuments to his worth are rising in every land on the globe, the idea of Burns forlorn and starving in Scotland blinds us with tears. He died a hundred and ten years ago in a storm of trouble and pain, full of despairing care about his wife and bairns, deserted by his canny fault-counting friends. But in the midst of it all he knew something of the worth of his short life's work. . . . When lying forsaken in the shadow of death, he said to his despairing wife, 'Never mind, I'll be more respected a hundred years after I am dead than I am now.' How gloriously this prophecy has been fulfilled! His fame began to grow from the day of his death, and year by year it has grown higher and brighter, cheering and enriching all mankind. In the halls of fame there is none like his. 'The birthday of no other human being is so universally celebrated'; and, as Lord Rosebery well says, 'He reigns over a greater dominion than any empire the world has yet seen, and his name excites a more enthusiastic worship than that of any saint in the calendar.' And this marvelous ever-growing admiring devotion is perfectly natural. Could Burns have seen it, how glad he would have been! What is the secret of it all? It is his inspiring genius derived from heaven, glowing with all-embracing sympathy. The man of science, the naturalist, too often loses sight of the essential oneness of all living beings in seeking to classify them in kingdoms, orders, families, genera, species, etc., taking note of the kind and arrangement of limbs, teeth, toes, scales, hair, feathers, etc., measured and set forth in meters, centimeters, and millimeters, while the eye of the Poet, the Seer, never closes on the kinship of all God's creatures, and his heart ever beats in sympathy with great and small alike as 'earth-born companions and fellow mortals' equally dependent on Heaven's eternal love. As far as I know, none in all the world so clearly recognized the loving fatherhood of God as our ain Robert Burns, and there has been none in whose heart there flowed so quick and kind and universal a sympathy. One calls to mind his field mouse, 'Wee, sleekit, cowrin', tim'rous beastie,' turned out of house and home, its store of food scattered and cold winter coming on; the tender pity for silly sheep and cattle, and ilk hopping bird, 'wee helpless thing' shelterless in a winter snowstorm; the wounded hare crying like a child; the unfortunate daisy, 'wee, modest, crimson-tippèd flower' crushed amang the stoure. He extended pity and sympathy even to the deil, entering into his feelings and hoping he might perhaps be able to repent and escape from his gloomy den.

> 'Hear me, Auld Hangie, for a wee.
> An' let poor damnèd bodies be;

> I'm sure sma' pleasure it can gie,
>   Ev'n to a deil,
> To skelp an' scaud poor dogs like me
>   An' hear us squeel....
>
> But fare-you-weel, Auld Nickie-Ben!
> O, wad ye tak a thought an' men'!
> Ye aiblins might — I dinna ken —
>   Still hae a stake:
> I'm wae to think upo' yon den,
>   Ev'n for your sake!

Many a song he sang in the few troubled years allotted him, and he made all the world his debtor. But Scotland's debt is in several ways peculiar. He brought her forward into a bright light and made her great among the nations, and he saved the grand Scottish language when it was in danger of sinking into English. Though unfit for science it is wonderfully rich in love-words for telling 'a' the pleasure o' the heart, the lover and the friend.' And since Burns's poems are enshrined in gude braid Scots, the world will never allow it to perish.

None in this land of plenty can realize the hardships under which Burns's immortal work was accomplished. Of what we call education he had almost nothing — he was brought up on the Bible in his father's auld clay biggin. This was his school and college, his poor neighbors and the fields and the sky his university. He sang untrained like a stream or a bird, while under the crushing weight of doure unchangeable poverty — a kind of poverty unknown in America, where doors open everywhere to affluence and ease. When he was in the fullness and strength of early manhood, standing five feet ten, his great eyes flashing, such eyes as Walter Scott said he had never seen in any other countenance, as bold and brave and bonnie a chiel as ever trod yird, he toiled from daybreak till dark, digging, plowing, reaping, thrashing for three dollars a month!

On my lonely walks I have often thought how fine it would be to have the company of Burns. And indeed he was always with me, for I had him by heart. On my first long walk from Indiana to the Gulf of Mexico I carried a copy of Burns's poems and sang them all the way. The whole country and the people, beasts and birds, seemed to like them. In the Sierra I sang and whistled them to the squirrels and birds, and they were charmed out of fear and gathered close about me. So real was his companionship, he oftentimes seemed to be with me in the flesh, however wild and strange the places where I wandered — the Arctic tundras so like the heathery muirlands of Scotland, the leafy Alleghanies, icy Alps and Himalayas, Manchuria, Siberia, Australia, New Zealand — everywhere Burns seemed at home and his poems fitted everybody.

Wherever a Scotsman goes, there goes Burns. His grand whole, catholic soul squares with the good of all; therefore we find him in everything everywhere. Throughout these last hundred and ten years, thousands of good men have been telling God's love; but the man who has done most to warm human hearts and bring to light the kinship of the world, is Burns, Robert Burns, the Scotsman.

4. *Yesterday, Today, and Tomorrow*

Undated.

Emerson [1] was the most serene, majestic, sequoia-like soul I ever met. His smile was as sweet and calm as morning light on mountains. There was a wonderful charm in his presence; his smile, serene eye, his voice, his manner, were all sensed at once by everybody.

I felt here was the man I had been seeking. The Sierra, I was sure, wanted to see him, and he must not go before granting them an interview! A tremendous sincerity was his. He was as sincere as the trees, his eye sincere as the sun.

Undated.

It is my faith that every flower enjoys the air it breathes. Wordsworth, Professors Wagner, French, and Darwin claim that

[1] For an account of Emerson's friendship with John Muir, see Badè's *Life and Letters of John Muir*, vol. I, pp. 252–62.

plants have minds, are conscious of their existence, feel pain and
have memories.

September 4, 1908.

Happy the man to whom every tree is a friend — who loves
them, sympathizes with them in their lives in mountain and plain,
in their brave struggles on barren rocks and windswept ridges,
and in joyous, triumphant exuberance in fertile ravines and valleys sheltered, waving their friendly branches, while we, fondling
their shining plumage, rejoice with and feel the beauty and
strength of their every attitude and gesture, the swirling surging
of their life blood in every vein and cell. Great as they are and
widespread their forests over the earth's continents and islands,
we may love them all and carry them about with us in our hearts.
And so with the smaller flower people that dwell beneath and
around them, looking up with admiring faces, or down in thoughtful poise, making all the land or garden instinct with God.

1913 (?)

Damming Hetch Hetchy. A great political miracle this of
'improving' the beauty of the most beautiful of all mountain
parks by cutting down its groves, and burying all the thickets of
azalea and wild rose, lily gardens, and ferneries two or three
hundred feet deep. After this is done we are promised a road
blasted on the slope of the north wall, where nature-lovers may
sit on rustic stools, or rocks, like frogs on logs, to admire the
sham dam lake, the grave of Hetch Hetchy. This Yosemite Park
fight began a dozen years ago. Never for a moment have I
believed that the American people would fail to defend it for the
welfare of themselves and all the world. The people are now
aroused. Tidings from far and near show that almost every good
man and woman is with us. Therefore be of good cheer, watch,
and pray and fight!

Everything is Poetry. A bird is not feathers of certain colors,
or members of a certain length — toes, claws, bill, gape, culmen

— and of group so and so. This is not the bird that at heaven's gate sings, any more than man is a vertical vertebrate, five and one half feet long, with so many teeth and bones forming a right angle with the ground, though even all this is good in its way. On the rim of the Yosemite I once heard a man say: 'How was this tremendous old rocky gorge formed?' 'Oh, stop your science,' said another of the party. 'Hush! stand still and behold the glory of God!'

I suppose silent wonder would have been better, more natural at first. Still, as the warmth and beauty of fire is more enjoyed by those who, knowing something of the origin of wood and coal, see the dancing flames and are able to contemplate the grand show as having come from the sun ages ago, and slowly garnered in cells, so also are those Yosemite temples the more enjoyed by those who have traced, however dimly, the working of the Divine Mind in their making, who know why domes are here, and how sheer precipitous walls like El Capitan were predetermined by the crystallization of the granite in the dark, thousands of centuries before development, and who know how in the fullness of time the sun was called to lift water out of the sea in vapor which was carried by the winds to the mountains, crystallized into snow among the clouds, to fall on the summits, form glaciers, and bring Yosemite Valley and all the other Sierra features to the light. In offering us such vistas, thereby increasing our pleasure and admiration, Science is divine!

Hunting. Making some bird or beast go lame the rest of its life is a sore thing on one's conscience, at least nothing to boast of, and it has no religion in it.

This grand show is eternal. It is always sunrise somewhere; the dew is never all dried at once; a shower is forever falling; vapor is ever rising. Eternal sunrise, eternal sunset, eternal dawn and gloaming, on sea and continents and islands, each in its turn, as the round earth rolls.

Clouds. Think of the cooling shadows of summer clouds which benevolent Nature spreads over her darling forests and gardens — summer shadows of wonderful depth and brilliancy like the wings of a mother bird over her young.

The touch of invisible things is in snow, the lightest, tenderest of all material. I have lain in the calm deeps of woods with my face to the snowflakes falling like the touch of fingertips upon my eyes.

All for Each. The winds wander, the snow and rain and dew fall, the earth whirls — all but to prosper a poor lush violet!

There is a musical idea in every form. See, hear, how sharp, loud, and clear-ringing are the tones of the sky-piercing peaks and spires; and how deep and smooth and massive those of the swelling domes and round-backed ridge-waves; and how quickly the multitude of small features in a landscape suggest hurrying trills and ripples and waves of melody. We not only see the forms and colors of the mountains, but hear them. Plants and animals also seem to be music both in form and color. Everything breaks forth into form, color, song, and fragrance — an eternal chorus of praise going up from every garden and grove, a wide range of harmonies leading into the inner harmonies that are eternal.

Not like my taking the veil — no solemn abjuration of the world. I only went out for a walk, and finally concluded to stay out till sundown, for going out, I found, was really going in.

Death. The rugged old Norsemen spoke of death as *Heimgang* — home-going. So the snow-flowers go home when they melt and flow to the sea, and the rock ferns, after unrolling their fronds to the light and beautifying the rocks, roll them up close again in the autumn and blend with the soil.

Myriads of rejoicing living creatures, daily, hourly, perhaps every moment sink into death's arms, dust to dust, spirit to spirit — waited on, watched over, noticed only by their Maker, each arriving at its own heaven-dealt destiny.

All the merry dwellers of the trees and streams, and the myriad swarms of the air, called into life by the sunbeam of a summer morning, go home through death, wings folded perhaps in the last red rays of sunset of the day they were first tried. Trees towering in the sky, braving storms of centuries, flowers turning faces to the light for a single day or hour, having enjoyed their share of life's feast — all alike pass on and away under the law of death and love. Yet all are our brothers and they enjoy life as we do, share heaven's blessings with us, die and are buried in hallowed ground, come with us out of eternity and return into eternity. 'Our little lives are rounded with a sleep.' ...

Undated.

Death is a kind nurse saying, 'Come, children, to bed and get up in the morning' — a gracious Mother calling her children home.

Index

INDEX

Abbott, Gen. Henry L., 356 and n., 357, 360, 363, 364
Abies amabilis, 183 and n., 286, 381
Abies balsamea, 383
Abies bracteata, 360, 361
Abies concolor, 78, 285 n., 286 n., 343, 362. See also Fir, silver
Abies grandis, 183 and n., 285, 286, 358
Abies magnifica, 36, 78, 286 n. See also Fir, silver
Abronia, 363
Adenostema, 361, 362
Adiantum. See Fern, maidenhair
Afognak Island, 400
Agassiz, Alexander, 356 n.
Agassiz, Louis, 266
Agrostis, 89, 212
Airbergs, 128
Akenia, 146
Alaska, first journey to, 245–75; second journey to, 245, 275–80; cruise of the Corwin to, 245; islands on coast, 249–51, 268; long days of, 255; rain in, 255, 256; fisheries, 259, 260; agriculture, 260, 261; fourth journey to, 298–322; writing of, 335–37; Harriman Expedition to, 378–426
Alaska Peninsula, 399–401
Albatross, sooty, 394
Alder, 45, 88, 285, 286, 293, 317, 330, 348, 356, 390, 394, 395, 416
Aleutian Islands, 401, 404, 405, 411
Alexander Archipelago, 307
Alfilaria, 22
Alhambra Valley, 244, 245; life at the ranch in, 281, 282, 334–42, 354, 355
All for each, 439
Allen, Francis H., 366
Allium, 338
Allosorus, 286, 391
Amazon River, 428
Amelanchier. See Shad-bush
Andromeda, 411
Andromeda polifolia, 381, 382
Anemone, 400, 405, 410
Anemone nuttalliana, 146
Anemone occidentalis, 145–47
Annette Island, 381, 382

Antelope, 194
Ant-lions, 107, 210
Ants, 25, 45
Apache War, 433
Aplopappus, 106, 112
Aralia, 263
Arborvitæ (Thuja), 290–93, 358, 419, 421. See also Thuja gigantea
Arbutus, 358
Arcata, 358
Archer, Fla., 375
'Arches, royal,' 75
Arctostaphylos, 396, 400
Arizona, with the National Forest Commission in, 362, 363; later trips to, 427, 428, 431–33; characteristics of, 431, 432
Arnica, 410
Ash, 290, 356
Ash, mountain. See Mountain-ash
Ashland, Ore., 356
Aspen, 185, 316. See also Poplars
Aspidium, 263, 291, 394
Asplenium, 348
Asters, 106, 146, 176, 395, 397
Astragalus, 410
Atlantic Monthly, 365 n., 366 n.
Atmospheric effects, 12
Auk Glacier, 383
Auklets, 407
Aurora borealis, 321, 350
Avalanche Canyon, 325, 326, 328
Avalanche Cascades, 325
Avalanches, in the Yosemite, 123; scouring action of, 123; appearance and sound of, 124; breaking off trees, 142
Averell, Elizabeth, 384 n., 391
Averell, W. H., 420
Azaleas, 62, 76, 79, 91, 104, 209, 309, 344, 348

Baccharis, 23
Badè, Dr. William Frederic, 428
Baeria, 238
Bahia, 112
Baird Glacier, 382
Bald Mountain Dome, 229
Balm of Gilead, 91, 348

Balsam, 394
Banner Peak, 155, **156**
Banyan, 372, 373
Barometer, a mountain, 37, 38
Bath, a cascade, 121
Bats, 210
Bear, black, a boy and his first bear, 328
Bear, grizzly, 56, 328
Bear, Kodiak, 402 and n.
Bear Cascade, 325
Bear Meadow, 327
Bears, 85, 193, 328, 329, 344; their wallows, 73; thoughts on Sierra bears, 82, 83; roaring, 92
Bearskin Meadow, 323
Beauty, as a synonym for God, 208
Beaver, 242
Beetles, 210
Behring Sea, 405-11
Bell Springs, 359
Berkshire Hills, Mass., 368
Bidwell, Gen. and Mrs. John, 190, 237
Bidwell Landing, 190, 236
Big Creek, 174, 217
Big Meadows, 179
Big Smoky Mountains, 365
Big Trees, 174
Big Tuolumne Meadows, 180
Birch, 188, 317, 383, 400-02, 409, 414
Birds, 56, 81, 85, 120, 158, 210, 217, 225, 227, 237, 239, 270, 285, 315; a flock of small waders, 28; two splendid waders, 30; a sermon from a bird, 58; in winter, 87; cans for, 116; 'blue loons,' 239; the real bird, 437, 438
Black, A. G., 172, 191 n., 349
Black, Mrs. A. G., 172, 191 n.
Black Meadow, 331
Black Peak, 81
Black Springs, 349
Blackberries, 289
Blackbird, 287
Black's, 106
Black's Hotel, 191 n., 192
Black's Ranch, 191
Bloody Canyon, 93, 195
Bluebells, 317, 401
Bluebird, 104
Bluebird, arctic, 323
Blue-curls, 216
Boat, the Spoonbill *alias* Snagjumper, 237, 240
Bobolink, 165
Bogoslof Island, 406
Books, 94, 95
Booming Cascades, 326
Boston, 364, 366
Boulder Camp, 345

Boulder Creek, 333
Boulder Valley, 175
Bower Cave, 191 n.
Bracken, 54, 185. *See also* Pteris
Brady Glacier, 389, 394
Brady, Gov. John Green, 387
Brandywine Creek, 371
Brant, 237
Brewer, William H., 356 n., 380, 401
Bridal Veil Fall, 63
Bridal Veil meads, 349
Brigham, Dr., 287
British Columbia, coast of, 247 ff., 380, 381
Brodiæa, 32, 33, 133, 289, 338
Brooklime, 216
Brown, an Englishman, 193-200
Brown, Mr., 420
Brownie, a mule, 190
Brownlee, Mr., 394
Bryanthus, 79, 158, 236, **296**, 316, 394, 395
Buckbean, 381
Buckeye, 182, 355
Bumblebee, 94
Bunting, McKay's snow, 410 and n.
Bunting, snow, 386 and n.
Burns, Robert, 320; quoted, 24, 434, 435; thoughts written on birthday of, 433-36
Burroughs, John, on the Harriman Expedition, 378, 380, 400, 405 and n., 406 n., 407, 411, 413, 420, 421; his 'Narrative of the Expedition' quoted, 382 n.; kept aboard ship by Muir, 405 n., 406 n.; Muir's rhymes about, 422-26
Butte Creek, 194, 238
Butterflies, 2, 46, 65, 103, 112, 204, 207
Buzzard, California, 211
Buzzards, 212, 239

Calamagrostis, 216, 219
California, flora of, 58
California, S.S., 246, 247
Calochortus, 32
Caltha, 395, 397
Calycanthus, 56
Calypso, 388
Camassia, 216
Cambridge, Mass., 366
Campanula, 414
Campfire, a mountain, 218, **219**
Canary (goldfinch?), 28
Canby, Miss, 366
Canby, William Marriott, 334, 364 n., 366, 371
Cape Flattery, 421

Cape Spencer, 417
Capitan, El, 59, 60, 90, 137; smoking, 39; a climb up and down, 40, 41
Carduaceæ, 238
Carex, 42, 66, 70, 72, 73, 88, 146, 158, 162, 173, 330, 400. *See also* Sedges
Carr, Mrs. Ezra S., letter to, 1, 2
Carroll Glacier, 385
Casa Nevada, 172 and n., 173, 283 n.
Cascade Creek, 10, 204, 286
Cascade Lake, 287
Cascade Mountains, 290, 291
Cascade Reservation, 380
Cascades, small, 121, 171, 202
Cassiar, S.S., 262
Cassiope, 67, 79, 80, 167, 314, 316, 389, 394–98
Castilleia, 214, 286, 355, 390, 400
Castle Creek, 26
Castle Rock, 166
Cataract Glacier, 398 and n.
Cathedral Creek, 164
Cathedral Meadows, 89
Cathedral Peak, 96, 164
Cathedral Rocks, 60, 104, 105, 109, 116, 166, 324
Cattle, drowned, 244
Cat, 'Tom,' 342 and n.
Cave Dome, 325
Ceanothus, 38, 52, 55, 79, 214, 231, 322, 343, 360, 361, 379, 421
Cedar, black, 416
Cedar, incense (*Libocedrus decurrens*). *See* Libocedrus
Cedar, yellow. *See* Cypress, yellow
Cedar Keys, 374, 375
Cedars, 291, 419
Celery, wild, 387
Century Magazine, The, 337 and n., 367 n., 369
Cephalanthus, 239, 242
Chænactis, 141
Chamæbatia, 230
Chamæcistus, 410
Charpentier Glacier, 385
Chatham Straits, 262
Checan, 257
Cheilanthes, 286
Cheilanthes gracilis, 106
Cherry, wild, 49, 52, 173, 174, 209, 231, 290, 322, 344
Chestnut, 218, 359
Chickadee, chestnut-backed *or* red-poll, 400 and n.
Chickadees, 107
Chico, 236
Chilcat, 260, 262
Chinamen, 13, 394

Chinquapin, 209, 421
Chipmunk (*Tamias*), 143, 144, 207 and n., 211, 287
Chiquito Buttes, 174
Chiquito Creek, 174
Chiquito River, 222
Choate, Joseph H., 369
Chukchis, 407, 408
City of Pueblo, S.S., 302, 308
Civilization, 317; evils of, 85, 191, 192
Clark, Galen, 173 n., 174, 176, 178, 179, 349
Clark Range, 81
Clark's Meadows, 173
Clark's Station, 173
Claytonia, 407, 411
Clematis, 356, 363
Climbing, difficult, 73, 74, 77, 78, 149–51, 296; joy of, 329
Clitoria, 372
Clouds, imposing, 64, 297; shadows of, 439
Clouds' Rest, 59, 86, 102, 107
Clove, alpine, 216
Clover, 32
Coast Range, 15, 23, 24, 32, 55
Cocoanut Grove, 372, 373
Colby, Judge, 55
Colby, Gilbert Winslow, 55 n.
Colby, William E., 55 n., 299
Coleman, Mr., 342
Collier's, 242
Columbia Glacier, 394
Columbia River, 246, 247, 421
Columbines, 73, 76, 179
Colusa, 237
Compositæ, 19, 21–23, 25, 27, 30, 31, 55, 56, 76, 79, 80, 146, 179, 372, 376, 390, 391
Composite Island, 387
Connel, John ('Smoky Jack'), 1, 19
Cony, 207 and n.
Cook Inlet, 401, 402, 414, 415
Copper Creek, 327
Copper River, 393
Coptis, 259, 263, 381, 388
Cornel, dwarf, 263, 291, 293, 388
Cornel, red-stemmed, 214
Cornus, 56, 183. *See also* Dogwood
Corwin, S.S., 411 and n.
Cottonwood, 238, 330, 331, 383, 390
Coulterville, 36
Coville, Frederick V., 388 and n., 420
Coyotes, 14, 18
Cranberry, 259, 316, 388, 395, 400, 409
Cranberry, N.C., 365
Crane, sandhill, 240
Cranes, 92, 170, 224, 237; blue, 11, 56, 240. *See also* Heron, blue

Crater Lake, 356, 357
Crescent City, 357, 358
Cresses, 18, 28, 31
Crillon Glacier, 418
Crocker's, 347
Cross Sound, 418
Crow, Clark, 92, 207
Crowberry (*Empetrum*), 388, 400, 401, 405
Crows, 382
Crucifers, 49
Crystal Bay, 288
Crystal soil, 163
Cuba, 57
Cumberland Mountains, 57, 58
Cupressus nootkatensis, 380–82. *See also* Cypress, yellow
Curlews, 237, 240
Currant, 148, 263
Currant, black, 104
Currant, mountain, 209
Cypress, 419
Cypress, Lawson, 358 and n.
Cypress, yellow, *or* yellow cedar, 258 and n., 263, 388. *See also Cupressus nootkatensis*
Cypripedium, 400, 405
Cystopteris, 179

Daisies, 400, 401
Daisy, blue, 209, 348
Daisy, the, lines on, 314
Dall, Dr. William H., 391, 404, 414
Dalton Glacier, 392
Dandelions, 216
Danger, its effect on the mind, 296
Darwin, Charles, 118, 436
Davidson Glacier, 383
Dawn, 81, 82; a forest, 217, 218
Dead Lake, 85
Death, a life for every, 168; beauty of, 222; home-going, 439, 440; a kind nurse, 440
Deer, 179, 211, 348, 388; an unsuspicious doe, 143, 144; fawns, 144; two large bucks, 175; four confiding deer, 231–33
Deer, mule, 196
Deer Grove, 343
Delany, Patrick, 3, 34
Delaware, 366, 371
Dellenbaugh, Fred S., 391
Departure Bay, 248, 249
Devil-fish, 412
Devil's-club (*Echinopanax horridum*), 381, 394
Devil's Thumb, 382
Dicksonia, 57
Diller, Mr., 357

Dipper, American. *See* Water-ouzel
Disenchantment Bay, 390 and n., 391, 392
Dodecatheon, 26, 179, 216, 338, 388
Dogs, sheep, 7, 14, 23, 27, 34; Stickeen, 245, 246, 275–80, 418; death of a neighbor's dog, 340
Dogwood, 79, 81, 185, 218, 263, 286, 290, 293, 348. *See also* Cornus
Dome Cliff, 344
Dome Creek, 58
Don Pedro, 432, 433
Doran, Capt. Peter A., 378, 401, 419, 420
Douglas, David, 338
Douglas City, 419
Douglas Island, 382, 383
Doves, 329
Drabas, 409
Draper, Dorothea, 384 n.
Dream, a, 125
Dreaming, transcendental, 103
Dry Creek, 3–5, 10, 11, 14, 18
Dryas, 385, 389
Duck, wood, 237, 240
Ducks, 104, 175, 205, 225, 243, 286, 385; with young, 287
Dunbar, Scotland, 8
Duncan, William, 381
Dutch Harbor, 404, 405, 411
Dyerville, 359

Eagle, bald, 243
Eagle Cliff, 86, 90
Eagle Glacier, 383
Eagle Peak, 115
Eagle Point, 122
Eagles, 382; watching a hare, 9; and dog, 27; in a snowstorm, 114
Earthquake, Inyo, 330
Echeveria, 134
Echinopanax horridum. *See* Devil's-club
Egg Island, 390–93
Egrets, 243 n.
El Capitan. *See* Capitan
Electra Peak, 157 n.
Elliot, Daniel Giraud, 418
Emerald Bay, 286
Emerald Pool, 284
Emerson, Prof. Benjamin R., 421
Emerson, Ralph Waldo, 100; memories of, 436
Emmons, Lieut., 387, 388
Empetrum. *See* Crowberry
Enchantment Bay, 390–92
Engraving, 11
Ephemera, 339
Epilobium, 143, 162, 179, 185, 188, 216, 391

Epilobium latifolium, 390
Erigeron, 73, 80
Eriodictyon, 106
Eriogonum, 80, 81, 163, 231, 238, 295
Erysimum asperum, 208 n.
Eschscholtzia. *See* Poppy
Eskimos, 408, 409
Eunanus, 217
Eureka, 358, 359
Exercise, 98

Fairweather Range, 416, 417
Falls. *See* Waterfalls
Farmer, the, 67
Farquhar, Francis B., his 'Story of Mount Whitney,' 187 n.
Fay, Jerome, 193, 194, 196, 197
Feather River, 240
Fern, maidenhair *(Adiantum),* 63, 112, 128, 165
Fern, rock, 10, 330. *See also Gymnogramme*
Fern Ledge, 128
Fernow, Prof. B. E., 381, 401, 420
Ferns, 286. *See also individual species*
Ficus, 372, 373
Finch, house *(Carpodacus mexicanus),* 32
Finch, rosy, 406
Fiords, formation of, 267, 268
Fir, Douglas. *See* Spruce, Douglas
Fir, silver, 56, 204, 209, 219, 231, 323, 344. *See also Abies concolor* and *Abies magnifica*
Fir, white, 358 n.
Fire, a sequoia forest, 230, 231; forest fires, 351, 352
Firs, 181, 183, 211, 212, 217, 218, 221, 395
Fisher, Mrs., 402
Fisherman's Peak, 188 and n.
Flagstaff, Ariz., 362
Flicker, 85 and n., 215, 225
Florida, flora of, 58
Flowers, on the coral shores of Cuba, 57; and the poor, 352–54
Floy, Mr., 349
Flycatchers, 215, 329
Foerster Peak, 157 n.
Foggy Cape, 413
Forests, management of, 351, 352; fires in, 351, 352
Fort Wrangell, 257, 258, 260, 262, 271–73, 382
Fox, the hunter, 327, 328, 332
Fox, blue, 411
Fox Camp, 323, 330
Fox farm, 414
Fragrance, in the woods, 219

French Bar, 36
Fresno Basin, 222, 227
Fresno Dome, 213
Fresno Valley, 215
Fringe-pod, 28 n.
Fritillaria, 395, 400
Frogs, 21, 46, 49, 87, 335
Frost, crystals of, 67
Fulmar, 394

Galium, 216
Gannett, Henry, 390, 391, 395–97, 420, 421
Garden, a cliff, 133, 134
Gaultheria shallon, 291. *See also* Salal
Gayophytum, 216
Geese, wild, 130, 170, 237, 240, 243
Geikie Inlet, 387
Gentiana frigida, 154
Gentians, 142, 145, 148, 175, 176, 204, 348, 401, 409
Gentry's Station, 201
George, Judge, 379
George W. Elder, S.S., 378, 380
Geranium, 400, 401, 405, 414
Geum, 400
Gibbs, Herbert R., 366
Gifford, R. Swain, 400
Gilbert, C. K., 385 n., 387, 389, 391, 395, 414
Gilder, Richard Watson, 367, 368, 370
Gilder, Mrs. R. W., 368, 370
Gilia, 28, 30, 182, 216, 238, 327, 338
Ginkgo, 372
Girl mountaineers, 347
Glacial Period, 108
Glacier Bay, 245, 298, 302, 311–21, 384–87, 389
Glacier Cliff, 344
Glacier Monument, 327
Glacier Point, 51, 64, 86, 107, 110, 122, 123, 131, 172, 284
Glacier Point Glacier, 111
Glacier Point Mountain, 110
Glaciers, their work in Yosemite Valley, 35, 59, 60, 108–12; their work in Hetch Hetchy (Great Tuolumne Canyon), 68–72, 75, 76, 168, 345; make their own channels, 76; their work about Mt. Clark, 80; rivers of ice, 83; as writers of history, 88, 95, 96; offspring of the snow, 90; their mission, 90; eat their own offspring, 94; controlled by shadows, 108–10; and fissures in the rock, 112; of the San Joaquin, 148–51; their work in the Columbia River Valley, 246, 247; in Alaska, 258–60, 265–70, 275, 278–80, 382–87,

380–99, 402, 404, 415–19; Taylor Glacier, 275, 278–80; the Muir, 298, 302, 313–21, 385; talk with an old sea-captain on, 308, 309; their work in and about Puget Sound, 310; their food and drink, 317; music of, 317, 318; made of frost-flowers, 318; night thoughts on a glacier, 319, 320; their work in Massachusetts, 365, 368, 369; their work in New York, 370; Patterson, 382; Baird, 382; Auk, or Mendenhall, 383; Eagle, 383; Davidson, 383; discharging bergs, 384, 389, 394, 395, 399, 418; Pacific, 385, 386; Hugh Miller, 385; Charpentier, 385; Carroll, 385; Harriman, 386, 387; Brady, 389, 394; Yakutat, 389, 417; La Pérouse, 389; Malaspina, 390, 415, 417; Hidden, 391, 392; Nunatak, 391, 392; Hubbard, 391, 392; Dalton, 391, 392; Columbia, 394; Port Wells, 395, 396; Harriman (Prince William Sound), 396 and n.; Cataract, 398 and n.; Surprise, 398 and n.; Crillon, 418; Taku, 419

Glenbrook, 287, 288
Gnaphalium, 31, 112
'Go east,' 99
Goat, mountain, 316
God, in Nature, 138; never shouts, 153, 154
Goldenrods, 146, 162, 176, 231, 372
Goodrich, Capt., 387
Goodyera, 214
Gooseberries, 162
Gooseberry, purple, 47
Go-quicks. *See* Lizards
Gordon, a Scotchman, 357
Grand Canyon of the Colorado, 362, 363
Grand Cascade, 343
Grand Sentinel, 325, 330
Granite, 177
Grant Park, 322
Grant's Pass, 357
Grapevines, 63, 356; the Mission vine, 337
Grasses, 231, 263, 348; Agrostis, 89, 212; Calamagrostis, 216, 219
Grasshopper, 93
Gray, Asa, 190
Gray, Mrs. Asa, 366
Gray Mountain, 80, 139
Graysonville, 243
Great Basin, 158, 185
Great Smoky Mountains, 365
Great Tuolumne Canyon. *See* Tuolumne Canyon
Greenhorn Mountains, 185
Grenville Channel, 419

Grinnell, Dr. George Bird, 420
Grouse, 179, 287, 296; with young, 152, 153
Grouse, Canada, 414
Grub Gulch, 343
Gull, glaucous, 411 and n.
Gull, Point Barrow, 411 and n.
Gulls, 206, 243, 253, 286, 385, 394; eggs, 386, 387, 396
Guyot, Arnold, 266
Gymnogramme, 63, 106, 134, 263
Gymnogramme triangularis, 10

Haenke, Island, 390, 392
Hague, Arnold, 356 n., 357, 363
Hague, Mrs., 363
Half Dome. *See* South Dome
Hall Island, 410
Hare, and eagle, 9; habits, 10; and sheep, 21; a young, 28
Hare, little chief, 207 and n.
Harriman, Carol, 396 and n.
Harriman, Cornelia, 384 n., 391
Harriman, Edward H., 378, 387, 395, 396, 399, 402, 415, 416, 421
Harriman, Mrs. Edward H., 380, 401, 421
Harriman, Mary, 384 n., 391
Harriman, Roland, 396 and n.
Harriman Alaska Expedition, 378–426
Harriman Alaska Expedition, The, 413 and n.
Harriman Fiord, 397
Harriman Glacier, Glacier Bay, 386 and n., 387
Harriman Glacier, Prince William Sound, 386 n., 396 and n.
Harris, 359
Harwell, C. A., 172 n.
Hawk, fish, 243
Hawkins, Mr., 380
Hawks, 80, 170, 212, 216, 221, 224, 354
Hazel, 290, 293
Helenium, 216
Heliotrope, 30
Hemlock, 258, 291, 292, 357, 358, 419
Hemlock, mountain (Williamson spruce), (*Tsuga mertensiana* or *pattoniana*), 69 and n., 70, 78, 141 and n., 143, 146, 147, 154, 328, 394, 395, 397, 398
Hemlock, Western (*Tsuga heterophylla* or *mertensiana*), 263, 293, 381–83, 393, 416
Hemlock Mountain, 313
Hepaticæ, 13. *See also* Liverworts
Hepburn, a Scotchman, 193–200
Hermit, a, 223
Hermit Canyon, 325

Hermit Towers, 325
Heron, blue, 337. *See also* Cranes
Herons, 170, 240 and n., 243 and n., 342, 354
Hetch Hetchy. *See* Tuolumne Canyon
Hickory, 374, 376
Hidden Glacier, 391 and n., 392
Hills Ferry, 243
Hobson, Richmond Pearson, 369, 370
Hodgson, Mr. and Mrs., 375
Hoffman Glacier, 172
Hoffman Range, 59, 164
Hogs, 244
Holkam Bay, 419
Homer, Alaska, 399, 414, 415
Honeysuckles, 79, 107, 112, 293, 330
Hooker, Sir Joseph, 190
Hopeton, 1, 33, 241
Hornet (yellow-jacket), 364
Horse Corral, 333
Horses, 356, 359; a horse and a hornet, 364
Hosackia, 209
Hosackia grandiflora, 216
Hot Springs, Alaska, 388
Houghton, Mifflin & Co., 366 n.
Housatonic River, 367
Howling Valley, 384
Hubbard Glacier, 391
Huckleberries, 226, 289
Huckleberry, 381, 382, 386, 388, 389
Huckleberry, black, 291
Huckleberry, red, 291, 293, 358, 419
Hudson's Bay Company, 248, 307
Hugh Miller Glacier, 385
Hummingbird, rufous, 382 n., 398 and n.
Hummingbirds, 209, 225, 330
Humphrey, Capt. O. J., 396, 407, 411
Hunting, love of, 199; no religion in, 438
Hunt's, 356
Huntsville, Tenn., 365
Hutchings, J. M., 34, 35, 37, 349
Hyde's Mill, 229
Hydrophyllum, 400
Hypnum, 293

Icebergs, 311, 312, 384-87, 389, 394-96, 416, 418, 419
Icy Bay, 399, 415
Idleness, enjoyment of, 288
Iluiliuk, 405 n., 406 n.
Illilouette Basin, 36
Illilouette Canyon, 64, 104, 109
Illilouette Fall, 55, 58
Illilouette Glacier, 111, 172
Illilouette Valley, 172
Imagination, power of, 226
Immortality, 89

Independence, Cal., 187
Indian Canyon, 49, 86, 103, 104, 110, 111, 114, 122, 123
Indian Jim, 395-98
Indian Point, Siberia, 408
Indian summer, 97
Indians, 94, 184, 207, 227, 291, 359, 364; Modoc, 195; in Alaska, 257, 259, 262, 264, 270-75, 381, 386, 387, 392, 393; a Stickeen village, 270-72; Washoe, 288; Indian Henry, 296; and nature, 315; Alaskan missions to, 381; hunting seals, 392, 393; Indian Jim, 395-98; Aleuts, 397
Indio, 376
Ingraham, Prof., 289 n.
Insects, the air full of, 162
Inverarity, D. G., 396, 397
Inyo Mountains, 150, 185
Irwin, Mr., 338
Islet Lake, 157-59
Ivesia, 163, 179, 219, 323

Jackson, Rev. Sheldon, D.D., 262, 269
Jaeger, long-tailed, 409
Jamestown, S.S., 273, 274
Jay, Steller, 76, 217
Jays, 104, 107, 114, 122, 209, 215, 225, 286
Jepson, Dr. Willis Linn, 299
Johnson, Robert Underwood, 282, 367, 369-71
Jumbo-limbo, 373
Junco, Oregon, 382, 394
Juneau, 382, 383, 419
Jungermannia, 293
Juniper, 176, 196, 328, 376; damaged by an avalanche, 143
Juniperus occidentalis, 78

Kachemac Bay, 399
Kalmia, 79, 368, 389, 395
Kalmia glauca, 381
Kaweah River, 183, 229-31
Kearney, Thomas H., Jr., 390
Kearsarge Pass, 186
Keeler, Charles A., 379, 380, 382 n., 401, 413, 419; his *Friends Bearing Torches* quoted, 405 n., 406 n.; his 'Days among Alaska Birds,' 413 n.
Keith, William, 189, 201, 202, 282, 289 n., 296, 338, 341
Kellogg, Dr. Albert, 173 n., 176, 185, 186
Kelly, Capt., 396-98, 405, 412, 419
Kendel, Dr., 262
Kern River, 185
Key West, 373
Kichamak Bay, 414, 415
Killdeer, 237

Killing, in nature, 93
Kincaid, Prof. Trevor, 412-14
King, Clarence, 187 n.
King Island, 408, 409
King Mountain, 329
Kingfisher, 102, 382 n.
King's City, 360
Kings River, 100, 180-84, 226-29, 240, 241, 322, 330-32
Kings River Cañyon, 182
Kings River Forest, 176
Kings River Yosemite, 323-32
Kirby, Mr. and Mrs., 286
Kittiwake, 394
Klamath, 356
Klamath Lake, 195
Klamath Lake, Lower, 196
Klanack, 259
Knight's Landing, 240
Knowles, Frank F., 187 n.
Kodiak Island, 399-403, 414
Kukak Bay, 399, 400

La Grange, Cal., 1
Lace Cascade, 126
Ladybirds, 233
Lake, Mr., 331, 332
Lake Millar, 179
Lake Nevada, 79 and n.; nights and mornings on, 83-85
Lake Tahoe, 100; camping at, 282, 285-88
Lake Tenaya, 163-65, 204, 235, 236, 347
Lake Washburn, 79 n.
Lakes, as checks against floods, 208; as heaters and coolers, 208
Lambert, Mr., 349
Lamon's, 122, 349
La Pérouse Glacier, 389
Larks. *See* Meadowlarks
Larkspur, 33, 73, 133, 179, 214, 323, 338
Laurel, 106, 107, 112, 128, 130
Lava Beds, 189, 195, 197
Laws, 8
Leaning Dome, 325
Le Conte, Joseph, 299; extract from diary of, 35
Ledum, 79, 259, 263, 400
Lee, Mass., 368
Leersia, 214
Leidig's winter house, 118
Lemmings, 409
Leptosiphon, 28
Leuchera, 63
Leucosticte, *or* rosy finch, 406
Liatris, 372
Liberty Cap Cascades, 283
Libocedrus (*L. decurrens*) or incense

cedar, 38, 61, 66, 74, 79, 88, 111, 115, 135, 209, 221, 285, 286, 323, 343, 344, 348
Lichens, 148
Life-everlasting, 217, 323
Light, a word for, 67; morning, 81, 82; strands of, 92; broken and beaten to a foam, 127, 130; on a glacier, 319
Lilies, 30, 67, 73, 76, 128, 129, 171, 209, 327
Lilium, 32
Lily, small tiger, 286
Lily, soaproot, 133. *See also* Soaproot
Lily, Washington, 289
Lily, white, 286
Lily Hollow, 26
Lindley, Dr., 262
Linnæa borealis, 259, 293, 382
Linnæus, 153
Linnet, yellow, 286
Linnets, 215, 225
Linosyris, 175
Linum, 293
Lion Rock, 326
Liquidambar, 375
Listera, 382
Little Diamond Cascade, 126
Little Yosemite, 55, 172, 173
Lituya Bay, 402, 418
Live-oak. *See* Oak, live
Live-Oak, Fla., 375
Liverworts, 13 and n., 14, 91, 216
Lizards ('go-quicks'), 25, 26, 355
Logcock. *See* Woodpecker, pileated
Loneliness, 89, 319
Long Beach, Wash., 306
Longmire, James, 294 and n.
Longspur, Lapland, 406, 409
Lonicera, 419
Loomis, Mr., 309
Loon, 93, 270. *See also* Birds
Lost Arrow, 140, 141
Love, human and divine, 130; human, 138
Lowe Inlet; 381
Lower Klamath Lake, 196
Lukens, T. P., 347
Lupine, 162, 209, 214, 216, 327, 382, 390, 392, 410
Lycopodium, 259
Lyell Glacier, 162
Lyell group of mountains, 106
Lynn Canal, 383
Lysichiton kamtschatcense, 291 n.

McChesney, J. B., 101, 189, 201-03
McChesney, Mrs. J. B., 101
Madroña, 421

Magee, Thomas, 341 and n.
Magnolia, 376
Magnolia grandiflora, 374
Mahonia, 238
Maidenhair, 63, 112, 128, 165
Malaspina Glacier, 390, 415, 417
Mallards, 237, 240
Malva, 217
Mammoth Mountain, 157
Mangrove, 373
Manzanita, 38, 79, 103, 231, 343, 344, 360, 361
Maples, 36, 79, 173, 290, 293, 330, 356, 358, 421
Marchantia, 293
Mariposa Trees, 222
Marysville Buttes, 237, 238
Matterhorn, the, of the Sierra, 100, 159–61
Matterhorn Lake, 160–62
Mavis, 165
May, Albert, 349
Maynard, Mr., photographer, 257
Mazamas, the, 380
Meadowlarks, Western, 8 and n., 11, 13, 16, 17, 336 and n.
Meat-eating, 97
Mellichamp, Dr. J. H.. 376
Mendenhall Glacier, 383
Mentzelia, 286
Menyanthes, 397
Menyanthes trifoliata, 381
Menziesia, 382, 419
Merced Canyon, 56, 202
Merced Divide, 141, 147
Merced Glacier, 35
Merced Mountains, 59, 86, 102, 172
Merced Peak, 81
Merced River, 1, 35, 56, 169, 241, 242; a port, 57; its autumn pools, 90, 91; its young waters, 95; its meanders in the Yosemite Valley, 111
Merriam, Dr. C. Hart, 409, 413 n., 420
Meteor, 210
Metlakahtla, 381
Miami, Fla., 372, 373
Miami, S.S., 373, 374
Mice, collected in Alaska, 383, 394
Middleton Island, 415
Mifflin, George H., 366
Milk-drinking, 97
Mill Creek, 181
Mimulus, 131, 134, 293
Mimulus, musk, 209, 216
Mimulus, scarlet, 209
Minarets, the, 81, 100, 147; ascent of, 147–52
Mines, the Treadwell, 382 and n., 383; copper, 399

Mint, 216, 327
Mishacol, *or* Mishal, River, 297
Mistletoe, 45
Mobile, 375, 376
Mockingbird, 165
Modoc Indians, 195–97
Modoc Lava Beds, 189, 195, 197
Mohr, Dr. Charles T., 375, 376
Mojave Desert, 362
Monardella, 176, 216
Moneses, 401
Mono, volcanoes of, 150, 163
Mono Creek, 175
Mono Lake, 34, 205–07
Mono plains, 157
Monroe, Kirk, 373
Monroe, Mrs. Kirk, 373
Montague Island, 399
Moon, the, trees seen against, 49, 343
Moore's Mills, 322
Moose, 415
Morrill, Mr., 195
Morris, Dr. Lewis Rutherford, 412
Moses, Prof., 388
Mosquitoes, 409
Mosses, 13, 18, 165
Mt. Abbot, 174, 176
Mt. Adams, 290
Mt. Augusta, 417
Mt. Bremer, 195
Mt. Brewer, 184 n.
Mt. Clark, 59, 139; ascent of, 80
Mt. Clark Glacier, 79
Mt. Conners, 348
Mt. Cook, 416
Mt. Crillon, 417, 418
Mt. Dana, 59, 206
Mt. Darwin, 178 n.
Mt. Davis, 159 n.
Mt. Diablo, 336–38
Mt. Emerson, 179
Mt. Fairweather, 386, 417, 418
Mt. Gabb, 174, 176
Mt. Hoffman, 59, 96, 164
Mt. Hood, 290
Mt. Hubbard, 417
Mt. Humphreys, 178 and n.
Mt. Iliamna, 399, 414, 415
Mt. Isanotski, 404
Mt. Joaquin, 207
Mt. Langley, 186 n.
Mt. Lyell, 59, 102, 159, 162
Mt. Makushin, 411
Mt. Millar, 179
Mt. Muir, 187 n.
Mt. Pavloff, 403, 404
Mt. Pogromni, 404

Mt. Rainier, 282, 290, 291; ascent of, 289, 293–97
Mt. Rainier National Park, 379
Mt. Redoubt, 399
Mt. Ritter, 100, 147–50, 154–57, 176
Mt. St. Elias, 390, 415–17
Mt. St. Helens, 290
Mt. Shasta, 98, 190, 197, 281, 289, 290, 379, 421; first sight of, 189; ascent of, 189; hunting wild sheep on, 193–200
Mt. Shishaldin, 404
Mt. Starr King, 48, 50, 55, 63, 90, 105, 107, 122, 137
Mt. Tamalpais, 352
Mt. Tyndall, 100, 184 and n., 185
Mt. Vancouver, 416, 417
Mt. Wanda, 340, 355
Mt. Watkins, 141
Mt. Whitney, 100; ascent of, 186–88
Mountain-ash, 79, 185
Mountaineering, 77, 78
Mountains, as friends, 98; health and sanity in, 191, 192; man and, 215; color of, 314; fountains of men, 315
Mouse, wood, 211
Muir, David, 354 n.
Muir, Helen, 282, 335, 338, 355
Muir, John, his journals, 1; his coming to California, 1; shepherding for John Connel, 1–33; shepherding for Pat Delaney, 34; works for J. M. Hutchings in the Yosemite Valley, building and running a sawmill for fallen timber, 34, 35; exploring the Yosemite region, 35–173; down the Sierra to Mt. Whitney, 173–88; in the Yosemite Valley, 190–93; hunting mountain sheep on Mt. Shasta, 193–200; trip to the High Sierra with Swett, McChesney, and Keith, 1875, 201–07; more Sierra explorations, especially to study the sequoias, 207–35; boat-trip down the Sacramento River, 1877, 236–40; boat-trip on the Merced and San Joaquin Rivers, 241–44; first journey to Alaska, 1879, 245–75; marriage to Louie Strentzel, 245; second journey to Alaska, 1880, 275–80; trip to Alaska and the Arctic Ocean on the Corwin, 1881, 281; life on the fruit ranch in Alhambra Valley, 281, 282; ill health, 281; camping-trip to Lake Tahoe with C. C. Parry, 1888, 285–88; journey to Mt. Rainier with William Keith, 289–97; fourth journey to Alaska, 1890, 298–322; trip to Kings River Yosemite, 1891, 322–33; activities in cause of National Parks

and Forests, 334; honorary degrees from Harvard and Wisconsin, 334; life on the ranch, 1895, 335–42; writing 'The Discovery of Glacier Bay,' 335–37; slow composition, 337, 338; a mountain ramble, 342–50; at home on the ranch, 354, 355; in Oregon, California, and Arizona with the National Forest Commission, 1896, 356–64; journey East and South, 1898, 364–76; back in California, 376, 377; on the Harriman Expedition to Alaska and Behring Sea, 1899, 378–426; in the fight to save Hetch Hetchy, 427, 428, 437; publishes *Our National Parks*, 1901, 427; tour of Yosemite with Theodore Roosevelt, 1903, 427; world tour with Professor Sargent, 427; visit to Arizona broken by death of his wife, 1905, 427, 428; journeys to South America and Africa, 1911, 428; death, 1914, 428
WRITINGS
    Cruise of the Corwin, The, 245
    'Discovery of Glacier Bay, The,' 337
    'Explorations in the Great Tuolumne Canyon,' 68 n.
    Mountains of California, The, 299, 394
    My First Summer in the Sierra, 2
    'New Sequoia Forests of California, The,' 231 n.
    'Pacific Coast Glaciers,' 413 n.
    'Rival of the Yosemite,' 322 n.
    'Save the Meadowlarks,' 336 n.
    Steep Trails, 190, 193 n.
    Thousand-Mile Walk to the Gulf, A, 1, 2 n.
    Travels in Alaska, 245, 246, 298, 299 n.
    Verse in journal, 91, 314, 321, 353, 422
Muir, Mrs. John, 281, 282, 298, 383. *See also* Strentzel, Louie
Muir, Wanda, 337, 338, 355, 367 n., 369 n.
Muir Glacier, 298, 302, 313–21, 385
Muir Gorge, 345
Muir Inlet, 312–21, 387
Mule, a, 190; rescued from a river, 331, 332
Murrelet, marbled, 394
Murres, 406, 407, 411
Mushrooms, 13, 14
Music, of nature, 16; of the woods, 219; in form, 439
Myosotis, 27
Myrica, 358

Nacimiento River, 360
Nanaimo, 248, 380
National Forest Commission, 334; journey with, 356–64
National Parks, thoughts upon, 350–54
Nature, spirituality of, 137, 138; purity of, 222
Negroes, 373–75
Neil's Ranch, 242
Nemophila, 373
Nemophila maculata, 25
Nevada, 190
Nevada Cliff, 55
Nevada Fall, 55, 58, 68, 172, 173; work of Park Commission on, 283 and n; description of, 283–85
New Orleans, 376
New York, 367, 369–71
Newhall, Mr., 360
Niagara Creek, Cal., 22, 23, 26
Nisqually River, 296, 297
North Carolina, 365
North Dome, 326
North Ritter Glacier, 156, 157
North Tower, 327
Nunatak Glacier, 392
Nunatak Inlet, 391
Nuthatch, 215
Nyssa, 375, 376

Oak, black, 36, 115, 330
Oak, blue, 379. *See also* Oak, Douglas
Oak, California (*Quercus californica*), 359, 360
Oak, chestnut, 56
Oak, Douglas, 342, 360, 361
Oak, Garryana. *See Quercus garryana*
Oak, Kellogg, 323, 346, 349, 356
Oak, live, 45, 64, 111, 115, 123, 130, 182, 359, 360; character, 132, 133; in Florida, 372, 376
Oak, lobata. *See Quercus lobata*
Oak, red (*Q. rubra*), 371
Oak, scarlet (*Q. coccinea*), 371
Oak, white, 32, 330, 355, 356
Oakland, 100, 189, 192
Oaks, 51, 66, 106, 112, 180–82, 238, 288, 338, 343, 355, 360, 371
Old Mill Flat, 228
Olney, Warren, Sr., 299
Olympic Range, 248, 307, 308
Orca, 393, 394
Orchis, 401
Osborn, Henry Fairfield, 369, 370; his children, 370
Osborn, Mrs. Henry Fairfield, 369, 370
Otter, sea, 414
Ouzel. *See* Water-ouzel

Ouzel Creek, 343, 344
Overland Monthly, 68 n.
Owens River, 179
Owens Valley, 188 and n.
Owl, screech, 336
Owl, snowy, 411
Owls, 88, 192, 210, 215, 217
Oxytropis, 410

Pacheco Pass, 23
Pacific Glacier, 385
Page, Walter H., 365, 366
Page, Mrs. Walter H., 366
Paintbrush, 216
Palache, Dr. Charles, 385 n., 387, 412, 414, 421
Palatka, 374
Palemonium, 405
Palisades, 325
Palm, filifera, 376
Palmetto, 372
Park Commission, 283 and n.
Parkhurst, Ruth, 341
Parnassia, 323
Parry, Charles C., 282, 285–88
Partridge-berry, 259
Patou, Mrs., 257
Patterson Glacier, 382
Patterson Grant, 243
Pawpaw, 372
Pedicularis, 391
Pedro, Don, 432, 433
Pelican, white, 243
Pellæa, 134, 179, 286
Pellæa Brewerii, 188
Pentstemon, Brewer's, 330
Pentstemon confertus, 286
Pentstemons, 106, 112, 133, 203, 238
Pepper Flat, 359
Peril Straits, 387
Petrified Forest National Monument, 428
Phegopteris dryopteris, 400
Phegopteris polypodioides, 400
Phlox, 379
Phyllodoce, 81
Picea sitchensis. See Spruce, Sitka
Pigeons, 104, 106
Pika, 207 and n.
Pinchot, Gifford, 356 n., 357, 363
Pine, Frémont, 184 and n., 328
Pine, lodgepole *or* tamarack (*Pinus contorta*), 78, 85, 141, 174 and n., 175, 221, 226, 285, 356, 381, 383
Pine, long-leaf, 372
Pine, mountain, 153, 236, 328. *See also Pinus monticola*
Pine, muricata, 361

Pine, Murray, 356
Pine, nut or pinyon, 184 and n.
Pine, Sabine, (*Pinus sabiniana*), 360
Pine, sugar (*Pinus lambertiana*), 36, 78, 174, 183, 202, 212, 217, 218, 227, 322, 323, 330, 343, 344, 362, 421
Pine, yellow (*Pinus ponderosa*), 78, 135, 183, 188, 209, 212, 227, 287, 323, 343, 344, 356, 360, 362
Pines, 66, 96, 106, 107, 115, 118, 176, 180, 181, 221, 231; in winter storms, 117; two-leaved, 286
Pinus albicaulis, 78, 80 n., 141, 158, 159
Pinus contorta. See Pine, lodgepole
Pinus Coulteri, 361
Pinus flexilis, 80 and n., 185, 186, 188, 194, 328, 362
Pinus Jeffreyii, 285
Pinus lambertiana. See Pine, sugar
Pinus monticola, 78, 154, 381. See also Pine, mountain
Pinus muricata, 361
Pinus ponderosa. See Pine, yellow
Pinus tuberculata, 361
Plains, the, California, 5, 6
Plantago, 29
Plantago patagonica, 182
Plants, organs of, 30; as sentient beings, 436, 437
Plover, 385, 411
Plover Bay, 407, 408
Pohono Fall, 58, 59
Pohono Glacier, 109
Point Reyes, 361
Polemonium, 179, 407, 409, 414
Polygonum, 73, 216, 409
Polypodium, 63, 263
Polytrichum, 28
Poor family, a, 353
Poplars, 50, 106, 185, 394. See also Aspens, Balm of Gilead, Cottonwood, and Populus
Popof Island, 412
Poppies, 407
Poppy, California, (Eschscholtzia), 27, 30
Populus Fremontii, 361
Porpoises, 304
Port Blakely, 292
Port Clarence, 408, 409
Port Townsend, 248, 307, 309
Port Wells Glacier, 395, 396
Portland, Ore., 246, 290, 379, 380, 421
Portulaca, 27, 338
Potentilla, 217, 323, 390, 397
Poverty-grass, 342
Pratt, William, 55
Pribilof Islands, 406, 407, 411

Primula, 188, 401, 409, 410
Primula suffrutescens, 147, 148, 186
Prince of Wales Sound, 399
Prince William Sound, 393–98, 418
Princess Royal Island, 381
Ptarmigan, 316, 398, 416; eggs, 386
Pteris, 76, 106. *See also* Bracken
Puffins, 411
Puget Sound, 291
Pumice, 356
Pyramid, Kings River, 326
Pyrola, 259, 263, 264, 286, 289

Quail, 323
Quail, mountain, 223, 344
Quarry Mountain, 321
Quartzite, 259
Queen, S.S., 308
Quercus agrifolia, 360, 361
Quercus chrysolepis, 359 and n., 360
Quercus garryana, 359, 360
Quercus lobata, 360, 361, 379
Quercus vaccinifolium, 288 and n.

Rabbits, jack, 343
Rabe, Carl, 187 n.
Race Rocks, 305
Rain, 29, 42, 130, 131; the first, 6
Rainbows, 16, 18; lunar bows in spray, 61, 62
Rancho Chico, 190
Randall, Harry, 34; Yosemite explorations with, 36, 56
Ranunculus, 26, 323
Raspberry, 79, 289, 344
Raspberry Island, 400
Rat, wood, 207; a thieving, 141, 145
Rattlesnake Hill, 359
Rattlesnakes, 28, 343–46
Ravens, 319
Raymond, Cal., 342, 343
Reality, 85
Records, 88
Recreation, compulsory, 234
Red Mountain, 140, 141
Red Peak, 81
Red Slate Peak, 176
Redwood (*Sequoia sempervirens*), 228, 229, 303, 351, 358, 359, 420, 429
Reid, John, 354 and n.
Reid Inlet, 385–87
Rennie, Mr., 338
Rhamnus, 79, 112, 285, 361
Rhett Lake, 196, 197
Rhododendron, 358, 400
Rhododendron, Kamchatka, 403
Rhododendron catawbiense, 365
Rhus, 373

Ribbon Fall, 58
Ribes, 79, 388
Ridgway, Robert, 414
Ritter, Prof. William E., 401, 412
Ritter Cascade, 146, 154
Ritter Glacier, 155
Ritter Lake, 157
Rivers, wild energy of Sierra, 65
Riverside Press, 366
Roan High Bluff, 365
Roaring Falls, 324
Roaring River, 324, 332
Robins, 76, 197, 202, 209, 215, 287, 335, 337
Robinson, the artist, 330, 331
Rodgers Peak, 159 n.
Rogue River, 357, 421
Romanzoffia, 390
Roosevelt, Theodore, 427, 428
Rose, brier, 49, 76, 244, 327
Roses, planting, 339
Roses, wild, 79, 91, 175, 176, 285, 290, 293, 309, 317, 380, 414
Ross, Thomas, 338
Round Meadow, 323
Royal Arch Falls, 58
Royal Arches, 75, 110, 349
Rubus, 173, 183, 291, 293, 343, 388, 390, 400
Rubus chamæmorus, 381, 388
Rubus nutkanus. See Salmon-berry
Rue, 216
Runnel, 216
Rush Creek Canyon, 161
Rushes, 330, 348
Russell Fiord, 391

Sacramento, 240
Sacramento River, 190, 379; a boat trip down, 236-40
Safety Cove, 419, 420
Sage, 287, 376
Sage hen, 200, 207
Sailing, a fine mode of motion, 303, 304
St. Augustine, Fla., 372
St. Elias Alps, 416, 417
St. John's-wort, 216
St. Lawrence Island, 409, 410
St. Matthew Island, 410, 411
St. Paul Island, 406, 407
Salal (*Gaultheria shallon*), 291, 293
Saldovia, 414, 415
Salix Barclayi, 416
Salmon, 244, 260; canneries, 259
Salmon-berry (*Rubus nutkanus*), 217, 263, 344, 348, 386, 388, 400, 401, 419
Salvia, 238
San Bernardino, 362

San Francisco, 352, 353
San Francisco Call, 336 n.
San Gorgonio Pass, 376
San Joaquin Canyon, 175-77
San Joaquin Cascades, 177
San Joaquin River, 100, 174-77, 213, 218, 222, 224, 241-44; trip to upper canyons of, 141-63
San Joaquin Yosemite, 176
San Miguel Canyon, 361
San Miguelito Ranch, 360
San Rafael, 361
Sandpipers, 237, 411
Sand Point, 403, 404, 412
Sand-storm, a, 432
Santa Lucia Mountains, 360
Sarcodes. See Snowflower
Sargent, Prof. Charles S., 334; Western journey with, 356-64; in the East with, 364, 365, 367, 371; world tour with, 427
Sargent, Mrs. Charles S., 365
Sargent, Miss Mary, 367
Sargent, Robeson, 427
Saunders, Dr. Alton, 412
Saxifrage, 264, 293, 390, 391, 409
Saxifrage, giant (*Peltiphyllum peltatum*), 209 and n., 379
Scandinavian, an old, 308, 309
Scrophularia, 27 and n., 32
Sea, a gale at, 305, 306
Sea-captain, a, on glaciers, 308, 309
Seal business, 405
Sea-lions, 406, 407
Seals, 392, 393; fur seals, 406, 407
Seattle, 292, 293, 380, 420, 421
Second Sentinel, 330
Sedges, 323. See also Carex
Senecio, 162, 216
Sentinel Cascade, 325
Sentinel Dome, 35, 59, 63; view from, 55, 172, 173
Sentinel Meadow, 330
Sentinel Rock, 109-11, 123
Sequoia gigantea, 421; the Kings River Big Trees, 182, 183, 228, 229, 322, 323, 333, 430; 'General Grant,' 182; the Mariposa grove, 182, 183, 222, 430; study of in 1875, 209-35; seed, 226, 235; nobility of, 228; a fire among, 230, 231, 380; growth and manner of falling, 233, 234; form, 235; cones, 235; future of, 235; cutting of, 322, 323, 429-31; the Calaveras grove, 429, 430; protection of, 430, 431
Sequoia sempervirens. See Redwood
Seven Gables, 324, 330
Shad-bush (Amelanchier), 79, 144

Shadow Lake, 79 n.
Shadows, controlling the glaciers, 108–10; at Yosemite Fall, 132; of trees, 132; in the woods at night, 136, 137
Shakan, 257 n.
Sheep, domestic, 239, 244; devastation caused by, 173, 174, 327, 348, 351
Sheep, mountain, 94, 179; a strong skull, 158; a fine band, 177, 178; a sheep hunt, 193–200; in Alaska, 316
Sheep and shepherding, 1–9, 12–15, 17–23, 25, 26, 29
Sheep Rock, 193, 194
Shelikof Strait, 413
Shepherds, 3; Mike, 19, 20; unpoetical character of, 29; Billy, 34
Shumagin Islands, 402–04, 412
Siberia, a glimpse of, 407, 408
Sideroxylon, 373
Sierra Club, 299, 428
Sierra Club Bulletin, 187 n.
Sierra Nevada, life and exploration in, 1–193, 201–35, 283–88, 322–33, 342–50; its allurement, 92; happiness offered by, 99; seen from Alhambra Valley, 336
Silene, 391, 410
Sims, William, 173 n., 176, 185, 186
Siskiyou Mountains, 290
Sisson, Mr., 193, 199, 289
Sisson's Station, 189, 193, 289
Sitka, 257–61, 387, 388
Skunk-cabbage, 291
Slate, 177, 179, 397; and earthquake action, 149
Slate Ridge, 174
Slifer, R. G., 401
Smilax, 264, 293, 356
Smith, Sidney, 355
Smoke, nature's use of, 226, 227
Smoky Jack. *See* Connel, John
Smoky Jack's sheep camp, life at, 2–33
Snake, green water, 344
Snakes. *See* Rattlesnakes
Snelling's, 1, 5, 18, 32
Snoqualmie Falls, 293
Snow, 38, 39, 44, 47–49, 52, 54, 65, 66, 113–20, 123, 124, 439
Snow, F. A., 172 n., 283 n.
Snow, Mrs. F. A., 172 n.
Snow Dome, 319
Snowballs for coffee, 163
Snow-banners, 139, 140
Snowberry, 217
Snowflake, hyperborean, 410 and n.
Snowflower (Sarcodes), 67, 214
Snowstorms, 321
Soaproot, 355. *See also* Lily, soaproot

Soda Springs, Cal., 282, 349
Soda Springs, Wash., 297
Solitude, 295
Solomon's-seal, 263
South Canyon Fall, 55
South Dome, *or* Half Dome, *or* Tissiack, 45–51, 53, 59, 61, 86, 87, 90, 104, 105, 107, 113, 127, 137, 201; as a barometer, 37, 38; in snow mantle, 38–40, 101, 102, 115
South Lyell Glacier, 172
South Tower, 325
Sparrow, golden-crowned, 400 and n.
Sparrow, song, 170, 400
Sparrow, Townsend's fox, 400
Sperry, James L., 430
Sphinx, the, Kings River, 325
Spider, 26; gossamer and dew beads, 103, 104; cobweb streamers, 112, 196
Spiræa, 63, 79, 128, 179, 238, 286, 289, 409; white and red, 293
Spiranthes, 216
Spirit, 138
Spraguea, 80, 162, 327
Spring, 87, 97, 98
Springs, of the Sierra, 224
Sproat, Mrs., 35, 38
Spruce, Douglas, 66, 78, 290–93, 344, 351, 356, 358 and n., 362, 381, 419–21
Spruce, Sitka *or* Menzies (*Picea sitchensis*), 258 and n., 263, 291, 293, 358, 359, 382, 383, 388, 389, 393, 394, 395, 398, 400, 405, 416, 419
Spruce, Williamson. *See* Hemlock, mountain
Spruce Island, 400
Squirrel, Douglas, 225, 226, 419
Squirrels, 144, 204, 211–13, 217, 218, 225
Squirrels, gray, 367, 369, 370
Squirrels, ground, 27, 349
Squirrels, red, 369, 370
Stanislaus River, 244
Stars, 171
Starwort, 400
Steel, Mr., 379
Stellaria, 293, 410
Stevens, Gen. Hazard, 294
Stickeen, the dog, 245, 246, 418; memories of, 275–80
Stickeen River, 266
Storms, 93; a complicated storm, 45; of winter, 117, 118; harmony of, 312
Strait of Georgia, 311
Strait of Juan de Fuca, 306
Stratton, Mr., 367
Strawberry, wild, 285, 293, 388, 390
Streams, music of, 95, 96; a charming cliff stream, 133; checks on speed, 208

Strentzel, Dr. John, 244, 245, 281
Strentzel, Louie, marriage to John Muir, 245. *See also* Muir, Mrs. John
Streptopus, 382, 389
Sturgeon Bay, 413, 414
Suisun swamps, 355
Sukatasse River, 297
Sum Dum Bay, 245
Sunnyside Bench, 35, 61, 104, 108, 111, 112, 120, 131, 133, 135-37, 139; charms of, 120, 121, 127, 128
'Sunnyside Observations,' 101 ff.
Sunrise, eternal, 438
Sunsets, 12, 13, 15; in a sheep's eyes, 26; in Alaska, 253-55
Surprise Glacier, 398 and n.
Suwanee River, 375
Swans, 92, 205, 410
Swett, John, 189, 201-03, 341 and n.
Swett, Mary, 189
Sycamore, 182, 236, 237, 361
Symphoricarpos, 293
Syringa, 63

Taku Glacier, 419
Taku Indian Village, 383
Taku Inlet, 245
Tamias. *See* Chipmunk
Tanager, Louisiana *or* Western, 339
Taxodium, 371
Taylor Bay, 418
Taylor Glacier, 245; adventure on, 275, 278-80
Teal, blue-winged, 237
Tehachapi Mountains, 377
Tenaya Canyon, 58, 96, 172
Tenaya Creek, 348
Tenaya Falls, 133, 165
Tenaya Glacier, 172
Tennessee, 57, 58, 365
Terns, eggs, 396
Tesla, Nikola, 367, 370
Texas, 376
Thimbleberries, 344. *See also Rubus nutkanus*
Thistles, 154, 162
Thomas Mills, 181
Thoreau, Henry D., 97
Thousand Islet Lakes, *or* Islet Lake, 157, 158 and n., 159
Thousand-Mile branch of San Joaquin, 162
Thrasher, brown, 165
Three Brothers, 49, 59, 102
Three Domes, 345, 346
Three Hermits, 325
Thrush, 263
Thrush, hermit, 400

Thrush, water. *See* Water-ouzel
Thuja. *See* Arborvitæ
Thuja gigantea, 380-82
Thunder, avalanches of, 146
Thysanocarpus, 28 n.
Tides, 89
Tillandsia, 372
Tioga Road, 347
Tissiack. *See* South Dome
Tissiacks, 94; of Kings River, 184
Titmouse, 104
Toad, horned, 27
Tomales Point, 361
Tongass, 419
Tornado Meadows, 323
Torrey, Bradford, 366
Totem-poles, 257, 270, 271
Tracks, 264
Trees, contented, 313; burial of, 313; good from, 428, 429; friendship of, 437
Trelease, Dr. William, 400
Trientalis, 400
Trinidad, Cal., 358, 359
Trout, 328, 329
Trout, salmon, 243
Truckee, 285
Truckee River, 285
Tsuga mertensiana. See Hemlock, mountain
Tsuga mertensian = aheterophylla, 263, 293, 381-83, 393, 416
Tsuga pattoniana. See Hemlock, mountain
Tule River, 233
Tulip, Mariposa, 203
Tuolumne Canyon, *or* Hetch Hetchy, 36, 204, 205; explorations in, 68-79, 343-47; days and nights in, 166-71; losing battle to save, 427, 428, 437
Tuolumne Divide, 89
Tuolumne Glacier, 70, 72
Tuolumne Meadows, 282
Tuolumne River, 1, 70, 74, 77, 166, 167, 204, 244, 343-46
Tuolumne Valley, 162
Tupelo, 376. *See also* Nyssa
Turner's, 242
Tussilago, 410
Twenty Hill Hollow, 21, 23, 24, 26, 27
Tyringham, Mass., 367

Ukiah, 359, 360
Umbelliferæ, 18
Umpqua River, 421
Unalaska Island, 404, 405, 411
Unga Island, 402-04
Unimak Island, 404, 414
United States Coast and Geodetic Survey, 190

Unity, 63
Upper Fresno Basin, 211, 212
Upper Illilouette Basin, 81
Utah, 190
Uyak Bay, 399, 400, 403

Vaccinium, 79, 259, 263, 293, 388, 391, 397, 401
Vaccinium ovalifolium, 401
Vaccinium uliginosum, 409
Van Bremers, the, 194–99; their mountain, 194
Vancouver coal measures, 248
Vancouver Island, 266, 268, 291, 307, 380, 381
Vanderbilt, Mr., 262, 387 n.
Vanderbilt, Mrs., 387
Vanderbilt, Annie, 387 n.
Van Rensselaer, Mariana G., 371
Van Trump, P. B., 294 and n.
Venus, the planet, 117
Veratrum, 214, 397
Vernal Fall, 55, 58, 59, 172; character, 68
Victoria, B.C., 303, 306, 307, 309, 311, 380
Vines, 58
Violet, gold, 21 and n.
Violets, 48, 49, 216, 217, 317, 323, 391, 401, 407, 409, 410
Violets, white, 66, 145, 217, 317
Visalia, 240
Voices in the wilderness, 135, 136
Volcanoes, 399, 402–04, 411, 414, 415

Wade Meadows, 347
Wah Mello, 213, 227
Wallflower, 207, 208
Warbler, yellow, 386, 400
Warbler, yellow-crowned, 394
Washburn, Mrs., 402
Washington, State of, 292
Washington, D.C., 371, 372
Washington Column, 122
Washingtonia filifera, 376 and n.
Wasp, and horse, 364
Wasps' nests, 85
Water, differences in, 219, 220
Waterfalls, 50; after a storm, 43; a journey among, 44; various characters of those of Yosemite Valley, 58; without definite channels, 64, 65; settings of, 67; music of, 95, 96. *See also under individual falls*
Water-ouzel (American dipper), 56, 66, 76, 84 and n., 87 and n., 93, 101, 102, 112–14, 118, 119, 130, 148, 165, 166, 192, 193, 326

Wawona Falls, 209 and n., 350
Weasel, 152
Well, choke-damp in a, 192
Wellington coal mines, 248
West Hills, 335, 338, 355
West Point, 370
West Sentinel, 325, 330
Whales, 304, 305, 408, 414
Whaling, in Alaska, 408
White, Mrs., 402
White Pass, 383
Whitehead, Robert, 430 n.
Wilderness, delights of the, 299–301; fear of the, 314
Wildness, 90
Willamette River, 379
Willit's, 360
Willoughby Island, 385
Willow, arctic, 79
Willows, 49, 56, 79, 88, 103, 106, 183, 185, 188, 239, 286, 289, 293, 316, 330, 361, 385, 386, 390, 391, 394, 395, 400, 401, 409, 410, 416, 421; tangles of, 73. *See also Salix Barclayi*
Wilmington, Del., 366, 371
Wind, 98, 99; masses of, 128
Wistaria, 401
Wood, silicified, 23
Wood, Col., 362
Wood Island, 401
Woodbridge, Mr., 415
Woodchucks, 147, 387
Woodpecker, pileated, *or* logcock, 211, 215, 225 and n.
Woodpeckers, 104, 114, 210, 216, 217, 225
Woods, the, their uses, 220
Woods Hole, 364
Woodwardia, 121, 165
Wrangell Island, Alaska, 257, 258, 260, 262, 270, 382
Wrangell Narrows, 382, 419
Wren, Parkman's house, 340–42
Wrens, 104, 107, 217

Yakutat, 390, 393
Yakutat Bay, 392, 393, 415–18
Yakutat Glacier, 389, 417
Yarrow, 176, 216, 219, 293
Yelm Prairie, 293, 294
Yew, 293, 381
Yosemite Creek, 59, 110, 136, 204
Yosemite Creek Basin, 69
Yosemite Creek Glacier, 172, 203
Yosemite Falls, 50, 55, 61, 111, 113, 118; in the wind, 40, 113; rainbow colors on, 40, 123, 134; its glories, 58; lunar bows in, 61, 62; an adventure behind,

61, 62; the Lower Fall, 63, 64, 127–29; the Upper Fall, 106, 126, 132, 134, 135, 140, 201; ice-cone at foot of Upper Fall, 125; subsidiary falls, 129; shadows at, 132; music of, 140

Yosemite Flat, 344

Yosemite National Park, 282, 298; uses of, 350

Yosemite Pass, 345

Yosemite Point, 127

Yosemite Valley, Muir's explorations in and about, 34–173, 190–93, 201–36, 283–85, 342–50; M.'s early statement of his belief in the glacial origin of, 35; during and after snow, 38, 39; various characters of its waterfalls, 58, 59; work of glaciers in, 59, 60, 90, 108–12; its autumn river pools, 90, 91; its spring lakelets, 96; difference in climate of north and south sides, 120; pine groves and meadows of, 137; recession from the State to the Federal Government, 298, 299, 334, 427; under State management, 349; Roosevelt in, 427; the glory of God in, 438

Yosemites, on North Fork Kings River, 180; on the Tuolumne River, 343

Young, Rev. Samuel Hall, 245, 262, 269, 276 and n., 384 and n.

Yucca, 376

Zonotrichias, 335